AF358809

Advanced Numerical and Computer Methods in Civil Engineering

Advanced Numerical and Computer Methods in Civil Engineering

Editors

Dongming Li
Zechuan Yu

Basel • Beijing • Wuhan • Barcelona • Belgrade • Novi Sad • Cluj • Manchester

Editors

Dongming Li
School of Civil Engineering
and Architecture
Wuhan University of
Technology
Wuhan
China

Zechuan Yu
School of Civil Engineering
and Architecture
Wuhan University of
Technology
Wuhan
China

Editorial Office
MDPI
St. Alban-Anlage 66
4052 Basel, Switzerland

This is a reprint of articles from the Special Issue published online in the open access journal *Buildings* (ISSN 2075-5309) (available at: www.mdpi.com/journal/buildings/special_issues/R8XH61D33E).

For citation purposes, cite each article independently as indicated on the article page online and as indicated below:

Lastname, A.A.; Lastname, B.B. Article Title. *Journal Name* **Year**, *Volume Number*, Page Range.

ISBN 978-3-7258-1330-8 (Hbk)
ISBN 978-3-7258-1329-2 (PDF)
doi.org/10.3390/books978-3-7258-1329-2

Contents

About the Editors

Dongming Li

Dr. Dong-Ming Li received his Ph.D. in 2015 from the City University of Hong Kong with a postgraduate studentship funded by the UGC of Hong Kong S.A.R. From 2011 to 2012, he worked as a research assistant at the City University of Hong Kong for six months. From 2018 to 2020, he was a MSCA COFUND fellow at Cardiff University with the support of the Marie Sklodowska Curie Actions COFUND fellowship and the SÊR CYMRU II program. Dr. Li became an associate professor at Wuhan University of Technology in 2015. His main research interests include numerical methods, computational mechanics, applied mechanics, soft matters, composites, meta materials and structures, and self-healing materials. Dr. Li's research in these fields has resulted in 40+ journal papers in related, top international journals, which have received a total of more than 900 citations with a h-index of 14. One of his co-authored paper received the Best Paper Award 2018 of Engineering Analysis with Boundary Elements. One paper for which he was the corresponding author was selected as an excellent scientific and technological paper in Hubei Province (2021 to 2023). Dr. Li has received the Shanghai Postgraduate's Outstanding Achievements (Dissertation) Award and the CMMM 2023 Best Presentation Award. He has been invited to be the guest editor and a youth editor board member of several international journals.

Zechuan Yu

Dr. Zechuan Yu received his Bachelor's degree from the University of Science and Technology of China and his Ph.D. from the City University of Hong Kong. He is currently a faculty member in the School of Civil Engineering and Architecture at Wuhan University of Technology. As of 2023, Dr. Yu has published approximately 30 SCI articles, garnered over 1,200 citations, and achieved an h-index of 20. Dr. Yu's research encompasses a diverse range of topics, including multi-scale molecular dynamics, calcium–silicate–hydrate, cementitious materials, the cement–epoxy interface, cellulose nanocrystals, and the application of deep learning methods in structural health monitoring. Among his interdisciplinary projects, one particularly intriguing ongoing study involves developing a generative deep learning model for atomistic simulations.

Preface

The development and application of advanced numerical and computer methods in civil engineering have become increasingly more important for modern engineers and researchers in recent decades. Advanced numerical and computer methods can be used, from the material level to the structural level, for solving nearly all engineering problems, solely or in combination with experimental/theoretical studies. In this Special Issue, we have successfully collected 10 excellent papers that present recent progress either in the novel development or the novel application of advanced numerical and computer methods for solving problems in civil engineering.

Dongming Li and Zechuan Yu
Editors

Article

On Interpolative Meshless Analysis of Orthotropic Elasticity

You-Yun Zou [1,2,†], Yu-Cheng Tian [1,†], D. M. Li [1,3,*], Xu-Bao Luo [1] and Bin Liu [1]

1   School of Civil Engineering and Architecture, Wuhan University of Technology, Wuhan 430070, China
2   School of Aeronautics and Astronautics, Zhejiang University, Hangzhou 310013, China
3   School of Science, Wuhan University of Technology, Wuhan 430070, China
*   Correspondence: domili@whut.edu.cn or dongmli2-c@my.cityu.edu.hk
†   These authors contributed equally to this work.

Abstract: As one possible alternative to the finite element method, the interpolation characteristic is a key property that meshless shape functions aspire to. Meanwhile, the interpolation meshless method can directly impose essential boundary conditions, which is undoubtedly an advantage over other meshless methods. In this paper, the establishment, implementation, and horizontal comparison of interpolative meshless analyses of orthotropic elasticity were studied. In addition, the radial point interpolation method, the improved interpolative element-free Galerkin method and the interpolative element-free Galerkin method based on the non-singular weight function were applied to solve orthotropic beams and ring problems. Meanwhile, the direct method is used to apply the displacement boundary conditions for orthotropic elastic problems. Finally, a detailed convergence study of the numerical parameters and horizontal comparison of numerical accuracy and efficiency were carried out. The results indicate that the three kinds of interpolative meshless methods showed good numerical accuracy in modelling orthotropic elastic problems, and the accuracy of the radial point interpolation method is the highest.

Keywords: interpolative shape functions; meshless method; elastic mechanics; orthotropic elasticity

1. Introduction

Unlike finite element methods (FEM), which are currently the most widely used in engineering computational simulations, meshless/mesh-free methods employ a node-based rather than element-based approach when constructing shape functions discretizing the problem domain [1,2]. This strategy makes meshless approaches an increasingly effective substitute to finite element methods when dealing with problems involving large deformation [3–6] and crack propagation [7–10], where numerical implementations may be restricted by predefined meshes/elements. However, mainstream meshless shape functions, such as the most widely used moving least-squares approximation [1,3,4,6–8,10] and the reproducing kernel function [2,5,9] and their various modifications, do not have the same interpolation properties as finite element methods. For the approximate rather than interpolated meshless methods, certain kinds of additional techniques, such as the Lagrange multiplier method [1,2] and the penalty method [3–10], will necessarily be used to enforce essential boundary conditions, with some unwanted side effects in terms of computational efficiency or numerical accuracy.

Therefore, lots of research efforts have been devoted to the construction of meshless shape functions with the Kronecker delta interpolation property, so that the essential boundary conditions could be easily and directly imposed in large-scale engineering modelling. A point interpolation method whose shape function is interpolative but prone to singularity was proposed by Liu et al. [11]. Later, Wang et al. [12] proposed the radial point interpolation method (RPIM) to overcome the singularity problem of polynomial point interpolation shape functions. Ren et al. [13] proposed the interpolating moving least-squares method (IMLSM) by improving the Lancaster's interpolative approach [14]

Citation: Zou, Y.-Y.; Tian, Y.-C.; Li, D.M.; Luo, X.-B.; Liu, B. On Interpolative Meshless Analysis of Orthotropic Elasticity. *Buildings* **2023**, *13*, 387. https://doi.org/10.3390/buildings13020387

Academic Editor: Elena Ferretti

Received: 24 October 2022
Revised: 17 January 2023
Accepted: 20 January 2023
Published: 31 January 2023

with a singular weight function. Wang et al. [15] used the non-singular weight function to develop an improved interpolating moving least-squares method (IIMLSM) thereafter. An interpolative complex variable moving least-squares method (ICVMLSM) was developed by Deng et al. [16]. An interpolative variational multiscale element-free Galerkin method was presented by Zhang and Li [17] for convection-diffusion equations and Stokes problems. Wang et al. [18] introduced the weighted orthogonal basis in the IIMLSM to get a diagonal moment matrix. An interpolating meshless local Petrov-Galerkin method (IMLPGM) with an IMLS scheme for steady-state heat conduction problems was reported by Singh et al. [19]. Bourantas et al. [20] modified the error functional of the IMLSM to construct almost interpolating shape functions with ensured invertibility of the moment matrix. Wang et al. proposed the stable collocation method [21,22] and the gradient smooth collocation method [23].

The application of different interpolative meshless methods to analyze the mechanical response of various anisotropic elastic solids widely present in natural [24,25] and artificial [26,27] engineering materials has also received extensive attention. Dinis et al. [28] proposed the natural neighbor RPIM and used it to analyze the problems of thin plates and shells of composite materials. Njiwa et al. [29] combined isotropic boundary element and local point interpolation to solve the three-dimensional anisotropic elasticity problem. Bui and Nguyen [30] developed a novel moving Kriging interpolative scheme for efficient meshfree vibration and buckling analysis of orthotropic plates. Fallah et al. [31] used the Delaunay triangulation scheme to discretize arbitrarily distributed node sets in the domain and proposed a meshless finite volume formula to model cracks and fracture in orthotropic media. A modified interpolative element-free Galerkin method was applied to the modelling of orthotropic thermoelastic fracture by Lohit et al. [32]. Luo et al. [33] developed an efficient and stable nodal integration RPIM to evaluate the buckling performance of variable-stiffness composite plates with elliptical cutouts.

We already know that interpolation meshless methods can directly impose essential boundary conditions, which is undoubtedly an advantage over other meshless methods. Therefore, the purpose of this paper is to compare the proposed meshless interpolation methods horizontally, so as to find out which meshless interpolation method has better accuracy, which is meaningful. In this study, one mature and two relatively fresh interpolating meshless methods, namely the RPIM, the IMLSM and the IIMLSM, are employed to construct interpolative shape functions of the displacement field of orthotropic elastic solids. The corresponding formulas of the radial point interpolative meshless method (RPIM), the interpolative element-free Galerkin method (IEFGM) and the improved interpolative element-free Galerkin method (IIEFGM) for orthotropic elasticity are established and the computer programs are developed. In implementations of all three interpolative meshless orthotropic elastic analyses, the displacement boundary conditions are imposed by the direct method. Three typical numerical examples are analyzed for verification purpose and to compare the differences in numerical performance between the three methods. Finally, we find that the three meshless interpolation methods have good numerical accuracy in the modeling of orthotropic elastic problems, and the radial point interpolation method has the highest accuracy.

2. Basics of Three Interpolative Meshless Shape Functions

Here we briefly review the three schemes, i.e., the RPIM [12], the IMLSM [13] and the IIMLSM [15], to construct the interpolative meshless shape function $u^h(x)$ of the displacement field $u(x)$ in the local domain of point x with n nodes x_I. Hereinafter, p or $p_i(x)(i = 1, 2, 3, \cdots, m)$ are used to represent the vector of m-term polynomial basis.

2.1. Radial Point Interpolation Method (RPIM)

The interpolative function $u^h(x)$ in the local domain could be constructed by the linear superposition of the radial basis and the polynomial basis as follows [12]:

$$u^h(x) = \sum_{i=1}^{n} r_i(x)a_i(x) + \sum_{j=1}^{m} p_j(x)b_j(x) = ra + pb \tag{1}$$

where r is the vector of n-term radial basis, and a and b are the coefficients corresponding to two kinds of basis. The generally adopted multiquadric (MQ) basis [34] is also employed,

$$r_i(x) = (c^2 + \|x - x_I\|^2)^q \tag{2}$$

where q and c are two coefficients.

Let the interpolation function $u^h(x)$ take the value of nodal displacement at each node, i.e.,

$$u = Ra + Pb \tag{3}$$

where u is the nodal displacement vector, and $P = [p_1 \quad p_2 \quad \cdots \quad p_n]^T$ and $R = [r_1 \quad r_2 \quad \cdots \quad r_n]^T$ are nodal basis matrices. With an extra constraint $P^T a = 0$, the unknown coefficient vectors a and b can be written as

$$a = \left[R^{-1} - R^{-1}P\left(P^T R^{-1} P\right)^{-1} P^T R^{-1} \right] u \tag{4}$$

$$b = \left(P^T R^{-1} P \right)^{-1} P^T R^{-1} u \tag{5}$$

Substituting Equations (4) and (5) into Equation (1) can yield

$$u^h(x) = \Phi(x)u \tag{6}$$

where $\Phi(x)$ is the shape function of RPIM,

$$\Phi(x) = r[R^{-1} - R^{-1}P\left(P^T R^{-1} P\right)^{-1} P^T R^{-1}] + p\left(P^T R^{-1} P\right)^{-1} P^T R^{-1} \tag{7}$$

2.2. Improved Interpolating Moving Least-Squares Method (IMLSM)

To construct interpolative moving least-squares shape functions, Ren et al. [13] employed a strategy to rebuild basis functions with a singular weight function $\overline{\omega}(x, x_I)$. The local interpolating function $u^h(x)$ of the IMLSM is

$$u^h(x) = \overline{p}_1(x)\overline{a}_1(x) + \sum_{i=2}^{m} \overline{p}_i(x)\overline{a}_i(x) = \overline{p}_1(x)\overline{a}_1(x) + \overline{pa} \tag{8}$$

where $\overline{a}$ is the unknown coefficient vector, and the reformed basis function $\overline{p}_i(x)(i = 1, 2, 3, \cdots, m)$ is reconstructed from the corresponding polynomial basis $p_i(x)$ as follows:

$$\overline{p}_i(x) = \begin{cases} \dfrac{1}{\left[\sum\limits_{I=1}^{n} \overline{\omega}(x, x_I) \right]^{1/2}} & i = 1 \\ p_i(x) - \sum\limits_{I=1}^{n} \Gamma(x, x_I)p_i(x_I) & i = 2, 3, \cdots, m \end{cases} \tag{9}$$

where $\gamma(x, x_I) = \overline{\omega}(x, x_I) \Big/ \sum\limits_{J=1}^{n} \overline{\omega}(x, x_J)$. The singular weight function $\overline{\omega}(x, x_I)$ is

$$\overline{\omega}(x, x_I) = \begin{cases} \left\| \dfrac{x - x_I}{\rho_I} \right\|^{-\alpha} & \|x - x_I\| \le \rho_I \\ 0 & \text{others} \end{cases} \tag{10}$$

where $\rho_I = d_{\max}d_c$ is the characteristic radius of the local support domain of point x, d_{max} is the scale factor to control the size of the local domain, d_c is the characteristic node spacing, and the parameter α is an even number and takes a value larger than zero in general.

Minimize the weighted error functional $J = \sum\limits_{I=1}^{n} \overline{\omega}(x, x_I)[u^h(x_I) - u(x_I)]^2$, and consider additional constraint $\sum\limits_{I=1}^{n} \overline{\omega}(x, x_I)\Gamma(x, x_I)u_I\overline{p}_i = 0$ to establish a group of algebraic equations to obtain the unknown coefficient term $\overline{a}$ as

$$\overline{p}_1(x)\overline{a}_1(x) = \sum_{I=1}^{n}\Gamma(x, x_I)u_I \tag{11}$$

$$\overline{a} = A^{-1}Bu \tag{12}$$

where $A = B\overline{P}_0^{\mathrm{T}}$, and the elements of the matrix $B_{(m-1)\times n}$ and the matrix $\overline{P}_0$ are

$$B_{kJ} = \begin{cases} \overline{\omega}(x, x_J)\overline{p}_k(x_J) & x \neq x_J \\ \Gamma(x, x_J)\sum\limits_{I=1,I\neq J}^{n}\overline{\omega}(x, x_I)[p_k(x_J) - p_k(x_I)] & x = x_J \end{cases} \tag{13}$$

$$\overline{P}_0 = \begin{bmatrix} \overline{p}_2(x, x_1) & \overline{p}_2(x, x_2) & \cdots & \overline{p}_2(x, x_n) \\ \overline{p}_3(x, x_1) & \overline{p}_3(x, x_2) & \cdots & \overline{p}_3(x, x_n) \\ \vdots & \vdots & \ddots & \vdots \\ \overline{p}_m(x, x_1) & \overline{p}_m(x, x_2) & \cdots & \overline{p}_m(x, x_n) \end{bmatrix} \tag{14}$$

Substituting Equations (11) and (12) into Equation (8), the interpolative function can be rewritten as

$$u^h(x) = \sum_{I=1}^{n}\Phi_I(x)u_I = \Phi(x)u \tag{15}$$

where the shape function of IMLSM is

$$\Phi(x) = (\Phi_1(x), \Phi_2(x), \cdots, \Phi_n(x)) = \gamma + \overline{p}A^{-1}B \tag{16}$$

$$\gamma = (\gamma(x, x_1), \gamma(x, x_2), \cdots, \gamma(x, x_n)) \tag{17}$$

2.3. Improved Interpolating Moving Least-Squares Method (IIMLSM)

To remove the unwanted singular weight function in practice, some modifications are made to the field variable function $u(x)$ and the basis function in the IIMLSM by Wang et al. [15]. The transformed field variable function is written as

$$\widetilde{u}(x) = u(x) - \sum_{I=1}^{n}\widetilde{\gamma}(x, x_I)u(x_I) \tag{18}$$

where $\widetilde{\gamma}(x, x_I) = \zeta(x, x_I)\Big/\sum\limits_{J=1}^{n}\zeta(x, x_J)$ and $\zeta(x, x_I) = \prod\limits_{J\neq I}(\|x - x_I\|^2/\|x_I - x_J\|^2)$. Its local moving least-squares interpolation is defined as

$$\widetilde{u}_x^h(x) = \sum_{i=2}^{m}\widetilde{p}_i(x)\widetilde{a}_i(x) = \widetilde{p}\widetilde{a} \tag{19}$$

The modified basis function $\widetilde{p}$ is constructed from the original polynomial basis as

$$\widetilde{p}_i(x) = p_i(x) - \sum_{I=1}^{n}\widetilde{\gamma}(x, x_I)p_i(x_I) \tag{20}$$

where apparently $\widetilde{p}_1(x) = 0$.

To minimize the weighted discrete error norm $J = \sum\limits_{I=1}^{n} \omega(x - x_I)[\tilde{u}_x^h(x_I) - \tilde{u}(x_I)]^2$, the unknown coefficient $\tilde{a}$ could be expressed as

$$\tilde{a} = \tilde{A}^{-1}\tilde{B}u \tag{21}$$

where $\tilde{A} = \tilde{P}_0^{\mathrm{T}}\omega\tilde{P}_0$, $\tilde{B} = \tilde{P}_0^{\mathrm{T}}W(I - \tilde{Y})$, $\omega = \mathrm{dig}[\omega(x - x_1) \ \cdots \ \omega(x - x_n)]$, I is the identity matrix, $\omega(x - x_I)$ can take any non-singular weight function, $\tilde{P}_0$ is the $n \times (m - 1)$ matrix of the nodal basis row vector $\tilde{p}$, and $\tilde{Y}$ is the $n \times n$ matrix of the row vector $\tilde{\gamma}(x, x_I)$.

Substituting Equation (21) into Equation (18), the IIMLS interpolation $u^h(x)$ of $u(x)$ is

$$u^h(x) = \Phi(x)u \tag{22}$$

where the IIMLS shape function $\Phi(x)$ is

$$\Phi(x) = \tilde{\gamma} + \tilde{p}\tilde{A}^{-1}\tilde{B} \tag{23}$$

$$\tilde{\gamma} = (\tilde{\gamma}(x, x_1), \tilde{\gamma}(x, x_2), \cdots, \tilde{\gamma}(x, x_n)) \tag{24}$$

3. The Establishment of Discrete Equations

The three meshless interpolative methods, namely the RPIM, the IMLSM, and the IIMLSM, are used to approximate orthotropic elastic displacement fields and to discretize the Galerkin weak form of control equation. The corresponding radial point interpolative meshless method (RPIM), the interpolative element-free Galerkin method (IEFGM) and the improved interpolative element-free Galerkin method (IIEFGM) for orthotropic elastic problems are presented. The Galerkin weak form of control equation for orthotropic elasticity can be written as

$$\int_{\Omega} \delta(Lu)^{\mathrm{T}}D(Lu)d\Omega - \int_{\Omega} \delta u^{\mathrm{T}}bd\Omega - \int_{\Gamma_t} \delta u^{\mathrm{T}}\bar{t}d\Gamma = 0 \tag{25}$$

where L is the partial differential operator, D is the orthotropic elastic matrix, u is the column of the nodal displacement, b and $\bar{t}$ are the columns corresponding to the body force and the surface traction, respectively, and Ω and Γ_t represent the problem domain and the force boundary, respectively.

The constitutive matrix of orthotropic material is represented as

$$D = \begin{bmatrix} s_{11} & s_{12} & 0 \\ s_{21} & s_{22} & 0 \\ 0 & 0 & s_{66} \end{bmatrix}^{-1} \tag{26}$$

where s_{ij} is the orthotropic elastic compliance coefficient.

Discretizing Equation (25), we can get

$$KU = F \tag{27}$$

where U is the column of the total displacement, and the overall stiffness matrix K and the total external force vector F could be assembled from the nodal values, respectively, as follows:

$$K_{IJ} = \int_{\Omega} B_I^{\mathrm{T}}DB_J d\Omega \tag{28}$$

$$F_I = \int_{\Omega} \Phi_I^{\mathrm{T}}bd\Omega + \int_{\Gamma_t} \Phi_I^{\mathrm{T}}\bar{t}d\Gamma \tag{29}$$

where $\boldsymbol{\Phi}_I = \begin{bmatrix} \Phi_I & 0 \\ 0 & \Phi_I \end{bmatrix}, \boldsymbol{B}_I^{\mathrm{T}} = \begin{bmatrix} \Phi_{I.1} & 0 & \Phi_{I.2} \\ 0 & \Phi_{I.2} & \Phi_{I.1} \end{bmatrix}.$

The interpolative shape functions of the above three methods can be used in meshless methods to impose displacement boundary conditions as easily as in the finite element methods. In the following three numerical examples, the interpolative meshless methods adopt the direct method to apply the corresponding essential boundary conditions. The principle of the direct method is to recombine the equations according to the determined and undetermined nodal displacements:

$$\begin{bmatrix} K_{aa} & K_{ab} \\ K_{ba} & K_{bb} \end{bmatrix} \begin{pmatrix} u_a \\ u_b \end{pmatrix} = \begin{pmatrix} f_a \\ f_b \end{pmatrix} \tag{30}$$

where u_a and u_b are the unknown and the known nodal displacements, respectively. The overall stiffness matrix and the total external force vector are partitioned to K_{aa}, K_{ab}, K_{ba}, K_{ba}, f_a and f_b, respectively, according to the dividing of the displacement. Since u_b is known, u_a can be obtained as

$$u_a = K_{aa}^{-1}(f_a - K_{ab}u_b) \tag{31}$$

4. Numerical Examples

In this section, three numerical examples are analyzed by using the three developed interpolative meshless methods, namely the radial point interpolative meshless method (RPIM), the interpolative element-free Galerkin method (IEFGM) and the improved interpolative element-free Galerkin method (IIEFGM), to demonstrate their numerical performance. The corresponding numerical results are validated with both the analytical solutions and those obtained by the element-free Galerkin method (EFGM). In all meshless implementations, the rectangular local domain is used and the 4×4 Gauss integral is adopted. Relative errors e_r and energy norm errors e_e are defined to compare numerical accuracy:

$$e_r = \frac{\text{Numerial result} - \text{Exact solution}}{\text{Exact solution}} \times 100\% \tag{32}$$

$$e_e = \sqrt{\frac{1}{2} \int_{\Omega} \left(\varepsilon^h - \varepsilon\right)^{\mathrm{T}} D\left(\varepsilon^h - \varepsilon\right) d\Omega} \tag{33}$$

where ε^h and ε are the numerical results and exact solutions of the strain, respectively.

4.1. Clamped-Clamped Beam Subjected to Uniformly Distributed Load

The orthotropic material clamped-clamped beam is shown in Figure 1. Beam span l is 48 m, the depth h is 12 m, the upper boundary is under uniformly distributed load, and the load q is 60 KN/m. Regardless of the structure weight, the numerical example is modelled as a plane stress problem.

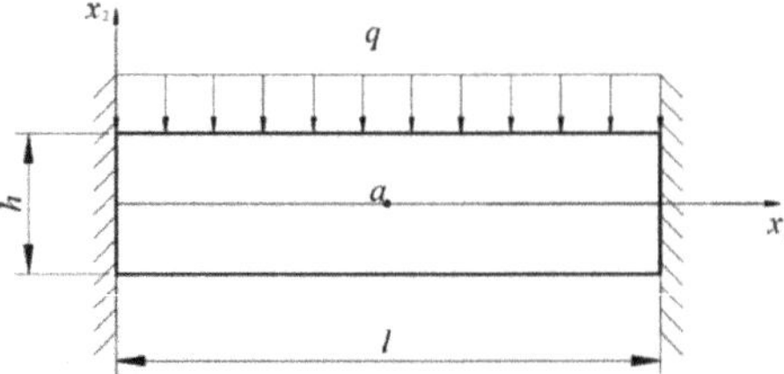

Figure 1. Clamped-clamped beam subjected to uniform load.

The material compliance coefficients (unit: m/KN) of the beam are $s_{11} = 0.078 \times 10^{-10}$, $s_{12} = -0.038 \times 10^{-10}$, $s_{21} = -0.038 \times 10^{-10}$, $s_{22} = 0.080 \times 10^{-10}$, and $s_{66} = 0.233 \times 10^{-10}$.

The exact solutions of the displacements and the stresses for this problem are expressed as [35]

$$u_1 = (2x_1 - l)(s_{12} + s_{66})qx_2^3/h^3 + (l - 2x_1)(x_1 - l)s_{11}qx_1x_2/h^3 \tag{34}$$

$$\begin{aligned} u_2 = &- \left[2s_{11}s_{12}(6x_1^2 - 6lx_1 + l^2) - 3s_{11}s_{22}h_2 + 3s_{12}^2h^2\right]qx_2^2/(4s_{11}h^3) + \\ &\left(2s_{12}^2 - s_{11}s_{22} + s_{12}s_{66}\right)qx_2^4/(2s_{11}h^3) - (s_{11}s_{22} - s_{12}^2)qx_2/(2s_{11}) + \\ &x_1(x_1 - l)(2s_{11}x_1^2 - 3s_{66}h^2 - 2s_{11}lx_1)q/(4h^3) \end{aligned} \tag{35}$$

$$\begin{aligned} \sigma_1 = &2(2s_{12} + s_{66})qx_2^3/(s_{11}h^3) - [2s_{11}(6x_1^2 - 6lx_1 + l^2) + \\ &3s_{12}h^2]qx_2/(2s_{11}h^3) + qs_{12}/(2s_{11}) \end{aligned} \tag{36}$$

$$\sigma_2 = -q\left(4x_2^3 - 3h^2x_2 + h^3\right)/2h^3 \tag{37}$$

$$\tau_{12} = 3q(l - 2x_1)\left(h^2/4 - x_2^2\right)/h^3 \tag{38}$$

The computing parameters of the four meshless schemes are obviously different. The optimal values for each method are surely not the same. To be able to compare the four methods horizontally, this paper adopts the best results with regard to the energy norm error of each method within a certain range of the computing parameters for accuracy comparison. The computing parameters in a certain range for all four methods are collected as: the number of nodes $n_1 = (19, 21, 23, 25)$ in the x_1 direction and the number of nodes $n_2 = (7, 9, 11, 13)$ in the x_2 direction for a uniform mesh, a fixed uniform background mesh with $l_1 \times l_2 = 12 \times 8$ for the Gauss quadrature, the scale factor $d_{max} = (1.5, 2.5, 3.5, 4.5)$ in the local domain, $q = (-0.5, 0.5, 1.5, 2.5)$ and $c = (1.0, 3.0, 5.0)$ for the MQ radial basis, the singular weight parameter $\alpha = (4, 6, 8)$ in IMLSM and the penalty factor $\beta = 3 \times 10^{14}$ in EFGM. Therefore, according to the different combinations of each computing parameter within the corresponding ranges, there are, in total, 64, 768, 192 and 64 groups of computing settings for the EFGM, the RPIM, the IEFGM, and the IIEFGM that need to be evaluated, respectively. We recode the optimal computing parameters of each method with the lowest energy norm error in all corresponding computing sets in Table 1. The numerical results according to these four computing settings are employed for accuracy comparison.

Table 1. Optimal computing parameters of each method within the tested range.

	$n_1 \times n_2$	$l_1 \times l_2$	d_{max}	Others
EFGM	19×11	12×8	3.5	$\beta = 3 \times 10^{14}$
RPIM	25×11	12×8	4.5	$q = 2.5, c = 1.0$
IEFGM	25×13	12×8	2.5	$\alpha = 6.0$
IIEFGM	19×13	12×8	1.5	—

To investigate the characteristics of the four shape functions, Table 2 shows the contour plots of the shape functions of the four methods with the parameters in Table 1 and their first-order derivatives at the beam center point a (24, 0). It can be seen from Table 2 that the contour plots of the shape functions of the three interpolative meshless methods are completely consistent. They all meet the property of the Kronecker delta function and are significantly different from the contour maps of the MLS. The difference between the four methods is mainly reflected in the contour maps of the first-order derivatives of the shape functions, which may be the main reason for the difference in numerical accuracy of the three meshless interpolative methods.

Table 2. Contour plot of the shape functions and their 1st order derivate at point (24, 0) by four different methods.

	MLS	RPIM	IMLS	IIMLS
Φ	0.000; 0.000; 0.036; 0.091; 0.054; 0.109; 0.073; 0.018; 0.000; 0.000	0.000; 0.000; 0.000; 0.125; 0.250; 0.375; 0.000; 0.000	0.000; 0.000; 0.000; 0.125; 0.250; 0.375; 0.000; 0.000	0.000; 0.000; 0.000; 0.125; 0.250; 0.375; 0.000; 0.000
$\Phi_{,x}$	0.000; 0.000; 0.000; -0.008; 0.008; -0.016; 0.016; -0.023; 0.023; 0.000; 0.000	0.000; 0.000; 0.000; 0.000; -0.097; 0.097; -0.048; 0.048; 0.000; 0.000	0.000; 0.000; -0.026; 0.026; -0.052; 0.052; 0.000	0.000; 0.000; 0.000; -0.007; -0.021; -0.014; 0.021; 0.014; 0.007; 0.000; 0.000
$\Phi_{,y}$	0.000; 0.000; 0.017; 0.035; 0.052; 0.000; -0.035; -0.052; -0.017; 0.000; 0.000	0.000; 0.000; 0.146; 0.292; 0.000; -0.292; -0.146; 0.000; 0.000	0.000; 0.058; 0.115; -0.058; -0.115; 0.000	0.000; 0.000; 0.019; 0.058; 0.039; 0.000; -0.019; -0.058; -0.039; 0.000; 0.000

The nodal deflection u_2 on the central axis $x_2 = 0$ of the beam and the corresponding relative errors are shown and are compared in Figure 2. It is obvious that the meshless displacement results of the four methods agree very well with the analytical solutions, and all the relative errors are below 1.16%. However, the accuracy of the IEFGM is relatively lower than those of the other three meshless methods. The numerical solutions of EFGM and RPIM have better stability, and their maximum relative errors do not exceed 0.06% and 0.36%, respectively. The maximum relative error of the IIEFGM is also below 0.59%.

Figure 3 shows the numerical stress results of σ_{22} on the central axis $x_2 = 0$ of the beam and the relative errors. The EFGM with relative errors below 0.38% showed the best numerical accuracy over all the four meshless schemes for this example. For the RPIM, the accuracy near the two fixed ends with a maximum relative error of 3.73%, which is higher than the relative error of 1.32% at the middle span of the beam, is relatively poor. The maximum relative errors of the IEGM and the IIEFGM are 3.51% and 4.27%, respectively. Figure 4 shows the cloud diagram of the Von Mises stress on the problem domain for the analytical solution and the four meshless numerical results. The numerical solutions of the Von Mises stress obtained by the four meshless methods showed good agreement to the exact ones.

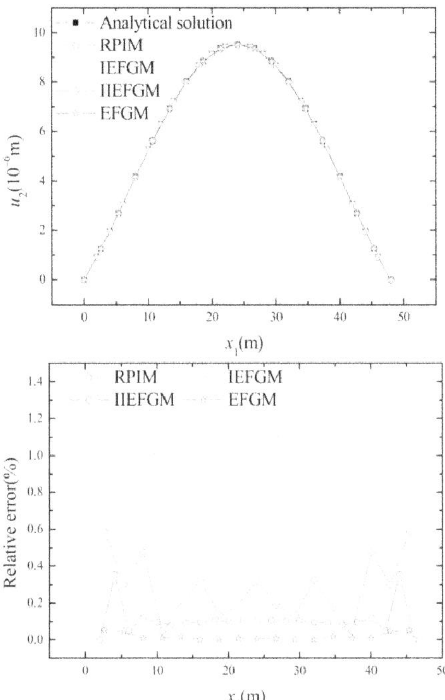

Figure 2. Deflection u_2 and its relative error at $x_2 = 0\ m$.

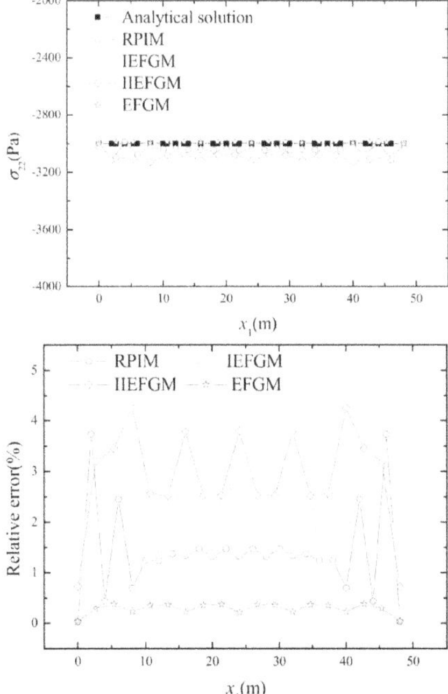

Figure 3. Stress σ_{22} and its relative error at $x_2 = 0\ m$.

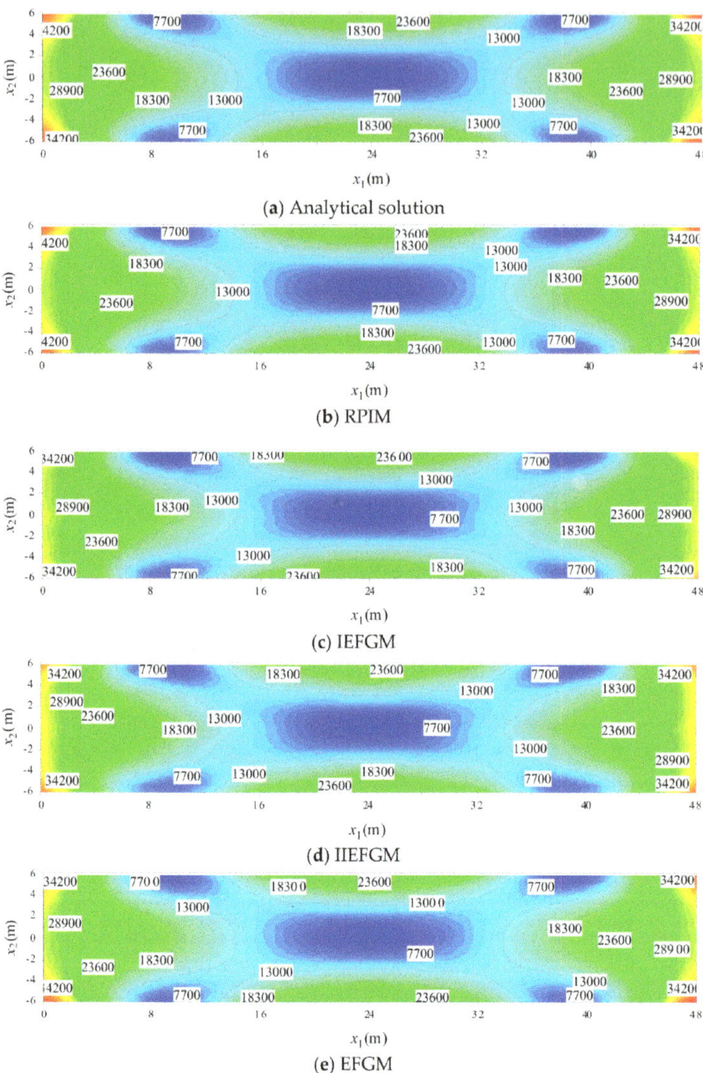

Figure 4. Contour of the Von Mises stress of the clamped-clamped beam.

4.2. Cantilever Beam Subjected to a Uniform Load

The orthotropic elastic cantilever beam shown in Figure 5 is considered in this example. Beam span l is 48 m, and depth h is 12 m. The upper boundary of the beam is subjected to a uniform load of $q = 1000$ N/m. The self-weight of the beam is ignored, and the structure is considered as in a plane stress state in the modelling.

The material coefficients of the beam are considered as $s_{11} = 0.0799 \times 10^{-10}$, $s_{12} = -0.0375 \times 10^{-10}$, $s_{21} = -0.0375 \times 10^{-10}$, $s_{22} = 0.0798 \times 10^{-10}$, and $s_{66} = 0.2326 \times 10^{-10}$. The analytical solutions of the displacements and the stresses of the beam are [35]

$$u_1 = \frac{2px_1x_2^3(s_{12}+s_{66})}{H^3} - \left[\frac{2s_{11}x_1^3}{H^3} - \frac{(9s_{12}-3s_{66})x_1}{10H} + \frac{(9s_{12}+12s_{66})L}{10H} - \frac{2s_{11}L^3}{h^3} \right] px_2 + \frac{s_{12}p(L-x_1)}{2} \tag{39}$$

$$u_2 = \frac{(s_{12}s_{66}-s_{11}s_{22}+2s_{12}^2)px_2^4}{2s_{11}H^3} - \left(\frac{s_{12}x_1^2}{H^2} + \frac{s_{12}s_{66}}{20s_{11}} + \frac{s_{12}^2}{10s_{11}} - \frac{s_{22}}{4}\right)\frac{3px_2^2}{H} - \frac{s_{22}px_2}{2}$$
$$+ \frac{p(L-x_1)^2}{20H^3}\left(10s_{11}x_1^2 + 20s_{11}Lx_1 - 12s_{66}H^2 - 9s_{12}H^2 + 30s_{11}L^2\right) \quad (40)$$

$$\sigma_1 = -\frac{6qx_1^2x_2}{H^3} + 2p\frac{2s_{12}+s_{66}}{4s_{11}}\left(\frac{4x_2^3}{H^3} - \frac{3x_2}{5H}\right) \quad (41)$$

$$\sigma_2 = -\frac{p(x_2+H)(H-2x_2)^2}{2H^3} \quad (42)$$

$$\tau_{12} = \frac{3px_1(4x_2^2 - H^2)}{2H^3} \quad (43)$$

As the geometry and the material in this example are similar to those in the example of Section 4.1, the trailed value range of the computing parameters in the meshless numerical modelling is chosen to be exactly the same as that in Section 4.1. Therefore, the optimal computing parameters for each meshless method are picked as the ones that get the lowest energy norm error among the tested value range and are listed in Table 3. The numerical results according to the optimal computing parameters in Table 3 are used for the comparison of the computational performance of the four meshless methods.

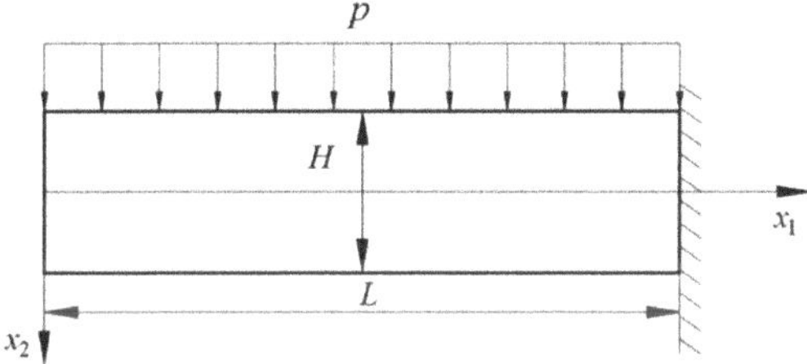

Figure 5. Cantilever beam under uniform load at its upper boundary.

Table 3. Optimal value of computing parameters in this example.

	$n_1 \times n_2$	$l_1 \times l_2$	d_{max}	Other Parameters
EFGM	19×13	12×8	3.5	$\beta = 3 \times 10^{14}$
RPIM	21×7	12×8	4.5	$q = 2.5, c = 1.0$
IEFGM	25×13	12×8	1.5	$\alpha = 6.0$
IIEFGM	25×13	12×8	1.5	—

Since there is no special difference in the accuracy of curves in each direction, this paper only provides curves in a certain direction for illustration. The meshless solutions for the deflections of the nodes on the central axis of the beam $x_2 = 0$ and the corresponding relative errors are presented in Figure 6. Generally, the numerical displacement results obtained by the four meshless methods show good accuracy and agree well with the exact solutions. The overall relative error values of RPIM, IEFGM and IIEFGM are about 0.07%, 1.1% and 1.6%, respectively, and the maximum relative error values are about 1.88%, 8.9% and 13.5%, respectively. The maximum relative error of EFGM is less than 0.27%. The relative errors of the three interpolative meshless methods, especially the IIEFGM and the IEFGM, show dramatic increase for the nodes at the fixed end of the beam.

In Figure 7, the numerical results of the Von Mises stress for the nodes on the central axis of the beam are compared to the values calculated from the analytical stress solutions. The corresponding relative errors of the Von Mises stress results are also presented in this figure. Unlike the case of the displacement solutions, the Von Mises stress solutions of the IEFGM and the IIEFGM show obvious deviation from the corresponding analytical solution at the fixed end of the beam. Generally, the relative errors of Von Mises stress calculated by the four meshless methods are relative larger at the free end of the beam and gradually decrease and become stabilized towards the fixed end. Meanwhile, the IEFGM and the IIEFGM show much poorer accuracy with very large relative error near the fixed end of the beam. The EFGM has the highest accuracy, with a nodal relative error around 0.72% with a maximum of 6.4%. The nodal relative errors of the Von Mises stress solutions for the RPIM, the IEFGM and the IIEFGM are mostly around 0.4%, 3% and 1%, respectively. When x_2 is between 10m and 40m, the variances of the RPIM, EFGM, IEFGM and IIEFGM are 0.0425, 0.1129, 0.797 and 0.546 respectively. Therefore, the Von Mises stress solutions of the RPIM are more accurate and stable than those of the other two interpolative approaches in this case.

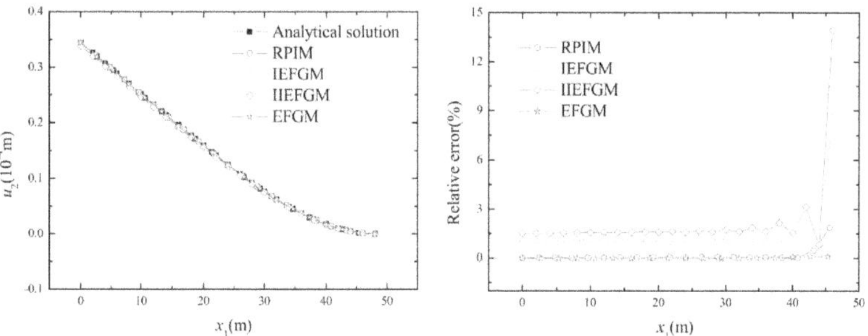

Figure 6. The nodes deflection and its relative error at central axis.

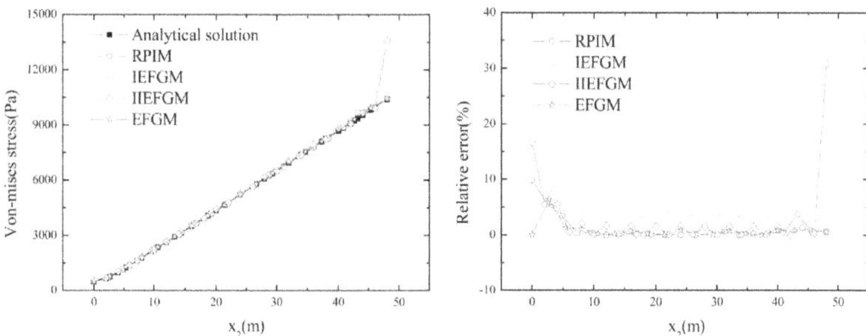

Figure 7. Solutions of the Von Mises stress and the relative error at $x_2 = 0$.

4.3. Ring under Pressures Applied Both Internally and Externally

To consider an orthotropic elastic ring under both internal and external compression, as shown in Figure 8, only one quarter of the structure needs to be modeled by symmetry. The inner and outer diameters of the ring are $a = 12$ m and $b = 20$ m, respectively. The pressures of $P_a = 100$ Pa and $P_b = 300$ Pa are applied both internally and externally. The numerical example is considered as in a plane stress state.

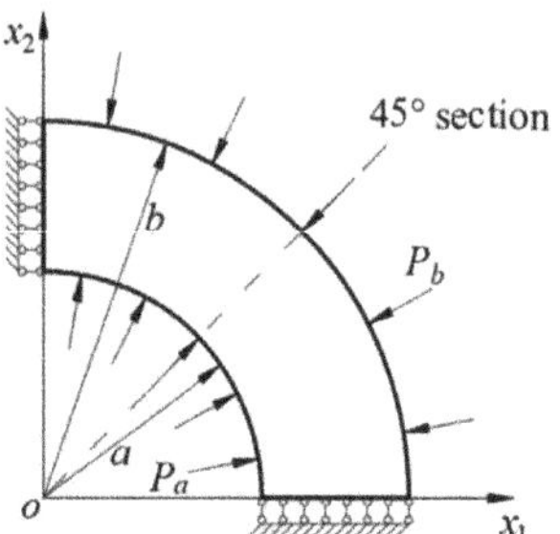

Figure 8. A quarter of the ring.

The ring is circular orthotropic, and the anisotropic pole is located at the center of the ring. The material compliance coefficients in a polar coordinate system are $s_{11} = 0.0799 \times 10^{-4}$, $s_{12} = -0.0375 \times 10^{-4}$, $s_{21} = -0.0375 \times 10^{-4}$, $s_{22} = 0.0798 \times 10^{-4}$ and $s_{66} = 0.2326 \times 10^{-4}$. The exact solutions of the radial displacement and the stresses are written as follows [35]:

$$u_r = \frac{b}{E_\theta(1 - c^{2k})}\left[(P_a c^{k+1} - P_b)(k - v_\theta)\left(\frac{r}{b}\right)^k + (P_a - P_b c^{k-1})c^{k+1}(k + v_\theta)\left(\frac{b}{r}\right)^k\right] \tag{44}$$

$$\sigma_r = \frac{P_a c^{k+1} - P_b}{1 - c^{2k}}\left(\frac{r}{b}\right)^{k-1} - \frac{P_a - P_b c^{k-1}}{1 - c^{2k}}c^{k+1}\left(\frac{b}{r}\right)^{k+1} \tag{45}$$

$$\sigma_\theta = \frac{P_a c^{k+1} - P_b}{1 - c^{2k}}k\left(\frac{r}{b}\right)^{k-1} + \frac{P_a - P_b c^{k-1}}{1 - c^{2k}}kc^{k+1}\left(\frac{b}{r}\right)^{k+1} \tag{46}$$

where $c = a/b, k = \sqrt{s_{11}/s_{22}}, E_\theta = 1/s_{22}, v_\theta = -s_{12}/s_{22}$.

In order to effectively compare the accuracy of the four meshless methods with completely different computing parameter groups, the following appropriate ranges are also selected for several kinds of the computational parameters in this example: for a uniform mesh of the nodes, the number of nodes in the radial direction $n_1 = (7, 9, 11, 13)$, the number of nodes in the circular direction $n_2 = (19, 21, 23, 25)$; for a fixed uniform mesh of $l_1 \times l_2 = 6 \times 24$ for the background cells of the Gauss quadrature, the dimensionless factors for scaling the influence domain $d_{max} = (1.5, 2.5, 3.5)$, $q = (-0.5, 0.5, 1.5, 2.5)$ and $c = (1.0, 3.0, 5.0)$; for the MQ radial basis in the RPIM, the singular weight parameter $\alpha = (4, 6, 8, 10)$ for the IEFGM, and the penalty factor $\beta = 3 \times 10^{14}$ for the EFGM to enforce displacement boundary conditions. Apparently, all 48, 576, 192 and 48 groups of the computing parameters need to be tested for the EFGM, the RPIM, the IEFGM and the IIEFGM, respectively, to determine the optimal settings for each method with regard to the relatively lower energy norm error. The corresponding optimal sections of the parameters for the four meshless schemes are presented in Table 4.

Table 4. Optimal parameters each method within the tested range.

	$n_1 \times n_2$	$l_1 \times l_2$	d_{max}	Other Parameters
EFGM	9×25	6×24	2.5	$\beta = 2 \times 10^8$
RPIM	11×25	6×24	3.5	$q = 1.5, c = 1.0$
IEFGM	13×25	6×24	1.5	$\alpha = 4.0$
IIEFGM	7×25	6×24	1.5	—

The nodal solutions of the radial displacement at section $\theta = 45°$ of the ring with the values of parameters in Table 4, the variances of the nodal relative errors in Table 5 and the relative errors are shown in Figure 9. It is obvious that the displacement results of the four meshless approaches are in generally good agreement with the analytical ones. The accuracy of the RPIM and the EFGM are almost the same and are higher and more stable than the other two methods. The maximum values of the nodal relative errors for the RPIM, IEFGM and the IIEFGM are about 0.0398%, 0.09% and 0.24%, respectively. The variances of the nodal relative errors for the RPIM, EFGM, IEFGM and the IIEFGM are about 0.0001, 0.000005, 0.00085 and 0.0043, respectively. The contour plots for the relative errors of the Von Mises stress obtained by the four meshless methods are presented in Figure 10. It could be found that the four meshless approaches can get good results of the Von Mises stress for this example, while the accuracy of each method in the different areas of the problem domain is not the same.

Table 5. The variances of the relative errors.

	EFGM	RPIM	IEFGM	IIEFGM
Radial displacement	0.0001	0.000005	0.00085	0.0043
Von Mises stress	0.07225	0.0545	0.391	0.9947

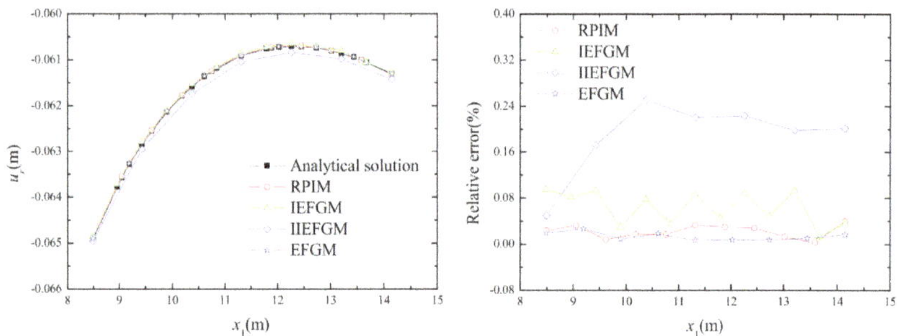

Figure 9. Radial displacement u_r and relative error at $\theta = 45°$ of the ring.

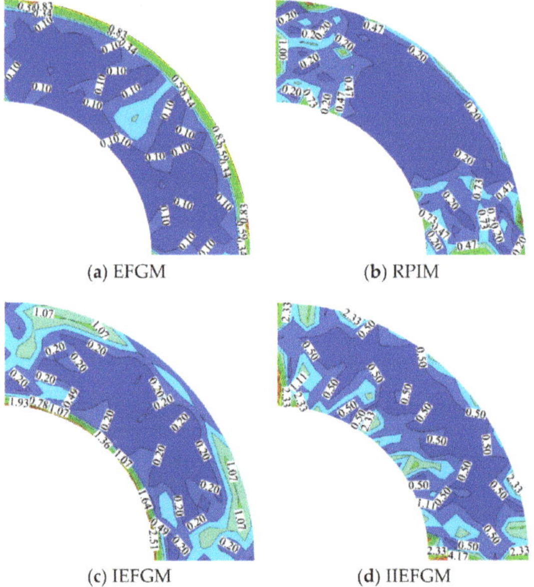

(a) EFGM

(b) RPIM

(c) IEFGM

(d) IIEFGM

Figure 10. Contour plots of the relative error for the Von Mises stress.

5. Conclusions

In this study, the RPIM, the IEFGM and the IIEFGM are applied to solve orthotropic elastic problems. The essential/displacement boundary conditions are applied by the direct method to calculate and analyze orthotropic clamped-clamped beams, orthotropic cantilever beams and orthotropic rings subjected to uniform loads. Since the principles and computing parameters of these three interpolative methods are very different, this paper uses the numerical solution with the optimal parameter group corresponding to the minimum energy norm error within a certain trailed range of each method to compare their accuracy. For example in Section 4.1, the maximum relative error of σ_{22} is 3.73%, which is higher than the relative error of 1.32% in the middle of the beam span. However, the maximum relative errors of the IEGM and the IIEFGM are 3.51% and 4.27%, respectively. For example in Section 4.2, the overall relative error values of nodal deformation for RPIM, IEFGM and IIEFGM are about 0.07%, 1.1% and 1.6%, respectively, and the maximum relative error values are about 1.88%, 8.9% and 13.5%, respectively. For example in Section 4.3, the accuracy of radial displacement for the RPIM and the EFGM are almost the same and are higher and more stable than the other two methods. The maximum values of the nodal relative errors for the RPIM, IEFGM and the IIEFGM are about 0.0398%, 0.09% and 0.24%, respectively. The results show that the three meshless interpolation methods have better numerical accuracy in the modeling of orthotropic elastic problems, and the radial point interpolation method (RPIM) has the highest accuracy. The research results of this paper can provide a certain reference value for future research on the selection of meshless form functions in interpolation. It is one of the regrets of this paper that the differences in numerical efficiency of these methods cannot be effectively investigated at the same time.

Author Contributions: Conceptualization, methodology, supervision, and funding acquisition, D.M.L.; investigation, validation, visualization and writing—original draft preparation, Y.-Y.Z. and Y.-C.T.; writing—review and editing, D.M.L., X.-B.L. and B.L.; Y.-Y.Z. and Y.-C.T. contributed equally to this work. All authors have read and agreed to the published version of the manuscript.

Funding: This work was supported by the National Natural Science Foundation of China (12002247), the Natural Science Foundation of Hubei Province (2018CFB129), Fundamental Research Funds for the Central Universities of China (WUT: 2016IVA022 and WUT: 20271IVB013) and the National innovation and entrepreneurship training program for college students of China (202210497088).

Data Availability Statement: Data will be made available on request.

Conflicts of Interest: The authors declare no conflict of interest.

References

1.   Belytschko, T.; Lu, Y.Y.; Gu, L. Element-free Galerkin methods. *Int. J. Numer. Methods Eng.* **2010**, *37*, 229–256. [CrossRef]
2.   Liu, W.K.; Jun, S.; Zhang, Y.F. Reproducing kernel particle methods. *Int. J. Numer. Methods Fluids* **1995**, *20*, 1081–1106. [CrossRef]
3.   Li, D.M.; Featherston, C.A.; Wu, Z.M. An element-free study of variable stiffness composite plates with cutouts for enhanced buckling and post-buckling performance. *Comput. Methods Appl. Mech. Eng.* **2020**, *371*, 113314. [CrossRef]
4.   Li, D.M.; Kong, L.H.; Liu, J.H. A generalized decoupling numerical framework for polymeric gels and its element-free implementation. *Int. J. Numer. Methods Eng.* **2020**, *121*, 2701–2726. [CrossRef]
5.   Chen, J.S.; Pan, C.H.; Wu, C.T.; Liu, W.K. Reproducing Kernel Particle Methods for large deformation analysis of non-linear structures. *Comput. Methods Appl. Mech. Eng.* **1996**, *139*, 195–227. [CrossRef]
6.   Li, D.M.; Tian, L.R. Large deformation analysis of gel using the complex variable element-free Galerkin method. *Appl. Math. Model.* **2018**, *61*, 484–497. [CrossRef]
7.   Li, D.M.; Liu, J.H.; Nie, F.H.; Featherston, C.A.; Wu, Z.M. On tracking arbitrary crack path with complex variable meshless methods. *Comput. Methods Appl. Mech. Eng.* **2022**, *399*, 115402. [CrossRef]
8.   Pan, J.H.; Li, D.M.; Luo, X.B.; Zhu, W. An enriched improved complex variable element-free Galerkin method for efficient fracture analysis of orthotropic materials. *Theor. Appl. Fract. Mech.* **2022**, *121*, 103488. [CrossRef]
9.   Liew, K.M.; Sun, Y.Z.; Kitipornchai, S. Boundary element-free method for fracture analysis of 2-D anisotropic piezoelectric solids. *Int. J. Numer. Methods Eng.* **2007**, *69*, 729–749. [CrossRef]
10.   Rabczuk, T.; Belytschko, T. Cracking particles: A simplified meshfree method for arbitrary evolving cracks. *Int. J. Numer. Methods Eng.* **2004**, *61*, 2316–2343. [CrossRef]
11.   Liu, G.R.; Gu, Y.T. A point interpolation method for two-dimensional solids. *Int. J. Numer. Methods Eng.* **2001**, *50*, 937–951. [CrossRef]
12.   Wang, J.G.; Liu, G.R. A point interpolation meshless method based on radial basis function. *Int. J. Numer. Methods Eng.* **2002**, *54*, 1623–1648. [CrossRef]
13.   Ren, H.P.; Cheng, Y.M. The interpolating element-free Galerkin method for two-dimensional elasticity problems. *Int. J. Appl. Mech.* **2011**, *3*, 735–758. [CrossRef]
14.   Lancaster, P.; Salkauskas, K. Surface generated by moving least squares methods. *Math. Comput.* **1981**, *37*, 141–158. [CrossRef]
15.   Wang, J.F.; Sun, F.X.; Cheng, Y.M. An improved interpolating element-free Galerkin method with a nonsingular weight function for two-dimensional potential problems. *Chin. Phys. B* **2012**, *21*, 090204. [CrossRef]
16.   Deng, Y.J.; He, X.Q. An improved interpolating complex variable meshless method for bending problem of Kirchhoff plates. *Int. J. Appl. Mech.* **2017**, *9*, 1750089. [CrossRef]

17. Zhang, T.; Li, X.L. A variational multiscale interpolating element-free Galerkin method for convection-diffusion and Stokes problems. *Eng. Anal. Bound. Elem.* **2017**, *82*, 185–193. [CrossRef]
18. Wang, Q.; Zhou, W.; Feng, Y.T.; Ma, G.; Cheng, Y.G.; Chang, X.L. An adaptive orthogonal improved interpolating moving least-square method and a new boundary element-free method. *Appl. Math. Comput.* **2019**, *353*, 347–370. [CrossRef]
19. Singh, R.; Singh, K.M. Interpolating meshless local Petrov-Galerkin method for steady state heat conduction problem. *Eng. Anal. Bound. Elem.* **2019**, *101*, 56–66. [CrossRef]
20. Bourantas, G.; Zwick, B.F.; Joldes, G.R.; Wittek, A.; Miller, K. Simple and robust element-free Galerkin method with almost interpolating shape functions for finite deformation elasticity. *Appl. Math. Model.* **2021**, *96*, 284–303. [CrossRef]
21. Wang, L.; Qian, Z. A meshfree stabilized collocation method (SCM) based on reproducing kernel approximation. *Comput. Methods Appl. Mech. Eng.* **2020**, *371*, 113303. [CrossRef]
22. Wang, L.; Liu, Y.; Zou, Y.; Yang, F. A gradient reproducing kernel based stabilized collocation method for the static and dynamic problems of thin elastic beams and plates. *Comput. Mech.* **2021**, *68*, 709–739. [CrossRef]
23. Qian, Z.; Wang, L.; Gu, Y.; Zhang, C. An efficient meshfree gradient smoothing collocation method (GSCM) using reproducing kernel approximation. *Comput. Methods Appl. Mech. Eng.* **2021**, *374*, 113573. [CrossRef]
24. Quintero MA, M.; Tam CP, T.; Li, H.T. Structural analysis of a Guadua bamboo bridge in Colombia. *Sustain. Struct.* **2022**, *2*, 000020.
25. Ponzo, F.C.; Antonio, D.C.; Nicla, L.; Nigro, D. Experimental estimation of energy dissipated by multistorey post-tensioned timber framed buildings with anti-seismic dissipative devices. *Sustain. Struct.* **2021**, *1*, 000007.
26. Liew, K.M.; Lei, Z.X.; Yu, J.L.; Zhang, L.W. Postbuckling of carbon nanotube-reinforced functionally graded cylindrical panels under axial compression using a meshless approach. *Comput. Methods Appl. Mech. Eng.* **2014**, *268*, 1–17. [CrossRef]
27. Wu, Z.M.; Raju, G.; Weaver, P.M. Optimization of postbuckling behaviour of variable thickness composite panels with Variable Angle Tows: Towards 'buckle-free' design concept. *Int. J. Solids Struct.* **2018**, *132–133*, 66–79. [CrossRef]
28. Dinis LM, J.S.; Jorge RM, N.; Belinha, J. A 3D shell-like approach using a natural neighbour meshless method: Isotropic and orthotropic thin structures. *Compos. Struct.* **2010**, *92*, 1132–1142. [CrossRef]
29. Njiwa, R.K. Isotropic-BEM coupled with a local point interpolation method for the solution of 3D-anisotropic elasticity problems. *Eng. Anal. Bound. Elem.* **2011**, *35*, 611–615. [CrossRef]
30. Bui, T.Q.; Nguyen, M.N. A novel meshfree model for buckling and vibration analysis of rectangular orthotropic plates. *Struct. Eng. Mech.* **2011**, *39*, 579–598. [CrossRef]
31. Fallah, N.; Nikrafter, N. Meshless finite volume method for the analysis of fracture problems in orthotropic media. *Eng. Fract. Mech.* **2018**, *204*, 46–62. [CrossRef]
32. Lohit, S.K.; Gaonkar, A.K.; Gotkhindi, T.P. Interpolating Modified Moving Least Squares based element free Galerkin method for fracture mechanics problems. *Theor. Appl. Fract. Mech.* **2022**, *122*, 103569. [CrossRef]
33. Luo, X.B.; Li, D.M.; Liu, C.L.; Pan, J.H. Buckling analysis of variable stiffness composite plates with elliptical cutouts using an efficient RPIM based on naturally stabilized nodal integration scheme. *Compos. Struct.* **2022**, *302*, 116243. [CrossRef]
34. Hardy, R.L. Theory and applications of the multiquadrics-Biharmonic method (20 years of discovery 1968–1988). *Comput. Math. Appl.* **1990**, *19*, 163–208. [CrossRef]
35. Lekhnitskiĭ, S.G. *Anisotropic Plates*; Gordon and Breach: Philadelphia, PA, USA, 1968.

 *buildings*

Article

Two-Step Identification Method and Experimental Verification of Weld Damage at Joints in Spatial Grid Structures

Hui Liu [1,2], Jianwei Huang [1], Xueliang Wang [1,2,*] and Xiuwen Lv [1]

[1] School of Civil Engineering and Architecture, Wuhan University of Technology, Wuhan 430070, China; drliuh@263.net (H.L.); 334142@whut.edu.cn (J.H.); lv553326@whut.edu.cn (X.L.)

[2] Hainan Institute of Wuhan University of Technology, Sanya 572025, China

* Correspondence: wxllhb@163.com

Citation: Liu, H.; Huang, J.; Wang, X.; Lv, X. Two-Step Identification Method and Experimental Verification of Weld Damage at Joints in Spatial Grid Structures. *Buildings* 2023, 13, 2141. https://doi.org/10.3390/buildings13092141

Academic Editor: Francisco López-Almansa

Received: 16 July 2023
Revised: 12 August 2023
Accepted: 18 August 2023
Published: 23 August 2023

Abstract: Welded joints in grid structures are susceptible to damage and destruction when exposed to random excitation. The complexity of the grid structure poses challenges for realizing the damage recognition of welded joints. In this study, a two-step method is proposed specifically for damage identification of welded joints in grid structures, combining wavelet analysis and fuzzy pattern recognition to accurately identify the location and extent of damage in welded joints. Firstly, the structure is divided based on the analysis of the influence range of joint damage. Key joints are selected within the sub-regions where sensors are installed, and the acceleration response of these key joints is measured. Wavelet analysis is then applied to identify the sub-regions where weld damage occurs. Secondly, an equivalent finite element model is established for joints with varying degrees of damage. The damage index, calculated as the ratio of the absolute value of the difference in the first-order element strain mode of the members, increases with the degree of damage during joint weld damage. By monitoring the changes in the damage index of sensitive members, which exhibit significant changes with varying weld damage degrees, a damage pattern database is constructed for each sub-region. The membership degree between joint damage and the patterns in the pattern database is then calculated to determine the location and degree of weld damage. To validate the effectiveness of the proposed method, an experiment was conducted using a grid structure model with replaceable members. Highly sensitive FBG sensors were designed to measure the acceleration response of the joints, resulting in accurate identification of damaged sub-regions solely through the measurement of key joint acceleration responses. Furthermore, within the damaged sub-regions, the fuzzy pattern recognition method precisely determined the location and degree of weld damage in the joints. The experimental results demonstrate that the proposed method effectively reduces the complexity of the structure by dividing the grid structure into sub-regions, and enables the two-step identification method to achieve successful damage identification for the joints in the grid structure with high efficiency and accuracy.

Keywords: grid structure; joint weld damage; wavelet transform; fuzzy pattern recognition; strain mode difference ratio; experiment verification; FBG sensors

1. Introduction

Welded spatial grid structures are extensively utilized in civil engineering, particularly in densely populated public areas like exhibition halls, stadiums, and theaters, which often serve as prominent architectural landmarks in cities. The occurrence of engineering accidents in these structures can result in significant economic losses and have severe social implications. Among the various components of welded grid structures, welded hollow spherical joints are particularly prone to damage due to the complex multi-directional loading conditions they experience. Moreover, Adin analyzed the mechanical properties of welded joints produced by different welding methods and pointed out that the welding methods have great influence on the mechanical properties of welded joints [1]. Wu et al., indicated that due to the welding process and environmental influences, the welding zone tended to lead to stress concentration, making the welded joints vulnerable locations for structural damage [2]. Therefore, it is essential to accurately identify any damage that may occur in these joints to ensure the safety of such structures in service.

The dynamic responses of engineering structures often contain valuable information that can provide insights into their working performance and condition. Analyzing the dynamic response enables us to understand dynamic characteristics and assess structural integrity to a certain extent. Several studies have focused on structural damage identification based on modal parameters extracted from dynamic response. Wei et al., conducted a comprehensive review of methods for damage identification in beams or plates, considering modal parameters such as natural frequencies, modal shapes, curved modal shapes, and a combination of modal shapes and frequencies [3]. Similarly, Zhu et al., developed an effective damage detection method for shear wall structures by analyzing variations in first-mode amplitudes [4]. Yin et al., proposed a practical method for detecting damage in bolted joints using noisy incomplete modal parameters by only limited data acquisition, and demonstrated the effectiveness of the combination of numerical simulations and experimental validations [5]. Ditommaso et al., proposed a methodology for damage localization in frame structures subjected to strong ground motion through monitoring modal curvature variations [6]. Zhang et al., introduced a displacement modal shape processing method based on difference accumulation, which was used as a damage characterization parameter to detect damage in composite materials [7]. Zhou et al., investigated modal flexibility extraction and damage identification using multi-reference hammering in reinforced concrete (RC) beams, and performed static and dynamic experiments on simply supported RC beams to validate this method [8]. Chang et al., demonstrated the effectiveness of modal parameter identification and vibration-based damage detection through field experiments on a simply supported steel truss bridge [9]. Fang et al., employed a substructure-based damage identification method utilizing the acceleration frequency response function (FRF) to identify damage in a six-story steel frame structure [10]. These studies highlight the significance and potential of using dynamic response analysis for structural damage identification. However, as the complexity of the structure increases, the sensitivity of damage identification methods that rely on changes in structural dynamic characteristics as damage indicators becomes less robust. This makes it difficult to apply these damage identification methods to practical structures.

The modal strain energy (MSE), derived from the structural modal shape and stiffness matrix, has been recognized as a sensitive physical property that undergoes changes before and after structural damage. As a result, several damage identification methods utilize modal strain energy as a damage indicator in structural health monitoring. Cha et al., proposed a novel damage detection method that employed a hybrid multi-objective optimization algorithm based on MSE to detect damages in various three-dimensional steel structures [11]. Li et al., developed an improved modal strain energy (IMSE) method for detecting damage in offshore platform structures, utilizing modal frequencies as the basis for the approach. Numerical and experimental studies both demonstrated the effectiveness and practicality of the IMSE method [12]. Arefi et al., employed MSE and modal shapes reconstructed by the Guyan reduction method (GRM) as damage indices for identifying structural damage [13]. Huang et al., addressed computational efficiency and the lack of high-sensitivity damage indices in structural damage identification by proposing a framework based on the modal frequency strain energy assurance criterion (MFSEAC), modal flexibility, and an enhanced moth–flame optimization algorithm [14]. These studies highlight the significance of modal strain energy as a valuable parameter for structural damage identification based on dynamic characteristics. However, when the scale of the structure increases, the high dimensionality of the stiffness matrix and the truncation effects of modal shapes lead to a decrease in the ability of modal strain energy to characterize structural damage, potentially resulting in the inability to detect structural damage.

Spatial grid structures, with their complex shapes and large scale, exhibit complex stiffness characteristics, resulting in a dense distribution of natural frequencies. In the case of damage to multiple members within the structure, even complete loss of their bearing capacity, the resulting changes in modal parameters, such as natural frequencies, tend to be minimal. Consequently, damage indicator methods relying solely on the magnitude of modal parameter changes are often inadequate for diagnosing damage in these structures. To overcome this limitation, damage identification methods based on direct structural response information have been proposed. Sohn et al., utilized time series methods to identify structural damage by analyzing the dynamic response of the structure [15]. Lam et al., developed a Bayesian method to assess the damage status of railway ballast under a concrete sleeper using vibration data from in situ sleepers [16]. Salehi et al., introduced a novel structural damage detection technique based on multi-channel empirical mode decomposition (MEMD) of vibrational response data [17]. Zhu et al., used a combination of statistical regression and deep learning methods to predict the deformation of a dam based on multiple measuring points in different sections of the dam [18]. Xiao et al., proposed a methodology for identifying damage in semi-rigid frames with slender beams, which was applied on semi-rigid frame structures with different cross-sectional shapes by formulating an objective function based on minimizing the difference between the analyzed and measured joint displacements [19]. However, the practical application of these methods to grid structures is often hindered by the large-scale nature of the structures and the significant number of connections and members involved. Consequently, the installation of a large number of sensors on the structure to obtain the required response information for identifying damaged spherical weld joints becomes challenging and restricts the feasibility of such approaches in grid structures.

Xiao et al., proposed a stiffness separation method for damage identification in large-scale space truss structures that simplified the high-dimensional structural damage identification problem [20]. Therefore, dividing the grid structure into several simpler substructures provides a practical method for identifying joint damage within each substructure, which is analogous to dimensionality reduction processing. This concept forms the basis for the development of the two-step method presented in this paper, specially designed to identify weld damage in the welded joints of grid structures. The first step involves partitioning the grid structure into sub-regions and utilizing data acquired from a limited number of measurement key joints to identify the sub-regions where joint weld damage has occurred. In the second step, the focus is on recognizing the precise location and extent of joint weld damage within the identified sub-regions. The key to implementing the two-step approach for identifying weld damage at the joints of grid structures lies in the selection of appropriate identification methods for each step. These identification methods should meet criteria such as requiring minimal information, while also ensuring high accuracy and ease of implementation.

Wavelet analysis has gained attention in the field of structural damage detection due to its ability to capture local characteristics of signals in both time and frequency domains, leading to significant achievements. Kim et al., utilized continuous and discrete wavelet transforms in structural health monitoring (SHM) to investigate damage identification in beam structures [21]. Zhe F. et al., introduced a novel transmissibility concept based on wavelet transform to detect slight structural damage at its early stage [22]. Janeliukstis et al., employed a two-dimensional wavelet transform algorithm with isotropic Pet Hat wavelet to locate areas of damage in a numerically simulated aluminum plate model, finding that the lowest scale of the selected wavelet function yielded the best results for damage identification [23]. Zhu et al., proposed a new damage index for crack identification in functionally graded material (FGM) beams using wavelet analysis, defining the index based on the position of the maximum value of the wavelet coefficient modulus in the scale space [24]. Katunin et al., presented a novel damage identification approach using the 2D continuous wavelet transform-based algorithm to analyze differences in modal rotation fields obtained from shearographic measurements of structures [25]. Yazdanpanah et al., proposed an efficient wavelet-based refined damage-sensitive feature for nonlinear damage diagnosis in steel moment resisting frames (MRFs), which utilized acceleration responses extracted from the structures analyzed by incremental dynamic analysis (IDA) under various ground motion records [26]. Zhu et al., used the variational mode decomposition (VMD) wavelet packet denoising method to denoise the prototypical seepage pressure data of dams, and it was demonstrated through experiments that the method achieved high prediction accuracy and flexibility [27].

However, directly applying wavelet analysis to identify joint damage in grid structures may lead to incorrect results due to the complexity and symmetry of the structure. The presence of joint damage in the grid structure can cause significant response changes, especially at symmetrical locations. To overcome this challenge, partitioning the grid structure becomes essential as it effectively reduces complexity and breaks symmetry, rendering wavelet analysis advantageous for damage identification in sub-regions of the structure.

Furthermore, damage to a joint can result in amplitude changes in the singular values of the wavelet-transformed acceleration response of adjacent joints. Therefore, the wavelet analysis method can be employed not only for damage identification but also to determine the influence range of a damaged joint by analyzing the amplitude changes in the singular values of the wavelet transform of adjacent joints. This additional feature enhances the capabilities of the wavelet analysis method for accurately detecting and localizing joint damage within the grid structure.

Fuzzy pattern recognition, a method that involves extracting identification indicators and establishing membership functions, has been widely employed in structural damage identification. Wang et al., proposed a two-stage fuzzy pattern identification method based on fuzzy theory to establish a fuzzy pattern database for cable force, key point strain, and weld crack growth length. This method enabled the identification of earplate crack length in guyed mast structures under wind load [28]. Ren et al., extracted damage-sensitive features through statistical analysis of time history response data, forming statistical mode vectors that characterize the structural status. By comparing the distances between mode vectors in the feature set for different structural conditions, they were able to determine the damage [29]. Jiang et al., proposed a data fusion damage identification method based on fuzzy neural networks, which effectively utilized redundant and uncertain information for more accurate damage diagnosis [30].

However, extracting damage indices becomes challenging in grid structures with numerous joints and members. Moreover, pattern matching in a large pattern database based on established membership functions requires extensive analysis, which may result in low recognition accuracy or errors due to the complexity of the grid structure. Nevertheless, the implementation of the fuzzy pattern recognition method in sub-regions of the structure where joint damage has been identified presents a practical solution, which can significantly reduce the difficulty of identifying the location and extent of joint damage. By focusing pattern matching solely on the sub-region containing joint damage, the computational burden is substantially reduced, and potential identification errors caused by the complexity of the grid structure are effectively avoided.

Therefore, it is recommended to combine the wavelet analysis method in the first step and the fuzzy pattern recognition method in the second step to achieve joint weld damage identification in grid structures. In this way, the objective of the study was attained through the selective placement of acceleration sensors on key joints within the structural sub-region. By subjecting the response time histories collected from these sensors to wavelet analysis, the presence of joint damage within the sub-region can be identified. Subsequently, using this outcome, fuzzy pattern recognition can be applied specifically to the sub-region where the damage has been detected, allowing the determination of the location and severity of the joint damage. Furthermore, to validate the effectiveness of the proposed method, a physical grid structure model was fabricated and subjected to testing. The results of these experiments demonstrate that the proposed method can accurately identify the location and degree of damage in the structural joints. Importantly, this method requires minimal measurement information, making it practical and feasible for implementation in practical engineering applications.

2. The Methodology of Two-Step Identification Technology for Weld Damage of Welded Spherical Joints

The two-step damage recognition procedures for joints in grid structures are illustrated in Figure 1.

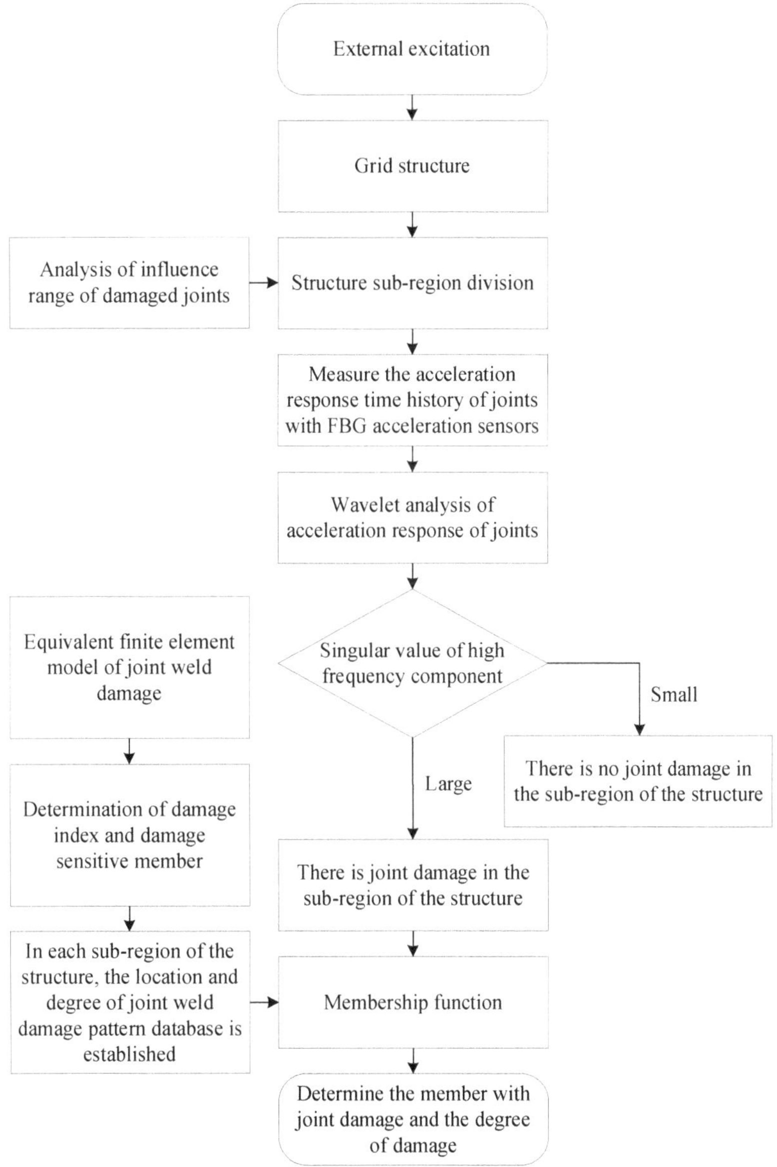

Figure 1. The detection framework of weld damage in joints.

According to Figure 1, the first step involves identifying the influence region of the damaged joints through wavelet transform analysis, arranging sensors in each sub-region, and identifying the damaged sub-region. The procedure is as followed: firstly, acquiring the time history of the acceleration response of both the damaged joint and neighboring joints when the grid structure is subjected to external excitation. Then, using the wavelet transform to analyze the acceleration response to obtain the singularity amplitude of the high-frequency signal, and determining the influence areas of the damaged joints, leading to the establishment of sub-regions within the structure. Additionally, the key joints in each sub-region were determined for deploying the fiber Bragg grating (FBG) acceleration sensors. In practical engineering, only the acceleration responses of the joints with FBG sensors in each sub-region are measured. The singular amplitude of the high-frequency signal of these acceleration responses, obtained through wavelet transform analysis, helps to judge whether weld damage in joints occurs in a sub-region.

The second step involves establishing a damage pattern database for welded joints in each sub-region, and identifying the location and degree of weld damage in the joints of sub-regions by adopting the fuzzy pattern recognition method. To ensure that the database covers various hollow spherical joints, three-dimensional solid weld connections between joints and members are modelled using the finite element method. These models consider different spherical wall thicknesses and pipe–sphere diameter ratios within the range of $0.3 \leq d/D \leq 0.5$, as specified by the JG/T11-2009 Standard for Welded Hollow Spherical Joints of Steel Grid Structures [31]. The axial stiffness coefficient and bending stiffness coefficient are obtained from these models and used in equivalent finite element models with varying crack sizes. Damage indices that characterize the location and extent of joint damage are extracted based on the damage equivalent model, and can be used to determine the members that are sensitive to joint damage. A vector composed of these damage indices from the sensitive members is used to characterize different patterns of joint damage. A pattern library of joint damage is then established, which contains information about the location and degree of joint damage for each sub-region.

Following the first step, if joint damage is identified to occur in a sub-region, the membership degree between the damage indices vector of each joint and each pattern in the pattern database is calculated only in this sub-region. Based on the membership values, the damaged joints can be located and the extent of damage can be assessed.

2.1. Damage Sub-Region Identification Based on Wavelet Analysis

The wavelet transform is a mathematical operation that involves taking the inner product between a basic wavelet function, denoted as $\psi(x)$, and the signal to be analyzed, denoted as $f(x)$, at different scales a and displacements b. It can be expressed as:

$$W_f(a,b) = \; < f(x), \psi_{a,b}(x) > \; = \frac{1}{\sqrt{a}} \int_{-\infty}^{+\infty} f(x)\psi\left(\frac{x-b}{a}\right)dx \tag{1}$$

where $\psi(x)$ represents the basic wavelet function, while a and b represent the scale factor and displacement factor, respectively.

The key characteristic of wavelet transform is its ability to provide variable time–frequency windows. The time–frequency window can be conveniently adjusted using an optimal base searching method. Wavelet analysis offers excellent time–frequency localization, allowing it to detect various frequency components of a signal by utilizing its adjustable time–frequency window. This property makes wavelet transform effective in detecting small-scale structural damage by analyzing signal singularities [32].

In this study, the acceleration response of joints was subjected to wavelet analysis. If a joint is damaged, not only that specific joint but also the neighboring joints exhibit singularities in the high-frequency signal. The magnitude of the singularity in the high-frequency signal of each joint determines the influence range of the joint's damage. By assessing the singularity values of each joint, the structure can be divided into sub-regions, and the key joints where sensors should be installed can be determined. Consequently, by measuring the response of these key joints, the identification of joint damage in the corresponding sub-region can be achieved.

2.2. Damage Location and Degree of Joints in the Damage Sub-Region Based on Fuzzy Pattern Recognition

2.2.1. Pattern Database

To establish the pattern database of weld damage in joints, the relationship between the degree of weld damage and the damage index was analyzed. The weld crack length was divided into intervals of 15 degrees, and different damage equivalent models were constructed and analyzed for each crack size. The first-order strain mode of each member was calculated, and the absolute difference in strain modes was obtained.

For each crack length, a set of absolute difference data of strain modes was generated. These data were used to determine the crack length of the joints. The absolute difference of strain modes was considered as the eigenvector to construct the pattern database. Each entry in the database corresponded to a specific crack length and its corresponding absolute difference data of strain modes.

To identify the damage in a joint with an unknown crack length, the absolute difference of strain modes for that joint was compared with the pattern database. By finding the closest match or determining the degree of similarity between the absolute difference of strain modes and the entries in the database, damage identification could be achieved.

The case study focused on a welded hollow sphere joint, specifically the WS1204 joint, which consisted of a hollow sphere with a diameter of 120 mm and a wall thickness of 4 mm, connected to a steel pipe with an external diameter of 48 mm and a wall thickness of 3.5 mm. To simplify the analysis, the calculation model considered only half of the sphere, as the stressed state of the welded hollow spherical joint under a unidirectional force was symmetrical. The fixed boundary conditions were applied along the hemisphere edge. The crack in the joint was assumed to be an opening penetrating crack located on the surface, which is a common type of crack in engineering. The crack section is shaded in Figure 2, where 2θ represents the crack angle. A three-dimensional solid model of the crack was established by dividing the structure into the crack body and non-crack body components, allowing accurate analysis of the crack behavior [33].

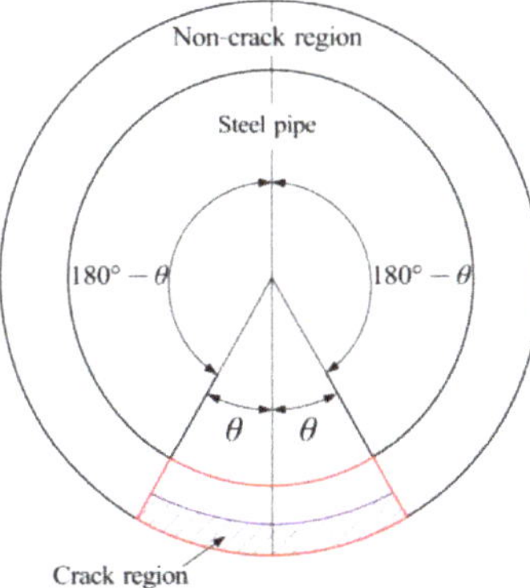

Figure 2. Location of cracks.

To accurately capture the singularity at the crack tip, the meshing process involved defining the singularity using the ANSYS pre-processing command KSCON. The parameters of the KSCON command were adjusted to ensure a uniform radiation pattern on the surface of the crack tip, as illustrated in Figure 3a. The mesh on the crack tip surface was then extended in the crack depth direction to form a solid mesh using a degenerated 20-node SOLID95 quadratic element, which is a singular element specifically designed to reflect the strain and stress singularity at the crack tip, as shown in Figure 3b.

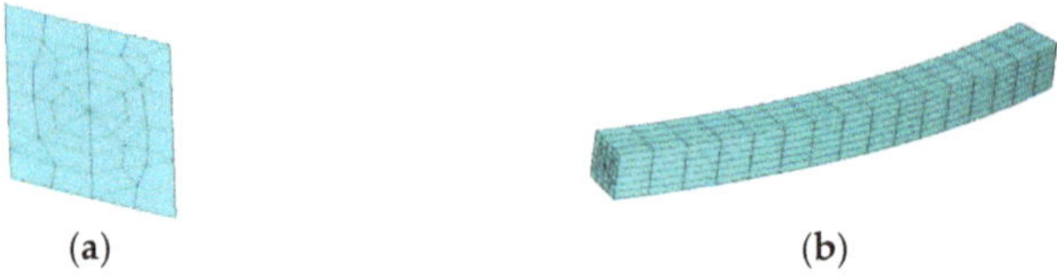

(a)   (b)

Figure 3. Mesh generation of cracked body, (**a**) element of the surface of crack front edge, (**b**) front body of crack.

For the transition region adjacent to the crack and the remaining portion of the structure, the element size was controlled using the LESIZE command, and the refinement of elements was controlled using the SMART command. This allowed for an appropriate mesh density in the vicinity of the crack and a coarser mesh in other regions. Finally, the complete mesh model of the joint was obtained, as shown in Figure 4.

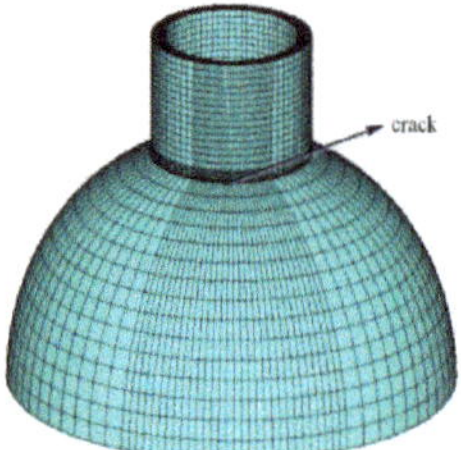

Figure 4. Model of welded hollow ball joint.

The crack shape of the weld joint was characterized by its depth and width, which had a notable impact on the stiffness of the welded hollow spherical joint. However, in this study, the influence of crack depth on the joint stiffness was neglected due to the observation that the axial and bending stiffness coefficients showed minimal variations when the crack depth changed within the small range of the pipe wall thickness [34]. Therefore, only the influence of crack width was considered.

To establish the relationship between crack width and axial and bending stiffness coefficients, an axial force F was applied to the welded hollow spherical joint model to obtain the average vertical displacement $\overline{\omega}$. The ratio of axial force to displacement yielded axial stiffness, expressed as $k = F/\overline{\omega}$. Additionally, a bending moment M was applied to the spherical joints, resulting in an angular displacement $\overline{\theta}$, and the bending stiffness $k_e = M/\overline{\theta}$, approximated as $k_e \approx M/(\overline{\omega_+} - \overline{\omega_-})/d$, where $\overline{\omega_+}$ was the mean displacement value of joints with positive vertical displacement, $\overline{\omega_-}$ was the mean displacement value of joints with negative vertical displacement, and d was the pipe diameter.

By assuming the crack depth to be half of the pipe wall thickness and varying the crack width, the axial stiffness coefficient and bending stiffness coefficient of the welded hollow spherical joint were calculated as shown in Figure 5.

To simulate the equivalent model of a joint with a member, the piecewise equivalent stiffness method is commonly used. In this method, the member is divided into three parts: the middle element and the fixed-length elements at both ends. The length of the elements at both ends is determined based on the size of the joint, and the rotation capacity of the joint is represented by reducing the moment of inertia of the beam element at both ends. This allows the continuous modification of the reduced coefficient of moment of inertia and the length of the member end to simulate different degrees of joint damage [35]. However, the length of the member end depends on the degree of joint damage, and it is not easy to determine the length in this way. Xiao et al., proposed a method for simultaneous damage identification of both section damage and joint damage in rigid frames, which furthermore can also be used to assess the rotational stiffness of semi-rigid connections, represented by fixed coefficients at the ends of components to characterize the rotational capacity between beams and columns [36]. In this study, the concept of adjustable stiffness is utilized to represent different connection cracks. The Matrix 27 element provided by ANSYS software was selected to simulate the stiffness of joints with cracks. The joint dimensions were ignored, and the effect of cracks on the stiffness of welded hollow spherical joints was considered. The connection stiffness of the joints was simulated by adjusting the spring stiffness, allowing consideration of changes in the axial, bending, or torsional stiffness. By adjusting the values of the corresponding elements in the Matrix 27 element's stiffness matrix, different crack-connected joints could be described.

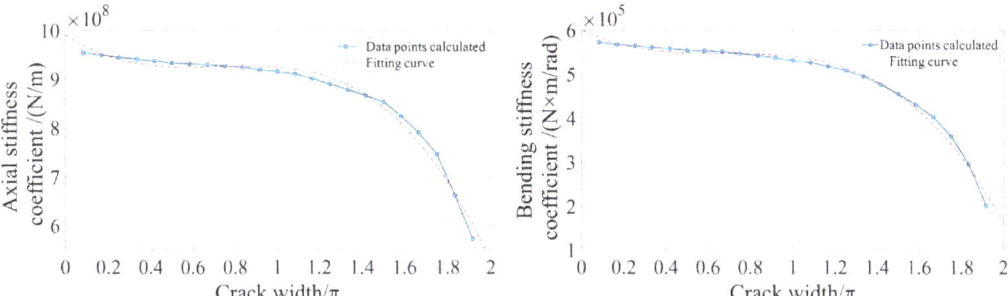

Figure 5. Fitting curves of stiffness coefficient.

When the structure is damaged, the stiffness in the damaged area decreases, leading to significant stress redistribution around the damage location. This, in turn, causes a substantial change in the strain mode at the damage location [37,38]. However, strain mode, as a relative quantity, can only be used for locating structural damage. Nevertheless, it is acknowledged that more severe damage results in a more significant change in strain modes. Therefore, in this paper, the ratio of the absolute value of the difference in the first-order element strain modes is proposed as the damage index to quantify and assess the extent of damage. By calculating this index, the damage location of the joint can be determined, and the severity of the damage can also be assessed.

The link element k in the structure is connected by nodes i, j. the strain ε_k which ignores the higher order terms is as follows:

$$\varepsilon_k = [(u_j - u_i)(x_j - x_i) + (v_j - v_i) \\ (y_j - y_i) + (w_j - w_i)(z_j - z_i)] / L^2 \tag{2}$$

where x_i, y_i, z_i and x_j, y_j, z_j are the coordinates of nodes j and j respectively, u_i, v_i, w_i and u_j, v_j, w_j are the displacements of nodes i and j respectively, and L is the length of the link element.

The absolute value of the element strain modal difference is sensitive to local damage and can be used to locate structural damage, which is expressed as follows:

$$\Delta\varepsilon(k) = \left| \varepsilon^{und}(k) - \varepsilon^{dam}(k) \right| \quad (k = 1, 2, \ldots, N) \tag{3}$$

where $\varepsilon^{und}(k)$, $\varepsilon^{dam}(k)$ are the strain mode of the kth element before and after structural damage, and N is the total number of elements.

Based on the analysis results of the equivalent model of welded hollow spherical joints with cracks, it was observed that the absolute value of the element strain mode difference increased with the damage aggravation. In particular, it was found that the absolute value of the element strain modal difference approached its maximum when the crack length was 345°. To quantify the damage degree of each element, the damage index of the kth element was defined as:

$$Id(k) = \frac{\Delta\varepsilon(k)}{\Delta\varepsilon(k)_{345°}} \tag{4}$$

where $\Delta\varepsilon(k)$ and $\Delta\varepsilon(k)_{345°}$ are the absolute differences of strain modes of the kth elements at arbitrary damage degree and crack length of $345°$, respectively. The $Id(k)$ increased with the damage degree monotonically. The crack length increased with the increase of $Id(k)$; when $Id(k) = 1$, the crack length was $345°$.

To determine the eigenvector that can effectively characterize the pattern of joint weld damage, it is important to identify the members that are most sensitive to the damage. When weld damage occurs at a particular joint, the corresponding member connected to that joint exhibits the most significant change in the damage index Id, and adjacent members may also experience considerable changes in their damage indices. These sensitive members are highly responsive to the joint damage, and their damage indices tend to increase as the degree of the joint damage worsens. Hence, these sensitive members can be selected as the key indicators (eigenvector) to characterize the specific damage pattern. The damage index Id values associated with these members are considered as the eigenvector components, representing the pattern of joint weld damage. Each set of damage indices Id corresponds to a specific damage pattern for a given joint, and together they form a pattern database.

To obtain the damage indices Id for each member, the strain modes are calculated using the vibration mode shapes of both the intact structure and various damage cases, based on the equivalent finite element model of the joint with weld damage. By analyzing the strain modes, the corresponding damage indices Id for each member can be determined and used to construct the pattern database, which facilitates the identification of the location and degree of joint weld damage in the structure.

2.2.2. Implementation of Fuzzy Pattern Recognition

The fuzzy pattern recognition process consists of the following steps. Firstly, the eigenvector is determined to establish the fuzzy pattern database. Then, the membership function is defined. Finally, the membership degree is calculated based on the membership principle to judge the membership pattern of the object and recognize it.

The membership function employs the distance formula to calculate the distance between the object to be identified, denoted as $b = (b_1, b_2, \cdots, b_m)$, and every pattern in the pattern database, represented as $a_i = (a_{i1}, a_{i2}, \cdots, a_{im})(i = 1, 2, \cdots, n)$. The distance formula is given by:

$$d_i(b, a_i) = \sqrt{\sum_{j=1}^{m}\left(\frac{b_j - a_{ij}}{a_{ij}}\right)^2} \tag{5}$$

where m and n are the dimension of the eigenvector and the number of patterns in the pattern database, respectively. The membership function is defined as:

$$A_{bi} = 1 - \frac{d_i(b, a_i)}{D} \tag{6}$$

where $D = \max(d_1(b, a_1), d_2(b, a_2), \ldots, d_n(b, a_n))$; if $A_{bl} = \max(A_{b1}, A_{b2}, \ldots, A_{bn})$, the object to be identified $b = (b_1, b_2, \cdots, b_m)$ belongs to the pattern $a_l = (a_{l1}, a_{l2}, \cdots, a_{lm})$.

In this study, $b = (b_1, b_2, \cdots, b_m)$ is the eigenvector composed of the Id value of the member, assuming the damage location and degree of joint are unknown, and $a_l = (a_{l1}, a_{l2}, \cdots, a_{lm})$ is the eigenvector composed of the Id of members connecting with different joints with different crack lengths.

3. Damage Identification of Weld Joints in Spatial Grid Structure

To investigate the applicability of the two-step method for identifying weld crack damage to joints, a welded space steel structure model was developed, as shown in Figure 6 (the numbers of the upper and lower chords are marked), which was used in the numerical analysis and experimental verification. Due to space limitation in the laboratory, the model dimensions were set to 3 m in length, 1.8 m in width, and 1.85 m in height. The overall height of 1.85 m was composed of a 1.5 m pillar and 0.35 m corresponding to the upper and lower chords of the grid structure. The hollow spheres utilized in the model were of WS1204 specification.

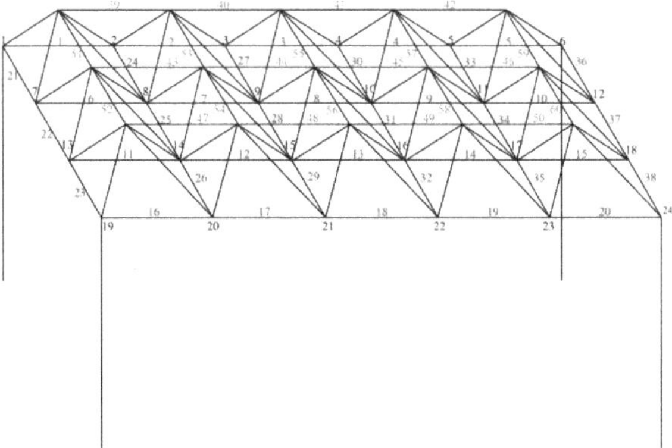

Figure 6. The numbers of nodes and chords of the structural finite element model.

The grid structure consisted of lower chords with a grid pattern of 5×3 and upper chords with a grid pattern of 4×2. All members in the structure were of type $\phi 48 \times 3.5$, with an outer diameter of 48 mm and a wall thickness of 3.5 mm. The structure was supported at all four corners by the pillars, constructed using circular steel tubes of type $\phi 60 \times 3.5$, with a net height of 1.5 m. The steel material in the structure had an elastic modulus of 206 GPa, Poisson's ratio of 0.3, and a density of 7850 kg/m^3.

The finite element model of the structure was created using the ANSYS (v. 12.0.) finite element analysis software. The member elements were simulated using the element type BEAM188. The mass of the spherical joints was represented by the MASS21. The welded spherical joints with cracks were simulated by the MATRIX27. The finite element model consisted of 43 nodes and 124 elements.

3.1. Damage Identification Sub-Region Division of Spatial Grid Structure

When weld damage occurs at a joint in the grid structure, it can lead to changes in the acceleration responses of this joint as well as the adjacent joints when subjected to external excitation. By performing wavelet transform on these acceleration responses, the singular values of the high-frequency components can be obtained. These singular values help determine the influence range of the damaged joint and guide the placement of sensors at specific locations within the structure.

In the analysis, joint 8 was chosen as an example. When a $90°$ weld damage occurred on the right side of joint 8, the singular values of the high-frequency components in the acceleration responses of joint 8 and its neighboring joints were examined using wavelet transform. The magnitude of these singular values was used to determine the range of influence caused by the weld damage.

To initiate the free vibration of the grid structure, a horizontal displacement of 2 cm was applied. At 20 s, the weld damage occurred on the right side of joint 8. The time–history acceleration responses of all joints in the structure were then calculated. Subsequently, the wavelet transform was performed on the acceleration response of each joint to obtain the singular value of the high-frequency component. Figure 7 illustrates the singular values of the high-frequency components for the lower chord joints 7, 8, 9, 10, and the upper chord joints 25, 31, 36, 37.

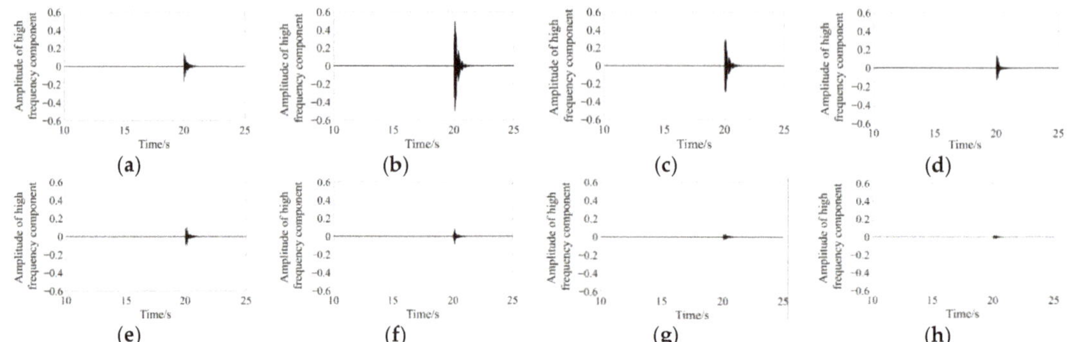

Figure 7. Wavelet transform of acceleration response of each joint, (**a**) No. 7 joint, (**b**) No. 8 joint, (**c**) No. 9 joint, (**d**) No. 10 joint, (**e**) No. 25 joint, (**f**) No. 31 joint, (**g**) No. 36 joint, (**h**) No. 37 joint.

From Figure 7, it is evident that when joint 8 is damaged, its acceleration response exhibits the highest singular value for the high-frequency component. As the distance from joint 8 increases, the singular values of the high-frequency components for the lower chord joints gradually decrease, indicating a diminishing influence caused by the damage of joint 8 with increasing distance. The distribution of singular values for the high-frequency components in the upper chord joints follows a similar pattern, and with a sufficient distance, the influence caused by joint 8 damage becomes negligible. By comparing the maximum singular value of the high-frequency component for each joint to that of joint 8, structural sub-regions can be identified, and the locations for sensor placement can be determined. In this case, the maximum singular value at joint 37 is significantly smaller than that at joint 8. Therefore, considering joint 37 as the boundary, the damage influence range of an individual joint is defined as a rectangle measuring 1.5 m × 1.8 m in the lower chord and a square measuring 1.2 m × 1.2 m in the upper chord. Consequently, the grid structure is divided into two sub-regions, labeled as A and B. An FBG accelerometer was mounted on joint 8 of sub-region A and joint 17 of sub-region B, as depicted in Figure 8.

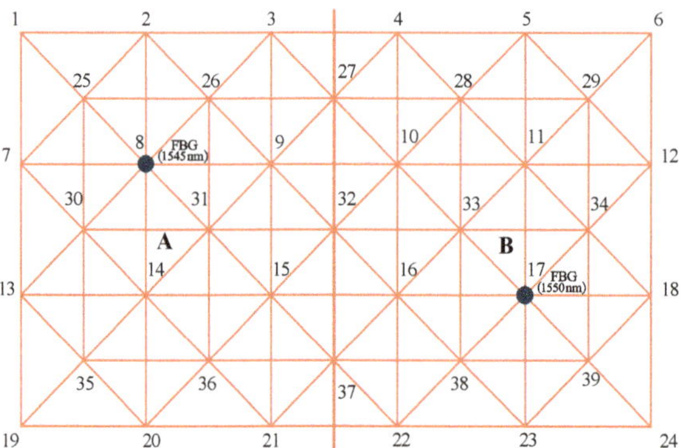

Figure 8. Sub-regional division of the structure and arrangement of joints for FBG accelerometers. A and B represent sub-region A and sub-region B.

3.2. Formation of Damage Recognition Pattern Database

The formation of an appropriate pattern database is crucial for diagnosing structural damage using the fuzzy pattern recognition method. To illustrate the process, the damage to joint 2 in the No. 24 member was taken as an example. In this case, the crack length range was assumed to be between $15\pi r/180$ and $345\pi r/180$ (where r was the outer radius of the tube), with an interval of $15\pi r/180$. Consequently, there were a total of 23 damage patterns for each weld damage scenario at joint 2.

Figure 9 displays the relationships between the *Id* values of the No. 24 and No. 25 members and the damage extent of joint 2. As the damage of joint 2 aggravated, the *Id* value of the No. 24 member connected to joint 2 exhibited the most significant change, while the *Id* value of the adjacent No. 25 member also experienced a noticeable change. Both *Id* values gradually increased with the aggravation of the damage to joint 2. Consequently, these sensitive members' *Id* values were selected as eigenvectors to effectively characterize this particular damage pattern.

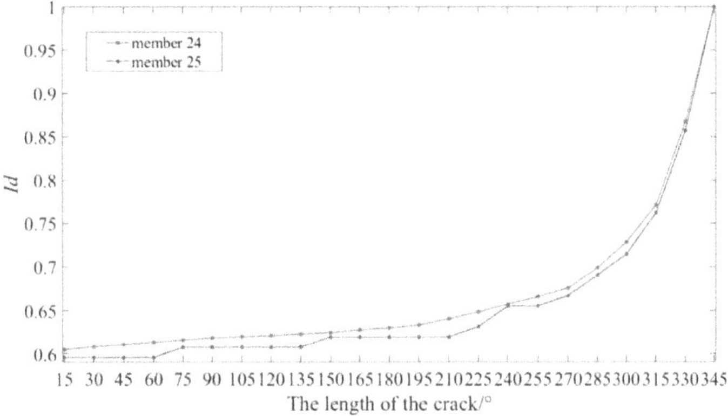

Figure 9. *Id* Changes with crack length.

Following a similar approach, the damage patterns for all joints corresponding to members were determined. In each sub-region, there were 60 pattern libraries, with each pattern library containing 23 damage patterns.

Then, a sinusoidal load close to the natural frequency of the structure was applied to induce forced vibrations. The displacement response of the structure during steady-state vibration was selected for analysis, allowing the determination of the strain mode of each member in the structure. In the sub-region where joint damage occurred, the strain mode of each member was compared with that observed during the 345° weld damage of the connecting joint, resulting in the acquisition of the *Id* for each member in the sub-region.

Subsequently, the membership degree between the eigenvectors representing each pattern library and the corresponding *Id* vectors of members in the damaged sub-region was calculated. The pattern associated with the highest membership degree represented the location and degree of the joint weld damage. To enhance accuracy, multiple peak displacement responses were selected to calculate the *Id* for each member, and the average value was employed for fuzzy pattern recognition.

3.3. Experimental Model and Sensor Placement for Grid Structures

In the aforementioned model, the members utilized were circular steel pipes with dimensions of $\phi48 \times 3.5$. The four pillars were constructed using $\phi60 \times 3.5$ circular steel pipes, which were welded onto 10mm thick steel plates measuring 0.5 m $\times$ 0.5 m. These steel plates were securely fastened to the concrete platform using high-strength bolts forming fixed supports. The upper ends of each pillar were connected to the upper grid via flanges. Each flange was fastened using six outer hexagon bolts with a diameter of 6 mm. The grid structure model was made by Q235; the yield strength of steel is 235 Mpa, as illustrated in Figure 10.

Figure 10. Structural model.

To ensure the accuracy of experimental verification, it is essential to produce both undamaged and damaged structures with identical quality. This is to prevent any identification errors caused by fabrication variations. Consequently, a model of the grid structure with replaceable rods was developed. To simulate joint weld damage, specific individual damaged members were designed. These damaged members could be substituted for the members within the test model, leaving the 90° angle at one end of the damaged member unwelded to the flange, as illustrated in Figure 11a. When weld damage occurred at a specific location in the model, the corresponding undamaged member could be replaced by the appropriate damaged member, resulting in a grid structure with a weld-damaged joint. The replaceable damaged members are presented in Figure 11b.

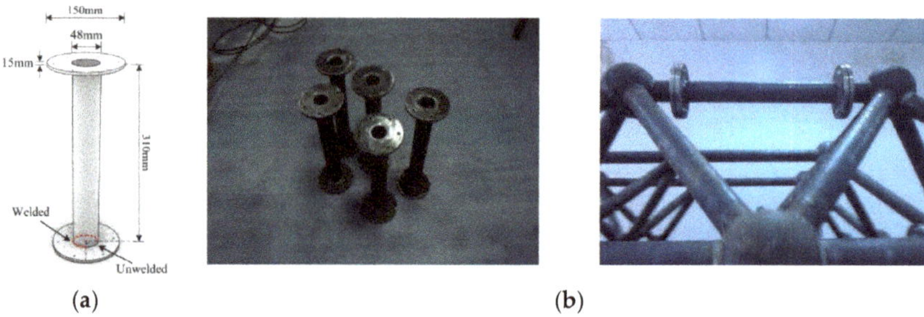

(a)   (b)

Figure 11. Replaceable members, (**a**) Three-dimensional schematic, (**b**) Actual object.

The FBG (fiber Bragg grating) sensor utilizes wavelength modulation as its sensing signal, ensuring that measurement signals remain unaffected by factors such as light source fluctuations, fiber bending losses, connection losses, and detector aging. Additionally, the use of wavelength division multiplexing enables the convenient connection of multiple FBGs in series on a single fiber, allowing distributed measurements. Due to these advantages, FBG sensors have found wide application in civil engineering monitoring and damage identification [39–41]. Malekzadeh et al., employed robust regression analysis (RRA) and cross-correlation analysis (CCA) techniques to locate structural damage using strain data collected by FBG sensors deployed on four-span bridge-type structures [42]. Elshafey designed and fabricated a fiber optic sensing array with eight sensing elements to measure time–history strain at various points on a simply supported beam subjected to random loading, by comparing the random decrements at different damage ratios to an intact case to identify the existence of damage [43].

In the whole length of FBG, there is such a relationship between the central wavelength, period, and effective refractive index:

$$\lambda_b = 2n\Lambda \tag{7}$$

where λ_b is the central wavelength, n is the effective refractive index of the core, Λ is the refractive index modulation period of the core.

When the strain ε of the optical fiber occurs, Λ becomes Λ':

$$\Lambda' = \Lambda(1 + \varepsilon) \tag{8}$$

According to photoelastic theory, the wavelength chan $\Delta\lambda$ is:

$$\Delta\lambda / \lambda_b = (1 - p)\varepsilon \tag{9}$$

where p is the effective photoelastic coefficient.

To avoid low measurement stability and noise disturbance in the existing methods, and consider the characteristics of low structural vibration frequency at the same time, a FBG (fiber Bragg grating) acceleration sensor was developed in this study based on the principle that strain and temperature variations affect the refractive index and period of the FBG, subsequently changing its reflection wavelength. In the configuration of the FBG sensor, a lever was located in the center of the sensor, the FBG was straightened above the lever, and a mass block located under the lever. The top of the mass block was connected to the bottom of the lever, and the bottom of the mass block was connected to the inside bottom of the sensor envelope by a spring. When the sensor was installed on the structure, the structure vibration drove the mass block to vibrate, and caused occurrence of tensile force in the FBG through the lever, changing the period and the refractive index of the FBG, and thus modifying the wavelength of its reflection. Notably, a linear relationship between the reflection wavelength and acceleration was observed, allowing acceleration measurement by monitoring the wavelength change. The design of the FBG acceleration sensor is shown in Figure 12.

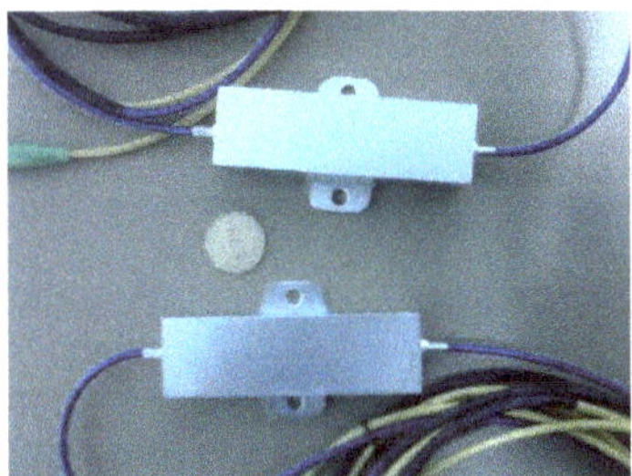

Figure 12. FBG accelerometers.

In order to achieve higher sensitivity, the measurement range of the FBG acceleration sensor was set to ± 1.0 g, taking into account the expected structural vibration amplitude. Two types of FBG acceleration sensors were designed, with central wavelengths of 1545 nm and 1550 nm, and tested to evaluate their performance, respectively. The test data are shown in Table 1. Additionally, the corresponding acceleration response curves are depicted in Figure 13.

Table 1. The performance test data of acceleration sensors.

Acceleration (m/s^2)	Amplitude of Acceleration Sensors (pm)	
	1545 nm	1550 nm
0.2	9.2	10
0.4	18.2	21
0.6	29	30
0.8	37	42

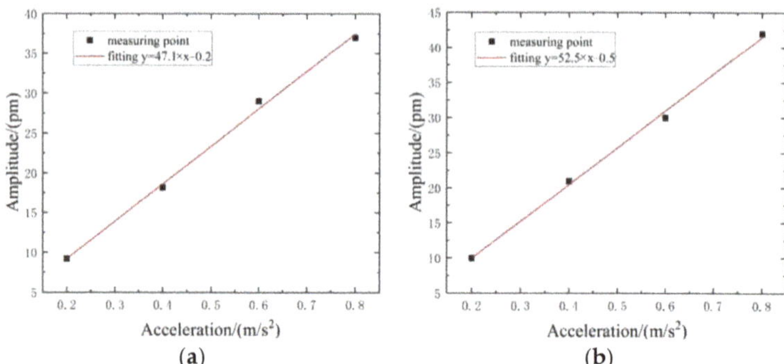

Figure 13. The performance curve of acceleration sensors, (**a**) 1545 nm, (**b**) 1550 nm.

Figure 13 illustrates the sensitivity of the FBG accelerometers with central wavelengths of 1545 nm and 1550 nm as 47.1 pm/(m/s^2) and 52.5 pm/(m/s^2), with measurement accuracies of 0.021 m/s^2 and 0.019 m/s^2, respectively. To monitor the structural vibrations, the two FBG accelerometers were installed on key joint 8 in sub-region A and key joint 17 in sub-region B, as shown in Figure 14.

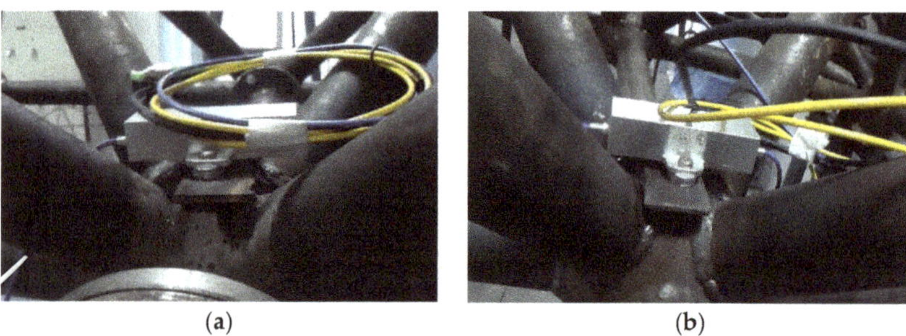

Figure 14. Acceleration sensors distribution, (**a**) No. 8 joint, (**b**) No. 17 joint.

The si425-500 FBG sensor demodulator, along with its supporting software, was employed to collect the testing data, which is a multi-channel, multi-sensor measurement system utilizing a calibrated wavelength scanning laser, operating at a sampling frequency of 250 Hz and offering a resolution of 1 pm.

3.4. Damage Identification of Weld Joints in Grid Structures

In order to further validate the feasibility and effectiveness of the proposed method, some damage cases were considered, as shown in Table 2.

Table 2. Damage Cases.

Damage Cases	Damaged Members	Damaged Joints of Member	Extent of Damage
Case1	Member 24	Joint 2	90°
Case2	Member 24	Joint 2	90°
	Member 33	Joint 5	90°

By applying an initial displacement on its top, the grid model was excited to vibrate freely, the acceleration responses of the representative joints 8 and 17 were measured with the FBG sensors and analyzed by wavelet transform. Subsequently, when weld damage occurred on a specific joint within this sub-region, the singular value of the high-frequency components of the acceleration response was obtained through wavelet transform, which was then used to determine whether weld damage had occurred in this particular sub-region.

To simulate the sudden occurrence of joint damage in the test and capture the time–history acceleration response, three steps of free vibrations were conducted. In the first step, a horizontal initial displacement of 2 cm in the X direction was applied to the intact structure, causing it to vibrate freely. The time–history acceleration responses of joint 8 and joint 17, referred to as data I, were collected during this vibration. In the second step, a 1 cm displacement in the X direction was applied to the intact structure, leading to another round of free vibration. The acceleration responses of the two joints, referred to as data II, were recorded. In the third step, the same 1 cm displacement in the X direction was applied to the damaged structure resulting in free vibration. The acceleration responses of the two joints were measured as data III.

The splice point of data I and data III was determined by overlapping the corresponding parts of data I and data II, on which data I and data III were spliced together. The combined dataset included the acceleration response information of the joints before and after the occurrence of structural damage.

Case 1, which serves as an example to demonstrate the steps of damage identification, involves damage solely at joint 2. The measured acceleration responses of joint 8 and joint 17 are presented in Figure 15.

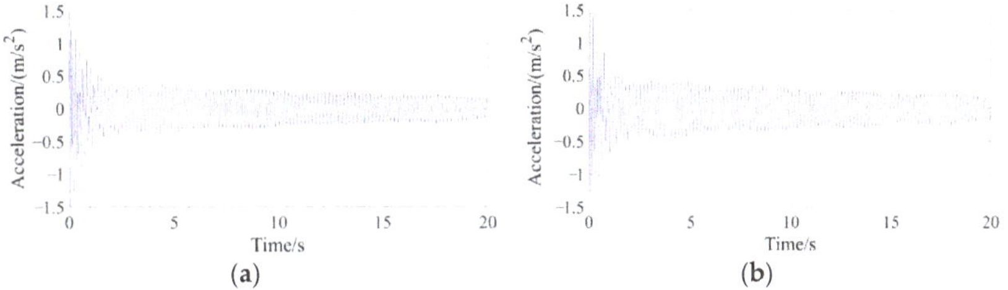

Figure 15. Acceleration response time history of two joints, (**a**) No. 8 joint, (**b**) No. 17 joint.

Upon observation, it becomes evident that during the stable free-vibration stage that the time–history curves obtained from the FBG accelerometers exhibit relatively smooth and sensitive behavior. This characteristic indicates that the designed accelerometer possesses excellent anti-noise performance and sensitivity, allowing precise measurement of the structural response.

The wavelet transform was performed on the acceleration responses of the joints to analyze and extract the singular value of the high-frequency component. The amplitude of this singular value can be utilized to determine whether weld damage has occurred at sub-region A or B within the structure. Figure 16 displays the amplitudes of the singular value of the high-frequency component for the two key joints.

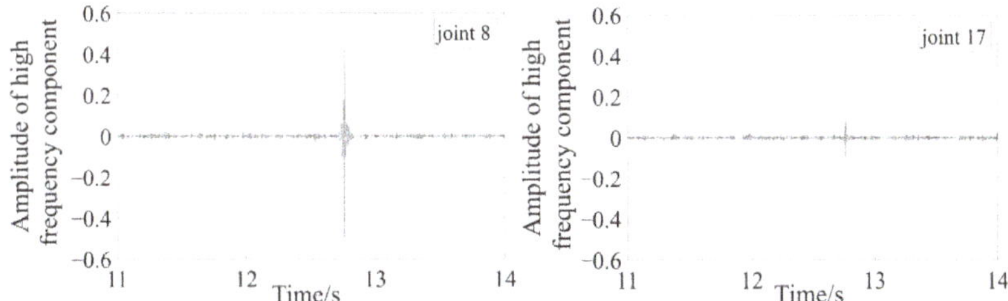

Figure 16. Case 1, the singular value amplitudes of high-frequency component of key joints.

Figure 16 reveals that the amplitude of the singular value of joint 8 is notably higher than that of joint 17. This stark difference indicates the presence of weld damage in structural sub-region A. Moreover, the results of Figure 16 also illustrate the feasibility of adopting the method proposed in this paper to simulate weld damage at the joints of the grid structure in the experiment, and the obtained acceleration response time histories of arrayed sensor joints also successfully characterize the occurrence of weld damage at the joints.

In the first step it was determined that the weld damage occurred in the sub-region A. In the second step, the membership degree between the eigenvectors of each pattern library and the corresponding *Id* vectors in this sub-region A was calculated using the strain mode difference. The strain mode difference of each member in sub-region A is shown in Figure 17.

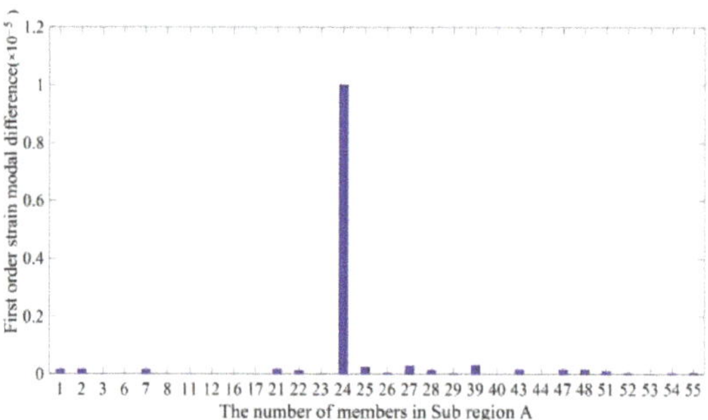

Figure 17. First order strain mode difference of elements in Sub-region A.

Subsequently, the damage index *Id* was determined by comparing the strain mode difference with the corresponding member with 345° joint damage in each pattern library. Finally, the membership degrees between the *Id* vector and all pattern libraries in sub-region A were calculated. The pattern with the highest membership degree represents the pattern of joint damage. Some membership degree diagrams are shown in Figure 18.

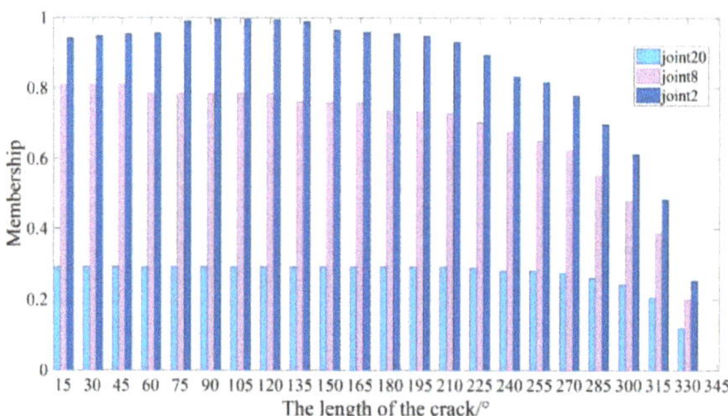

Figure 18. The membership degree between the damage pattern of case 1 and each typical pattern library.

Figure 18 depicts the membership degree between the damage pattern of Case 1 and the 23 pattern libraries of the damaged joint 2 of No. 24 member. It also shows the membership degree between Case 1 and the pattern libraries of the damaged joint 8 of No. 24 member, as well as the membership degree between Case 1 and the pattern libraries of the damaged joint 20 of No. 26 member.

Since joint 2 is adjacent to joint 8, and the sensitive members of the damage pattern for joint 8 are the No. 24 and No. 25 members, which are the same as those for joint 2, the membership degree is higher but still smaller than that of joint 2. On the other hand, joint 20 is farther away from joint 2, and its sensitive members are No. 25 and No. 26, which are not exactly the same as those for joint 2. As a result, the membership degree for joint 20 is much smaller compared with joint 2. Therefore, it is confirmed that the damage occurred on joint 2 of No. 24 member. The corresponding membership values are shown in Table 3.

Table 3. The Degree of Membership between the damage pattern of Case 1 and Pattern Library of Joint 2 Damage.

Extent of Damage	Degree of Membership	Extent of Damage	Degree of Membership	Extent of Damage	Degree of Membership
15°	0.9411	135°	0.9882	255°	0.8172
30°	0.9481	150°	0.9649	270°	0.7782
45°	0.9527	165°	0.9596	285°	0.6978
60°	0.9562	180°	0.9553	300°	0.6135
75°	0.9898	195°	0.9480	315°	0.4840
90°	0.9958	210°	0.9307	330°	0.2538
105°	0.9954	225°	0.8941	345°	0.00
120°	0.9931	240°	0.8325		

Based on the values of the membership degree, it is evident that the maximum membership degree is 0.9958 when the weld damage of joint 2 in No. 24 member is approximately 90°. This high membership degree indicates that the damage location is indeed at joint 2 of No. 24 member in sub-region A. Furthermore, the damage extent is most likely to be around 90°, which aligns with the findings of Case 1.

In Case 2, the identification process is similar to that of Case 1. Figures 19 and 20 present the results identifying the location and extent of weld damage for two joints. Table 4 provides further details and information related to the results.

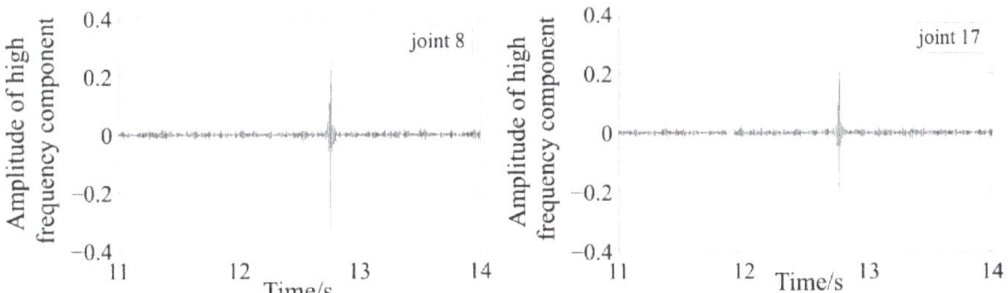

Figure 19. Case 2, the amplitude of singular value of high-frequency component of key joints.

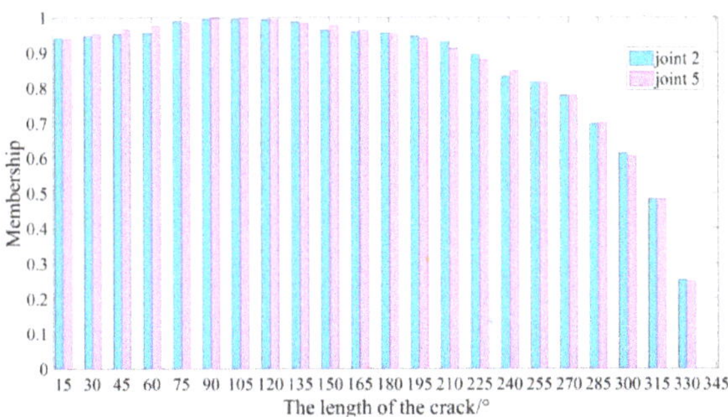

Figure 20. The membership degree between the damage pattern of Case 2 and corresponding pattern library.

Table 4. The Membership Degree between the damage pattern of Case 2 and Pattern Library of Joint 2 and Joint 5 Damage.

Extent of Damage	Degree of Membership		Extent of Damage	Degree of Membership	
	Joint 2	Joint 5		Joint 2	Joint 5
15°	0.9458	0.9453	195°	0.9345	0.9343
30°	0.9605	0.9601	210°	0.9056	0.9056
45°	0.9721	0.9727	225°	0.8746	0.8746
60°	0.9832	0.9822	240°	0.8418	0.8418
75°	0.9930	0.9915	255°	0.8096	0.8096
90°	**0.9954**	**0.9933**	270°	0.7742	0.7742
105°	0.9905	0.9893	285°	0.6943	0.6943
120°	0.9857	0.9848	300°	0.6011	0.6011
135°	0.9778	0.9773	315°	0.4790	0.4790
150°	0.9706	0.9703	330°	0.2465	0.2478
165°	0.9571	0.9569	345°	0.00	0.00
180°	0.9481	0.9479			

Based on Figure 19, it is apparent that there is minimal difference between the singular values of joint 8 and joint 17. This suggests that weld damage has occurred in both sub-regions A and B. Figure 20 and Table 4 provide additional insights into the identification process.

From Figure 20 and Table 4, it can be observed that the maximum membership degree for joint weld damage in the pattern library is 0.9954 for joint 2 connected with No. 24 member in sub-region A and 0.9933 for joint 5 connected with No. 33 member in sub-region B. Both cases indicate a possibility of 90° weld damage. These findings confirm that both joint 2 of No. 24 member in sub-region A and joint 5 of No. 33 member in sub-region B have the highest likelihood of experiencing 90° weld damage, which aligns with the damage scenario presented in Case 2.

3.5. Discussion

The experimental verification demonstrates that partitioning the grid structure allows it to fully leverage the benefits of the wavelet analysis method in capturing sudden shifts in joint response signals caused by structural damage, leading to a reduction in complexity and symmetry. As a result, it becomes feasible to identify the specific sub-region where joint damage occurs by solely analyzing the acceleration responses from sensors deployed on the joints.

By limiting the number of joints and members in each sub-region, the number of potential joint damage patterns per sub-region is significantly reduced. Consequently, a comprehensive pattern library can be constructed with minimal computational effort. Moreover, the adoption of the fuzzy pattern recognition method in the sub-regions with joint damage not only simplifies the recognition of structural damage but also enhances the efficiency and accuracy of the method's application.

Thus, the combination of these two methods enables the effective recognition of damage in large and intricate structures, such as grid structures, offering a promising approach for practical application.

While the experimental results have demonstrated the effectiveness of the method, its application to a practical engineering scenario poses certain challenges. One notable difficulty is the demanding service conditions of grid structures, leading to the inclusion of inherent noise in the original measurement information. Despite the recommendation to employ FBG acceleration sensors for enhanced measurement accuracy, the measured acceleration responses may still be affected by noise, potentially leading to an increase in the number of structural sub-regions. Consequently, this may result in reduced recognition efficiency and accuracy.

To address this issue, it is essential to process the measured raw information and apply noise reduction techniques. By doing so, the method can be effectively applied to accurately recognize damage in joints of grid structures in practical engineering.

4. Conclusions

Weld damage in joints is a prevalent issue in welded spatial grid structures. To address this problem, a two-step method for identifying weld damage in joints in such structures has been proposed and experimentally verified. The conclusions are as follows:

1. The structure is divided into sub-regions based on the influence range of joint weld damage, determined by analyzing the singular value of the high-frequency component of the acceleration response using wavelet transform. Thereby, reducing the complexity and breaking the symmetry of the grid structure transforms the structure into multiple simple structures. FBG accelerometers are installed only at representative joints within each sub-region. The wavelet transforms of the measured acceleration response at these very few joints facilitate the identification of sub-regions where weld damage has occurred, allowing subsequent damage location and extent identification limited to these sub-regions.

2. In the second step, a fuzzy pattern database is developed considering different damage extents and different damaged joints within each sub-region. The strain modal difference of all members, given the specific damaged location and extent, is used as a damage index to form a pattern. Since the number of joints in the sub-region of the grid structure is rather small, fewer calculations are required to build up the pattern library, and the scale of the pattern library is also smaller. Then, the membership degree between the measurement data and the pattern library enables easy identification of the location and extent of joint damage in the identified sub-region. The step-by-step application of wavelet analysis and fuzzy pattern recognition technology reduces calculation requirements and improves recognition efficiency.

3. The designed FBG accelerometer, which exhibits high sensitivity, is utilized in the identification technology. Performance tests and verification experiments demonstrate that data measured by the FBG accelerometer possess high accuracy and smoothness. By effectively reducing noise, recognition accuracy can be further improved. Consequently, both the proposed two-step identification method and the designed FBG accelerometer offer efficient solutions for damage identification in large-span grid structures, with potential applications in engineering.

These conclusions demonstrate that the proposed method addresses the barriers to damage identification in welded joints of grid structures, and in combination with FBG accelerometers, which have high measurement sensitivity and noise immunity, highlights its effectiveness, efficiency, and potential practical applications in the field of damage identification in welded spatial grid structures.

Although this paper has developed a two-step method for welded joint damage identification in grid structures, there are still some issues for further clarification. The use of crack-equivalent finite element model analysis to establish a library of welded joint damage patterns introduces a certain degree of error. Therefore, it is necessary to investigate a multiscale model capable of characterizing joint crack damage to provide more accurate damage patterns.

If this method is applied to damage identification in practical grid structures, in addition to processing original measurement information for noise reduction, it should also be considered that the damage index plays a crucial role in damage identification, and is the key to the success of damage identification [44]. The complexity of the grid structure may result in identification difficulties due to the high dimensionality of the damage index vectors. As a solution, more concise and sensitive damage indices should be developed to improve the applicability of the damage identification method.

Author Contributions: Supervision, H.L.; methodology, H.L. and X.W.; software, J.H.; test, X.W. and X.L.; writing—review and editing, H.L. and X.W.; funding acquisition, H.L. All authors have read and agreed to the published version of the manuscript.

Funding: The work described in this paper was funded by Hainan Provincial Joint Project of Sanya Yazhou Bay Science and Technology City (Grant No. 520LH058), Hainan Provincial Natural Science Foundation of China (Grant No. 522CXTD517, 522RC879).

Data Availability Statement: Not applicable.

Conflicts of Interest: The authors declare no conflict of interest.

References

1. Adin, M.S. A Parametric Study on the Mechanical Properties of MIG and TIG Welded Dissimilar Steel Joints. *J. Adhes. Sci. Technol.* **2023**, *37*, 1–24. [CrossRef]
2. Wu, Q.; Yu, A.; Kang, Z.; Kuang, C.; Liu, H. Molecular Dynamics Study on the Microscopic Mechanism of In-Service Welding Damage and Failure. *Eng. Fail. Anal.* **2022**, *137*, 106402. [CrossRef]
3. Wei, F.; Qiao, P. Vibration-Based Damage Identification Methods: A Review and Comparative Study. *Struct. Health Monit.* **2011**, *9*, 83–111.
4. Zhu, H.P.; Lin, L.; He, X.Q. Damage Detection Method for Shear Buildings Using the Changes in the First Mode Shape Slopes. *Comput. Struct.* **2011**, *89*, 733–743. [CrossRef]
5. Yin, T.; Jiang, Q.H.; Yuen, K.V. Vibration-Based Damage Detection for Structural Connections Using Incomplete Modal Data by Bayesian Approach and Model Reduction Technique. *Eng. Struct.* **2017**, *132*, 260–277. [CrossRef]
6. Ditommaso, R.; Ponzo, F.C.; Auletta, G. Damage Detection on Framed structures: Modal Curvature Evaluation Using Stockwell Transform under Seismic Excitation. *Earthq. Eng. Eng. Vib.* **2015**, *2*, 265–274. [CrossRef]
7. Zhang, Y.Z.; Yang, Y.; Kong, L.J.; Du, W.H. Research on Damage Identification Method of Composite Materials Based on Displacement Modal Shape Parameters. *J. Phys. Conf. Ser.* **2019**, *1168*, 52046. [CrossRef]
8. Zhou, Y.; Jiang, Y.Z.; Yi, W.J.; Xie, L.M.; Jia, F.D. Experimental Research on Structural Damage Detection Based on Modal Flexibility Theory. *J. Hunan Univ. Nat. Sci. Ed.* **2015**, *42*, 36–45.
9. Chang, K.C.; Kim, C.W. Modal-Parameter Identification and Vibration-Based Damage Detection of a Damaged Steel Truss Bridge. *Eng. Struct.* **2016**, *122*, 156–173. [CrossRef]
10. Fang, Y.L.; Su, P.R.; Shao, J.Y.; Lou, J.Q.; Zhang, Y. Substructure Damage Identification Based on Model Updating of Frequency Response Function. *Int. J. Struct. Stab. Dyn.* **2021**, *21*, 1–32. [CrossRef]
11. Cha, Y.J.; Buyukozturk, O. Structural Damage Detection Using Modal Strain Energy and Hybrid Multiobjective Optimization. *Comput. Aided Civil Infrastruct. Eng.* **2015**, *30*, 347–358. [CrossRef]
12. Li, Y.C.; Wang, S.Q.; Zhang, M.; Zheng, C.M. An Improved Modal Strain Energy Method for Damage Detection in Offshore Platform Structures. *J. Mar. Sci. Appl.* **2016**, *15*, 182–192. [CrossRef]
13. Arefi, S.L.; Gholizad, A.; Seyedpoor, S.M. Damage Detection of Structures Using Modal Strain Energy with Guyan Reduction Method. *J. Rehabil. Civ. Eng.* **2020**, *8*, 47–60.
14. Huang, M.; Li, X.; Lei, Y.; Gu, J. Structural Damage Identification Based on Modal Frequency Strain Energy Assurance Criterion and Flexibility Using Enhanced Moth-Flame Optimization. *Structures* **2020**, *28*, 1119–1136. [CrossRef]
15. Sohn, H.; Farrar, C.R. Damage Diagnosis Using Time Series Analysis of Vibration Signals. *Smart Mater. Struct.* **2001**, *10*, 446–451. [CrossRef]
16. Lam, H.F.; Yang, J.H.; Hu, Q.; Ng, C.T. Railway Ballast Damage Detection by Markov Chain Monte Carlo-based Bayesian Method. *Struct. Health Monit.* **2018**, *17*, 706–724. [CrossRef]
17. Salehi, M.; Azami, M. Structural Damage Localization through Multi-channel Empirical Mode Decomposition. *Int. J. Struct. Integr.* **2019**, *10*, 102–117. [CrossRef]
18. Zhu, Y.T.; Xie, M.X.; Zhang, K.; Li, Y.T. Dam Deformation Residual Correction Method for High Arch Dams Using Phase Space Reconstruction and an Optimized Long Short-Term Memory Network. *Mathematics* **2023**, *11*, 2010. [CrossRef]
19. Xiao, F.; Meng, X.W.; Zhu, W.W.; Chen, G.S.; Yu, Y. Combined Joint and Member Damage Identification of Semi-rigid Frames with Slender Beams Considering Shear Deformation. *Buildings* **2023**, *13*, 1631. [CrossRef]
20. Xiao, F.; Sun, H.M.; Mao, Y.X.; Chen, G.S. Damage Identification of Large-scale Space Truss Structures Based on Stiffness Separation Method. *Structures* **2023**, *53*, 109–118. [CrossRef]
21. Kim, H.; Melhem, H. Damage Detection of Structures by Wavelet Analysis. *Eng. Struct.* **2004**, *26*, 347–362. [CrossRef]
22. Zhe, F.; Xin, F.; Jing, Z. A Novel Transmissibility Concept Based on Wavelet Transform for Structural Damage Detection. *Smart Struct. Syst.* **2013**, *12*, 291–308.
23. Janeliukstis, R.; Rucevskis, S.; Akishin, P. Wavelet Transform Based Damage Detection in a Plate Structure. *Procedia Eng.* **2016**, *161*, 127–132. [CrossRef]
24. Zhu, L.F.; Ke, L.L.; Zhu, X.Q.; Xiang, Y.; Wang, Y.S. Crack Identification of Functionally Graded Beams Using Continuous Wavelet Transform. *Compos. Struct.* **2019**, *210*, 473–485. [CrossRef]

25. Katunin, A.; Santos, J.; Lopes, H. Damage Identification by Wavelet Analysis of Modal Rotation Differences. *Structures* **2021**, *30*, 1–10. [CrossRef]
26. Yazdanpanah, O.; Mohebi, B.; Yakhchalian, M. Selection of Optimal Wavelet-based Damage-sensitive Feature for Seismic Damage Diagnosis. *Measurement* **2020**, *154*, 107447. [CrossRef]
27. Zhu, Y.T.; Zhang, Z.D.; Gu, C.S.; Li, Y.T.; Zhang, K.; Xie, M.X. A Coupled Model for Dam Foundation Seepage Behavior Monitoring and Forecasting Based on Variational Mode Decomposition and Improved Temporal Convolutional Network. *Struct. Control Health Monit.* **2023**, *2023*, 3879096. [CrossRef]
28. Wang, X.L.; Qu, W.L.; Liu, H.; Caicedo, J.M.; Wang, X.X. Fuzzy Pattern Recognition Technique for Crack Propagation on Earplate Connection of Guyed Mast under Wind Load. *Struct. Control Health Monit.* **2017**, *24*, e2010. [CrossRef]
29. Ren, W.X.; Lin, Y.Q.; Fang, S.E. Structural Damage Detection Based on Stochastic Subspace Identification and Statistical Pattern Recognition: I. Theory. *Smart Mater. Struct.* **2011**, *20*, 115009. [CrossRef]
30. Jiang, S.F.; Zhang, S. Structural Damage Identification Method with Data Fusion on Fuzzy Neural Network. *Eng. Struct.* **2008**, *25*, 95–101.
31. *JG/T11-2009*; Welded Hollow Spherical Node of Space Grid Structures. China Building Industry Press: Beijing, China, 2009.
32. Feng, X.; Zhang, X.T.; Sun, C.S.; Motamedi, M.; Ansari, F. Stationary Wavelet Transform for Distributed Detection of Damage by Fiber-optic Sensors. *J. Eng. Mech.* **2014**, *140*, 04013004. [CrossRef]
33. Qu, W.L.; Lu, L.J.; Li, M. Solid Modeling Method for Structure with 3-D Straight Through Clack. *J. Wuhan Univ. Technol.* **2008**, *30*, 88–90. (In Chinese)
34. Ding, L.J. The Study of Equivalent Finite Element of Welded Hollow Ball with a Crack. Master's Thesis, Wuhan University of Technology, Wuhan, China, 2012. (In Chinese).
35. Li, L.J.; Che, W.X.; Yang, X.M.; Liu, F. Finite Element Analysis of Reticulated Shells with Semi-rigid Connections. *Spat. Struct.* **2007**, *39*, 43–49. (In Chinese)
36. Xiao, F.; Zhu, W.W.; Meng, X.W.; Chen, G.S. Parameter Identification of Structures with Different Connections Using Static Responses. *Appl. Sci.* **2022**, *12*, 5896. [CrossRef]
37. Li, Y.M.; Gao, X.Y.; Shi, Y.S.; Zhou, X.Y. Damage Diagnosis of Space Truss based on Change of Elemental Strain Modal. *J. Build Struct.* **2009**, *30*, 152–159. (In Chinese)
38. Yin, Z.X.; Xu, Z.M. Damage Identification of Cable Truss-cable Net Structure based on Strain Mode Difference. *Build Struct.* **2016**, *46*, 99–103. (In Chinese)
39. Zhang, X.L.; Liang, D.K.; Zeng, J.; Asundi, A. Genetic Algorithm-support Vector Regression for High Reliability SHM System based on FBG Sensor Network. *Opt. Lasers Eng.* **2012**, *50*, 148–153. [CrossRef]
40. Jang, B.W.; Lee, Y.G.; Kim, J.H.; Kim, Y.Y.; Kim, C.G. Real-time Impact Identification Algorithm for Composite Structures Using Fiber Bragg Grating Sensors. Struct. *Control Health Monit.* **2012**, *19*, 580–591. [CrossRef]
41. Kinet, D.; Megret, P.; Goossen, K.W.; Qiu, L.; Heider, D.; Caucheteur, C. Fiber Bragg Grating Sensors toward Structural Health Monitoring in Composite Materials: Challenges and Solutions. *Sensors* **2014**, *14*, 7394–7419. [CrossRef]
42. Malekzadeh, M.; Catbas, F.N. A Comparative Evaluation of Two Statistical Analysis Methods for Damage Detection Using Fibre Optic Sensor Data. *Int. J. Reliab. Qual. Saf. Eng.* **2014**, *8*, 135–155. [CrossRef]
43. Elshafey, A.; Marzouk, H.; Gu, X.; Haddara, M.; Morsy, R. Use of Fiber Bragg Grating Array and Random Decrement for Damage Detection in Steel Beam. *Eng. Struct.* **2016**, *106*, 348–359. [CrossRef]
44. Chen, Y.; Chen, W.S.; Hao, H.; Xia, Y. Damage Evaluation of a Welded Beam-column Joint with Surface Imperfections Subjected to Impact Loads. *Eng. Struct.* **2022**, *261*, 114276.1–114276.12. [CrossRef]

 *buildings*

Article

Application of a Modified Differential Quadrature Finite Element Method to Flexural Vibrations of Composite Laminates with Arbitrary Elastic Boundaries

Wei Xiang *, Xin Li and Lina He

School of Mechanical Engineering, Southwest Jiaotong University, Chengdu 610031, China
* Correspondence: xiangwei@swjtu.edu.cn

Abstract: This paper formulates a modified differential quadrature finite element method (DQFEM) by a combination of the standard DQFEM and the virtual boundary spring technique, which makes it easy to implement arbitrary elastic restraints by assigning reasonable values to the boundary spring stiffnesses. This new formulated method can offer a unified solution for flexural vibrations of composite laminates subjected to general elastic boundary combinations including all the classical cases. The influences of the number of Gauss–Lobatto nodes and the boundary spring stiffnesses on the convergence characteristics of natural frequencies are investigated, and some conclusions are drawn in terms of the minimum number of unilateral nodes required to generate convergent solutions and the optimal values of the boundary spring stiffnesses to simulate classical boundaries. Numerical examples are performed for composite laminates under various classical boundary conditions. Excellent accuracy, numerical stability, and reliability of the present method are demonstrated by comparisons with available exact and numerical solutions in open literatures. Additionally, for elastically constrained composite laminates, which are beyond the scope of most existing approaches, numerous new results obtained by the present method may serve as reference values for other research.

Keywords: differential quadrature finite element method (DQFEM); virtual boundary spring; composite laminates; arbitrary elastic boundary; flexural vibration

Citation: Xiang, W.; Li, X.; He, L. Application of a Modified Differential Quadrature Finite Element Method to Flexural Vibrations of Composite Laminates with Arbitrary Elastic Boundaries. *Buildings* **2022**, *12*, 1380. https://doi.org/10.3390/buildings12091380

Academic Editors: Zechuan Yu and Dongming Li

Received: 12 August 2022
Accepted: 1 September 2022
Published: 4 September 2022

Publisher's Note: MDPI stays neutral with regard to jurisdictional claims in published maps and institutional affiliations.

1. Introduction

Composite laminates are increasingly used in various engineering structures, such as space vehicles, aircraft, naval ships, and submarines, which are usually subjected to frequent dynamic loads. Hence, a thorough understanding of the vibration characteristics of laminates is critical for the design and analysis of composite structures.

In the past few years, many efforts have been devoted to developing accurate and efficient methods to determine the vibration behaviors of composite laminates. A comprehensive review of the recent works on this subject has been provided by Sayyad and Ghugal [1], covering both analytical and numerical methods. By contrast with analytical methods known to be limited to only a few cases, numerical methods are more effective in a wide range of cases involving various physical properties, arbitrary boundary conditions, and sophisticated loading configurations. Various numerical procedures are available for flexural vibrations of multi-layered composite plates [2–12], among which the most representative and widely used one is the finite element method (FEM) that has already been successfully incorporated into commercial software. However, FEM generally uses low-order approximating functions; consequently, a higher accuracy can only be achieved by mesh refinement, resulting in a higher computation cost. To tackle this problem, more intensive research activities are motivated, focusing on high-order schemes such as the mesh-free method [13,14] and differential quadrature (DQ) method [15,16], which tend to yield highly accurate solutions with far fewer degrees of freedom (DOFs) than low-order ones owing to the use of high-order basis functions.

Although increased interest is still in the extension of various numerical methods to the vibration analysis of composite laminates, most of the previous research was confined to the classical boundary conditions comprised of simply supported, clamped, and free boundaries. However, the boundary conditions of numerous engineering structures might not always be ideal essentially. Actually, elastic supports are more commonly seen in practice, and a relative lack of corresponding research still exists.

Motivated by the state-of-the-art, this paper aims to seek for an accurate, efficient, and reliable method for the flexural vibration analysis of composite laminates subjected to arbitrary elastic boundaries. The current work draws on the idea of a well-established high-order scheme referred to as the differential quadrature finite element method (DQFEM) [17,18], in which the DQ rule and the Gauss–Lobatto integration rule are utilized to discretize the energy functional of structures. Fast convergence, high precision, and efficiency, as well as remarkable versatility of the DQFEM, have been validated in the previous research [17,18]. The boundary conditions in both DQFEM and the standard FEM are implemented via the same way, that is, the elimination method. However, this classical approach is only limited to dealing with the classical boundaries but is unable to process general elastic boundaries.

On the purpose of extending the applicability of DQFEM to elastically restrained composite structures, a modified DQFEM is proposed by introducing the virtual boundary spring technique [19–28], in which general elastic restraints including several classical boundary conditions can be easily realized by assigning reasonable values to the virtual boundary spring stiffnesses.

It is well-known that equivalent single-layer laminate theories, which treat a laminated plate as an equivalent homogeneous and orthotropic single layer, are adequate to predict the global response behaviors of composite laminates. Therefore, in the present paper, the widely acknowledged first-order shear deformation theory (FSDT) [29,30] is adopted to model the flexural vibration behavior of composite laminates, since it affords the best compromise between accuracy and efficiency. A detailed formulation of this modified DQFEM is presented for flexural vibrations of rectangular laminates with general elastic restraints. Numerical examples are carried out to discuss the convergence characteristics and validate the accuracy of the present approach.

2. Modified DQFEM Formulation for Composite Laminates

2.1. Constitutive Relations for Composite Laminate

Figure 1 schematically shows a rectangular composite laminate (length a, width b, and thickness h) composed of multiple orthotropic layers with the same thickness and material properties. The xy plane of the Cartesian coordinate system is located on the mid-plane of the laminate, with the origin placed at one corner.

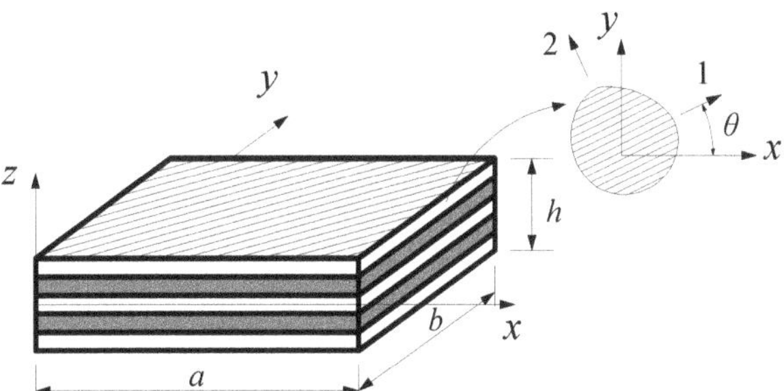

Figure 1. Schematic representation of a composite laminate.

Each single-layer of the laminate is usually assumed to be in the plane-stress state. As shown in Figure 2, σ_1 and σ_2 are the normal stress components in the principle directions of the single-layer, and τ_{12} represents the shear stress component. The constitutive equations relating in-plane stresses and strains for each layer are expressed in the material coordinate system as

$$
\begin{bmatrix} \sigma_1 \\ \sigma_2 \\ \tau_{12} \end{bmatrix} = \begin{bmatrix} Q_{11} & Q_{12} & 0 \\ Q_{12} & Q_{22} & 0 \\ 0 & 0 & Q_{66} \end{bmatrix} \begin{bmatrix} \varepsilon_1 \\ \varepsilon_2 \\ \gamma_{12} \end{bmatrix}, \quad \begin{bmatrix} \tau_{13} \\ \tau_{23} \end{bmatrix} = \begin{bmatrix} Q_{44} & 0 \\ 0 & Q_{55} \end{bmatrix} \begin{bmatrix} \gamma_{13} \\ \gamma_{23} \end{bmatrix}, \tag{1}
$$

in which ε_i ($i = 1, 2$) and γ_{ij} ($i, j = 1, 2, 3$) are strain components; Q_{ij} are modulus components with respect to the material coordinate system in the following form:

$$
Q_{11} = E_1/(1 - v_{12}v_{21}), Q_{12} = v_{12}E_2/(1 - v_{12}v_{21}), Q_{22} = E_2/(1 - v_{12}v_{21}) \\
Q_{44} = G_{13}, Q_{55} = G_{23}, Q_{66} = G_{12} \tag{2}
$$

where E_1 and E_2 are Young's moduli in prime material axes; v_{12} and v_{21} are Poisson's ratios.

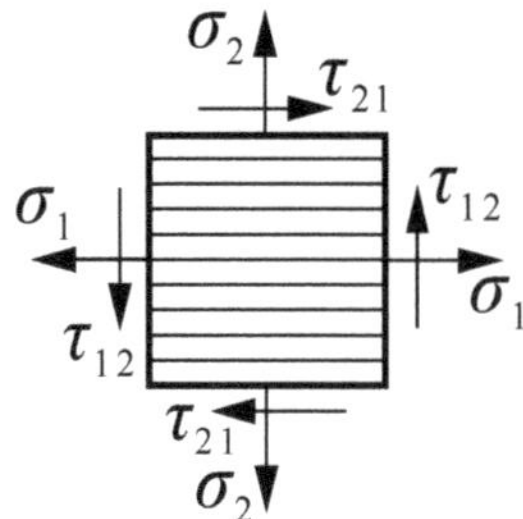

Figure 2. The plane-stress state.

For a unified formulation for each layer, Equation (1) should be transformed into the plate (laminate) coordinate system as

$$
\begin{bmatrix} \sigma_x \\ \sigma_y \\ \tau_{xy} \end{bmatrix} = \begin{bmatrix} \overline{Q}_{11} & \overline{Q}_{12} & \overline{Q}_{16} \\ \overline{Q}_{21} & \overline{Q}_{22} & \overline{Q}_{26} \\ \overline{Q}_{61} & \overline{Q}_{62} & \overline{Q}_{66} \end{bmatrix} \begin{bmatrix} \varepsilon_x \\ \varepsilon_y \\ \gamma_{xy} \end{bmatrix}, \quad \begin{bmatrix} \tau_{xz} \\ \tau_{yz} \end{bmatrix} = \begin{bmatrix} \overline{Q}_{44} & \overline{Q}_{45} \\ \overline{Q}_{54} & \overline{Q}_{55} \end{bmatrix} \begin{bmatrix} \gamma_{xz} \\ \gamma_{yz} \end{bmatrix}, \tag{3}
$$

in which $\overline{Q}_{ij}$ denotes the modulus component with respect to the plate (laminate) coordinate, and the following relationships are satisfied:

$$
\overline{Q}_{12} = \overline{Q}_{21}, \overline{Q}_{16} = \overline{Q}_{61}, \overline{Q}_{26} = \overline{Q}_{62}, \overline{Q}_{45} = \overline{Q}_{54}. \tag{4}
$$

2.2. Arrangement of Virtual Boundary Springs

To model the flexural vibration behavior of composite laminates, the first-order shear deformation laminate theory is adopted, and the displacement field is given by

$$
\begin{aligned}
u_1(x, y, z) &= -z\varphi_x(x, y) \\
u_2(x, y, z) &= -z\varphi_y(x, y) \\
u_3(x, y, z) &= w(x, y)
\end{aligned} \tag{5}
$$

where u_1, u_2, and u_3 are displacement components with respect to the three global axes x, y, z, respectively; w the deflection of a point on the middle surface. Based on the linear elastic theory, the strain components in terms of displacements can be defined as

$$
\begin{aligned}
\varepsilon_b &= \begin{bmatrix} \varepsilon_x & \varepsilon_y & \gamma_{xy} \end{bmatrix}^{\mathrm{T}} = -z \begin{bmatrix} \dfrac{\partial \varphi_x}{\partial x} & \dfrac{\partial \varphi_y}{\partial y} & \dfrac{\partial \varphi_x}{\partial y} + \dfrac{\partial \varphi_y}{\partial x} \end{bmatrix}^{\mathrm{T}} \\
\varepsilon_s &= \begin{bmatrix} \gamma_{xz} & \gamma_{yz} \end{bmatrix}^{\mathrm{T}} = \begin{bmatrix} \dfrac{\partial w}{\partial x} - \varphi_x & \dfrac{\partial w}{\partial y} - \varphi_y \end{bmatrix}^{\mathrm{T}}
\end{aligned} \tag{6}
$$

According to the basic assumptions of the virtual boundary spring technique, all the classical boundary conditions can be imposed by setting extremely large or small stiffnesses to the corresponding boundary springs, and any elastic boundary can be simulated by assigning reasonable and moderate values to the boundary spring stiffness. In FSDT, there are three generalized DOFs, namely, the deflection w and two rotations of the normal line φ_x and φ_y. Therefore, one line spring and two torsion springs linking the laminates with the foundation are arranged on each edge to restrain the three DOFs, as illustrated in Figure 3. The four edges $x = 0$, $y = 0$, $x = a$, and $y = b$ are numbered 1, 2, 3, and 4, respectively. For clarity, Figure 3 only gives a detailed illustration of boundary spring arrangements at sides 2 and 3.

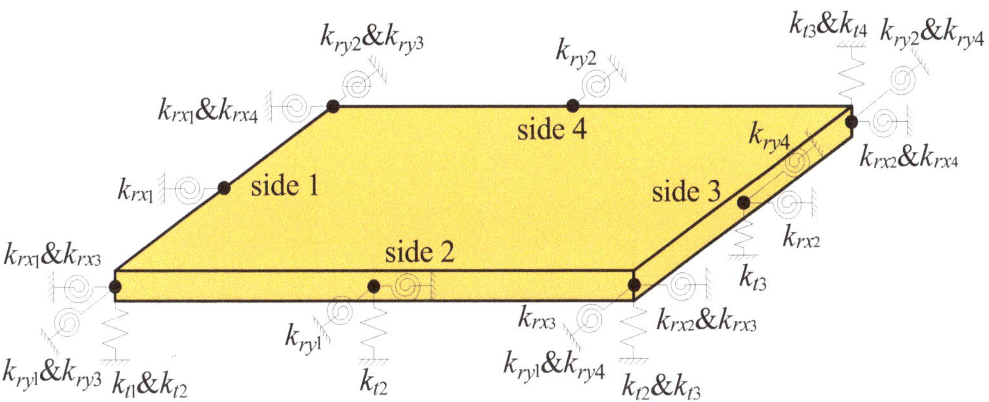

Figure 3. Arrangement of boundary springs.

The notations of all the boundary springs are explained in Table 1. The subscript i of line spring stiffness k_{ti} denotes the number of the side where the line spring is located; the subscripts 1 and 2 of torsion springs k_{rxi} and k_{ryi} denote restrictions on normal rotations, while 3 and 4 indicate constraints on tangent rotations.

Table 1. Notations and definitions of boundary springs.

Notations	Definitions
k_{ti} ($i = 1,2,3,4$)	Line spring of the i-th edge
k_{rxi} ($i = 1,2$)	Torsion springs restricting normal rotations of edges 1 and 3
k_{ryi} ($i = 1,2$)	Torsion springs restricting normal rotations of edges 2 and 4
k_{rxi} ($i = 3,4$)	Torsion springs restricting tangent rotations of edges 2 and 4
k_{ryi} ($i = 3,4$)	Torsion springs restricting tangent rotations of edges 1 and 3

2.3. Rectangular Plate Element

The previous studies [17,18] have shown that the DQFEM can afford highly accurate results even if the entire structure is modeled by very few elements, which is mainly attributed to the use of higher-order polynomials. In addition, the widely used Gauss–Lobatto points have proved to be better than the equally spaced Chebyshev and Legendre points [31–33] in boundary value problems. Therefore, in the present work, the whole plate is divided into just one element, with $M \times N$ Gauss–Lobatto nodes distributed in the domain, as shown in Figure 4.

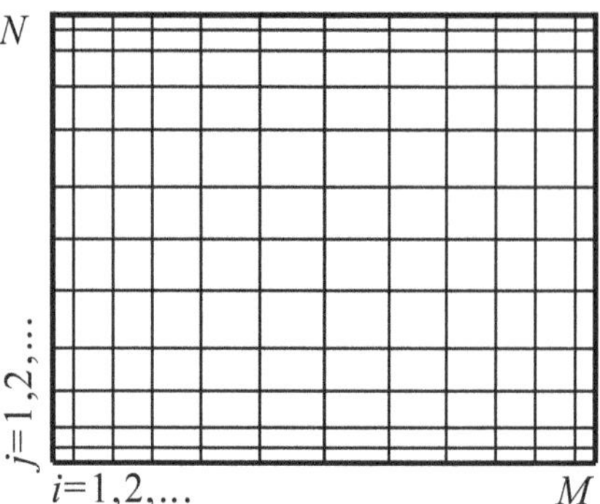

Figure 4. The distributions of grid nodes.

Introducing the Lagrange polynomials as the trial functions, the three generalized displacements can be expressed as

$$w(x,y) = \sum_{i=1}^{M} \sum_{j=1}^{N} l_i(x)l_j(y)w_{ij}$$
$$\varphi_x(x,y) = \sum_{i=1}^{M} \sum_{j=1}^{N} l_i(x)l_j(y)\varphi_{xij} \tag{7}$$
$$\varphi_y(x,y) = \sum_{i=1}^{M} \sum_{j=1}^{N} l_i(x)l_j(y)\varphi_{yij}$$

in which l_i and l_j are the Lagrange polynomials, and w_{ij}, φ_{xij}, and φ_{yij} are the deflections and rotations of the Gauss–Lobatto nodes (x_i, y_i).

To obtain the governing equations, Hamilton's principle is adopted:

$$\delta\Pi = \delta(U + V - T) = 0, \tag{8}$$

where δ is the symbol of variation, and the total potential energy Π consists of the strain energy U, the potential energy of boundary springs V, and the kinetic energy T.

For flexural vibrations of composite laminates, the strain energy can be expressed as the sum of each layer as

$$U = \frac{1}{2}\sum_{i=1}^{j} \int_{z_i}^{z_{i+1}} \int_A (\varepsilon_b^{\mathrm{T}} D_b^{(i)} \varepsilon_b + \kappa\varepsilon_s^{\mathrm{T}} D_s^{(i)} \varepsilon_s)\mathrm{d}A\mathrm{d}z, \tag{9}$$

in which j is the number of layers; κ is the shear correction factor; z_i and z_{i+1} denote the z coordinates of the top and bottom surfaces of the i-th layer in the Cartesian coordinate system; and $D_b^{(i)}$ and $D_s^{(i)}$ represent bending and shear rigidity matrices of the i-th layer in the forms of

$$D_b^{(i)} = \begin{bmatrix} \overline{Q}^{(i)}_{11} & \overline{Q}^{(i)}_{12} & \overline{Q}^{(i)}_{16} \\ \overline{Q}^{(i)}_{21} & \overline{Q}^{(i)}_{22} & \overline{Q}^{(i)}_{26} \\ \overline{Q}^{(i)}_{61} & \overline{Q}^{(i)}_{62} & \overline{Q}^{(i)}_{66} \end{bmatrix}, \quad D_s^{(i)} = \begin{bmatrix} \overline{Q}^{(i)}_{44} & \overline{Q}^{(i)}_{45} \\ \overline{Q}^{(i)}_{54} & \overline{Q}^{(i)}_{55} \end{bmatrix}. \tag{10}$$

Considering that the boundary springs are arranged continually on four edges, elastic potential energy stored in the boundary springs can be given in the integral form as

$$V = V_t + V_{rx} + V_{ry} = \frac{1}{2}\sum_{i=1}^{4} \int_0^{s_i} (k_{ti}w_i^2 + k_{rxi}\varphi_{xi}^2 + k_{ryi}\varphi_{yi}^2)\mathrm{d}s, \tag{11}$$

in which s_i denotes the length of the i-th side; w_i, φ_{xi}, and φ_{yi} represent the deflection and rotations of the i-th side.

Since the displacement field is continuous through the thickness, thus, the kinetic energy of the laminate can be written as

$$T = \frac{1}{2}\iint\limits_{A} \rho\omega^2(hw^2 + J\varphi_x^2 + J\varphi_y^2)\mathrm{d}x\mathrm{d}y, \tag{12}$$

in which ρ is the density of the laminate, and $J = h^3/12$ the axial moment of inertia; ω is the radial frequency of free vibration.

It needs to be pointed out that the potential energy stored in boundary springs is included in the total energy functional, and this special scheme has already taken boundary conditions into account; thus, during the subsequent solution procedures, no additional measures are required to process boundary conditions.

Three generalized node displacement vectors as defined as

$$\begin{aligned}
\overline{\varphi}_x^{\mathrm{T}} &= \begin{bmatrix} \varphi_{x11} & \cdots & \varphi_{xM1} & \varphi_{x12} & \cdots & \varphi_{xM2} & \cdots & \cdots & \varphi_{x1N} & \cdots & \varphi_{xMN} \end{bmatrix} \\
\overline{\varphi}_y^{\mathrm{T}} &= \begin{bmatrix} \varphi_{y11} & \cdots & \varphi_{yM1} & \varphi_{y12} & \cdots & \varphi_{yM2} & \cdots & \cdots & \varphi_{y1N} & \cdots & \varphi_{yMN} \end{bmatrix} \\
\overline{w}^{\mathrm{T}} &= \begin{bmatrix} w_{11} & \cdots & w_{M1} & w_{12} & \cdots & w_{M2} & \cdots & \cdots & w_{1N} & \cdots & w_{MN} \end{bmatrix}
\end{aligned} \tag{13}$$

Then, using the two-dimensional DQ rule in conjunction with the Gauss–Lobatto integration rule, the strain energy U, the potential energy of boundary springs V, and the kinetic energy T are further expressed in a simpler form as

$$U = \frac{1}{2}\sum_{i=1}^{j} \frac{z_{i+1}^3 - z_i^3}{3} \left\{ \begin{array}{l}
\overline{Q}^{(i)}_{11}\overline{\varphi}_x^{\mathrm{T}}\overline{A}^{(1)\mathrm{T}}C\overline{A}^{(1)}\overline{\varphi}_x + \overline{Q}^{(i)}_{12}\overline{\varphi}_x^{\mathrm{T}}\overline{A}^{(1)\mathrm{T}}C\overline{B}^{(1)}\overline{\varphi}_y \\
+\overline{Q}^{(i)}_{21}\overline{\varphi}_y^{\mathrm{T}}\overline{B}^{(1)\mathrm{T}}C\overline{A}^{(1)}\overline{\varphi}_x + \overline{Q}^{(i)}_{22}\overline{\varphi}_y^{\mathrm{T}}\overline{B}^{(1)\mathrm{T}}C\overline{B}^{(1)}\overline{\varphi}_y \\
+\overline{Q}^{(i)}_{16}\overline{\varphi}_x^{\mathrm{T}}\overline{A}^{(1)\mathrm{T}}C(\overline{B}^{(1)}\overline{\varphi}_x + \overline{A}^{(1)}\overline{\varphi}_y) \\
+\overline{Q}^{(i)}_{26}\overline{\varphi}_y^{\mathrm{T}}\overline{B}^{(1)\mathrm{T}}C(\overline{B}^{(1)}\overline{\varphi}_x + \overline{A}^{(1)}\overline{\varphi}_y) \\
+\overline{Q}^{(i)}_{61}(\overline{\varphi}_x^{\mathrm{T}}\overline{B}^{(1)\mathrm{T}}C + \overline{\varphi}_y^{\mathrm{T}}\overline{A}^{(1)\mathrm{T}}C)\overline{A}^{(1)}\overline{\varphi}_x \\
+\overline{Q}^{(i)}_{62}(\overline{\varphi}_x^{\mathrm{T}}\overline{B}^{(1)\mathrm{T}}C + \overline{\varphi}_y^{\mathrm{T}}\overline{A}^{(1)\mathrm{T}}C)\overline{B}^{(1)}\overline{\varphi}_y \\
+\overline{Q}^{(i)}_{66}(\overline{\varphi}_x^{\mathrm{T}}\overline{B}^{(1)\mathrm{T}} + \overline{\varphi}_y^{\mathrm{T}}\overline{A}^{(1)\mathrm{T}})C(\overline{B}^{(1)}\overline{\varphi}_x + \overline{A}^{(1)}\overline{\varphi}_y)
\end{array} \right\}$$
$$+ \frac{1}{2}\sum_{i=1}^{j} \frac{z_{i+1}-z_i}{3}\kappa \left\{ \begin{array}{l}
\overline{Q}^{(i)}_{44}(\overline{w}^{\mathrm{T}}\overline{A}^{(1)\mathrm{T}} - \overline{\varphi}_x^{\mathrm{T}})C(\overline{A}^{(1)}\overline{w} - \overline{\varphi}_x) \\
+\overline{Q}^{(i)}_{45}(\overline{w}^{\mathrm{T}}\overline{A}^{(1)\mathrm{T}} - \overline{\varphi}_x^{\mathrm{T}})C(\overline{B}^{(1)}\overline{w} - \overline{\varphi}_y) \\
+\overline{Q}^{(i)}_{54}(\overline{w}^{\mathrm{T}}\overline{B}^{(1)\mathrm{T}} - \overline{\varphi}_y^{\mathrm{T}})C(\overline{A}^{(1)}\overline{w} - \overline{\varphi}_x) \\
+\overline{Q}^{(i)}_{55}(\overline{w}^{\mathrm{T}}\overline{B}^{(1)\mathrm{T}} - \overline{\varphi}_y^{\mathrm{T}})C(\overline{B}^{(1)}\overline{w} - \overline{\varphi}_y)
\end{array} \right\}, \tag{14}$$

$$V = \frac{1}{2}\left(\sum_{i=1}^{4}\overline{w}^{\mathrm{T}}C\overline{K}_{ti}\overline{w} + \sum_{i=1}^{4}\overline{\varphi}_x^{\mathrm{T}}C\overline{K}_{rxi}\overline{\varphi}_x + \sum_{i=1}^{4}\overline{\varphi}_y^{\mathrm{T}}C\overline{K}_{ryi}\overline{\varphi}_y\right), \tag{15}$$

$$T = \frac{1}{2}\rho\omega^2(h\overline{w}^{\mathrm{T}}C\overline{w} + J\overline{\varphi}_x^{\mathrm{T}}C\overline{\varphi}_x + J\overline{\varphi}_y^{\mathrm{T}}C\overline{\varphi}_y), \tag{16}$$

in which $\overline{A}^{(1)}$ and $\overline{B}^{(1)}$ are weighting coefficient matrices given in Appendix A, and the matrices C, $\overline{K}_{ti}$, $\overline{K}_{rxi}$, and $\overline{K}_{ryi}$ ($i = 1, 2, 3, 4$) are defined as follows

$$C = \mathrm{diag}\left[C_1^x C_1^y, \cdots, C_M^x C_1^y, C_1^x C_2^y, \cdots, C_M^x C_2^y, \cdots, C_1^x C_N^y, \cdots, C_M^x C_N^y\right], \tag{17}$$

$$
\overline{K}_{t1} = \mathrm{diag}(\overbrace{K_{t1}, K_{t1}, \cdots, K_{t1}}^{N}), \ K_{t1} = \mathrm{diag}(k_{t1}, \overbrace{0, \cdots, 0}^{M-1})
$$
$$
\overline{K}_{t2} = \mathrm{diag}(\overbrace{k_{t2}, k_{t2}, \cdots, k_{t2}}^{M}, \overbrace{0, \cdots, 0}^{(N-1)M})
$$
$$
\overline{K}_{t3} = \mathrm{diag}(\overbrace{K_{t3}, K_{t3}, \cdots, K_{t3}}^{N}), \ K_{t3} = \mathrm{diag}(\overbrace{0, \cdots, 0}^{M-1}, k_{t3}) \tag{18}
$$
$$
\overline{K}_{t4} = \mathrm{diag}(\overbrace{0, \cdots, 0}^{(N-1)M}, \overbrace{k_{t4}, k_{t4}, \cdots, k_{t4}}^{M})
$$

$$
\overline{K}_{rx1} = \mathrm{diag}(\overbrace{K_{rx1}, K_{rx1}, \cdots, K_{rx1}}^{N}), \ K_{rx1} = \mathrm{diag}(k_{rx1}, \overbrace{0, \cdots, 0}^{M-1})
$$
$$
\overline{K}_{rx2} = \mathrm{diag}(\overbrace{K_{rx2}, K_{rx2}, \cdots, K_{rx2}}^{N}), \ K_{rx2} = \mathrm{diag}(\overbrace{0, \cdots, 0}^{M-1}, k_{rx2}) \tag{19}
$$
$$
\overline{K}_{ry1} = \mathrm{diag}(\overbrace{k_{ry1}, k_{ry1}, \cdots, k_{ry1}}^{M}, \overbrace{0, \cdots, 0}^{(N-1)M})
$$
$$
\overline{K}_{ry2} = \mathrm{diag}(\overbrace{0, \cdots, 0}^{(N-1)M}, \overbrace{k_{ry2}, k_{ry2}, \cdots, k_{ry2}}^{M})
$$

$$
\overline{K}_{ry3} = \mathrm{diag}(\overbrace{K_{ry3}, K_{ry3}, \cdots, K_{ry3}}^{N}), \ K_{ry3} = \mathrm{diag}(k_{ry3}, \overbrace{0, \cdots, 0}^{M-1})
$$
$$
\overline{K}_{ry4} = \mathrm{diag}(\overbrace{K_{ry4}, K_{ry4}, \cdots, K_{ry4}}^{N}), \ K_{ry4} = \mathrm{diag}(\overbrace{0, \cdots, 0}^{M-1}, k_{ry4}) \tag{20}
$$
$$
\overline{K}_{rx3} = \mathrm{diag}(\overbrace{k_{rx3}, k_{rx3}, \cdots, k_{rx3}}^{M}, \overbrace{0, \cdots, 0}^{(N-1)M})
$$
$$
\overline{K}_{rx4} = \mathrm{diag}(\overbrace{0, \cdots, 0}^{(N-1)M}, \overbrace{k_{rx4}, k_{rx4}, \cdots, k_{rx4}}^{M})
$$

where C_M^x and C_N^y given in Equation (A6) are the M-th and N-th Gauss–Lobatto weights with respect to x and y, respectively.

Define a displacement vector as

$$
w = \begin{bmatrix} \overline{\varphi}_x^{\mathrm{T}} & \overline{\varphi}_y^{\mathrm{T}} & \overline{w}^{\mathrm{T}} \end{bmatrix}^{\mathrm{T}}. \tag{21}
$$

Then the total potential energy can be further written in a compact form as

$$
\Pi = \frac{1}{2} w^{\mathrm{T}} K w - \frac{1}{2} \omega^2 w^{\mathrm{T}} M w, \tag{22}
$$

where the stiffness matrix K and mass matrix M are given by

$$
K = K_U + K_V, \ M = \rho\,\mathrm{diag}(JC, JC, hC), \tag{23}
$$

in which K_U and K_V account for the contributions of the strain energy and elastic potential energy of boundary springs, respectively, which are obtained as

$$
K_U = \begin{bmatrix} K_{11} & K_{12} & K_{13} \\ K_{21} & K_{22} & K_{23} \\ K_{31} & K_{32} & K_{33} \end{bmatrix}, \ K_V = \mathrm{diag}\left(\sum_{i=1}^{4} C\overline{K}_{rxi}, \sum_{i=1}^{4} C\overline{K}_{ryi}, \sum_{i=1}^{4} C\overline{K}_{ti} \right). \tag{24}
$$

The matrices in expressions of K_U are given below.

$$K_{11} = \frac{1}{2}\sum_{i=1}^{j} \frac{z_{i+1}^3 - z_i^3}{3}\left(\overline{Q}^{(i)}_{11}\overline{A}^{(1)\mathrm{T}}C\overline{A}^{(1)} + \overline{Q}^{(i)}_{16}\overline{A}^{(1)\mathrm{T}}C\overline{B}^{(1)} + \overline{Q}^{(i)}_{61}\overline{B}^{(1)\mathrm{T}}C\overline{A}^{(1)} + \overline{Q}^{(i)}_{66}\overline{B}^{(1)\mathrm{T}}C\overline{B}^{(1)} \right) + \frac{1}{2}\sum_{i=1}^{j} \frac{z_{i+1}-z_i}{3}\kappa\overline{Q}^{(i)}_{44}C$$

$$K_{12} = \frac{1}{2}\sum_{i=1}^{j} \frac{z_{i+1}^3 - z_i^3}{3}\left\{ \begin{array}{l} (\overline{Q}^{(i)}_{12} + \overline{Q}^{(i)}_{21} + \overline{Q}^{(i)}_{66})\overline{A}^{(1)\mathrm{T}}C\overline{B}^{(1)} \\ +(\overline{Q}^{(i)}_{61} + \overline{Q}^{(i)}_{16})\overline{A}^{(1)\mathrm{T}}C\overline{A}^{(1)} \\ +(\overline{Q}^{(i)}_{26} + \overline{Q}^{(i)}_{62})\overline{B}^{(1)\mathrm{T}}C\overline{B}^{(1)} \\ +\overline{Q}^{(i)}_{66}\overline{B}^{(1)\mathrm{T}}C\overline{A}^{(1)} \end{array} \right\} + \frac{1}{2}\sum_{i=1}^{j} \frac{z_{i+1}-z_i}{3}\kappa\overline{Q}^{(i)}_{54}C$$

$$K_{13} = -\frac{1}{2}\sum_{i=1}^{j} \frac{z_{i+1}-z_i}{3}\kappa(\overline{Q}^{(i)}_{44}C\overline{A}^{(1)} + \overline{Q}^{(i)}_{45}C\overline{B}^{(1)})$$

$$K_{21} = \frac{1}{2}\sum_{i=1}^{j} \frac{z_{i+1}^3 - z_i^3}{3}\left\{ \begin{array}{l} (\overline{Q}^{(i)}_{12} + \overline{Q}^{(i)}_{21} + \overline{Q}^{(i)}_{66})\overline{B}^{(1)\mathrm{T}}C\overline{A}^{(1)} \\ +(\overline{Q}^{(i)}_{61} + \overline{Q}^{(i)}_{16})\overline{A}^{(1)\mathrm{T}}C\overline{A}^{(1)} \\ +(\overline{Q}^{(i)}_{26} + \overline{Q}^{(i)}_{62})\overline{B}^{(1)\mathrm{T}}C\overline{B}^{(1)} \\ +\overline{Q}^{(i)}_{66}\overline{A}^{(1)\mathrm{T}}C\overline{B}^{(1)} \end{array} \right\} + \frac{1}{2}\sum_{i=1}^{j} \frac{z_{i+1}-z_i}{3}\kappa\overline{Q}^{(i)}_{54}C \qquad (25)$$

$$K_{22} = \frac{1}{2}\sum_{i=1}^{j} \frac{z_{i+1}^3 - z_i^3}{3}\left\{ \overline{Q}^{(i)}_{22}\overline{B}^{(1)\mathrm{T}}C\overline{B}^{(1)} + \overline{Q}^{(i)}_{26}\overline{B}^{(1)\mathrm{T}}C\overline{A}^{(1)} + \overline{Q}^{(i)}_{62}\overline{A}^{(1)\mathrm{T}}C\overline{B}^{(1)} + \overline{Q}^{(i)}_{66}\overline{A}^{(1)\mathrm{T}}C\overline{A}^{(1)} \right\} + \frac{1}{2}\sum_{i=1}^{j} \frac{z_{i+1}-z_i}{3}\kappa\overline{Q}^{(i)}_{55}C$$

$$K_{23} = -\frac{1}{2}\sum_{i=1}^{j} \frac{z_{i+1}-z_i}{3}\kappa(\overline{Q}^{(i)}_{54}C\overline{A}^{(1)} + \overline{Q}^{(i)}_{55}C\overline{B}^{(1)})$$

$$K_{31} = -\frac{1}{2}\sum_{i=1}^{j} \frac{z_{i+1}-z_i}{3}\kappa(\overline{Q}^{(i)}_{44}\overline{A}^{(1)\mathrm{T}}C + \overline{Q}^{(i)}_{54}\overline{B}^{(1)\mathrm{T}}C)$$

$$K_{32} = -\frac{1}{2}\sum_{i=1}^{j} \frac{z_{i+1}-z_i}{3}\kappa(\overline{Q}^{(i)}_{54}\overline{A}^{(1)\mathrm{T}}C + \overline{Q}^{(i)}_{55}\overline{B}^{(1)\mathrm{T}}C)$$

$$K_{33} = \frac{1}{2}\sum_{i=1}^{j} \frac{z_{i+1}-z_i}{3}\kappa\left(\overline{Q}^{(i)}_{44}\overline{A}^{(1)\mathrm{T}}C\overline{A}^{(1)} + \overline{Q}^{(i)}_{45}\overline{A}^{(1)\mathrm{T}}C\overline{B}^{(1)} + \overline{Q}^{(i)}_{54}\overline{B}^{(1)\mathrm{T}}C\overline{A}^{(1)} + \overline{Q}^{(i)}_{55}\overline{B}^{(1)\mathrm{T}}C\overline{B}^{(1)} \right)$$

The DQFEM formulation for free vibration analysis of composite laminate is eventually equivalent to an eigenvalue problem governed by a standard characteristic equation obtained from the Hamilton's principle as

$$(K - \omega^2 M)w = 0 \qquad (26)$$

It is noteworthy that the stiffness matrix K and mass matrix M are nonsingular due to the inclusion of boundary spring potential energy into the energy functional; thus, the characteristic Equation as (26) can be directly solved without reducing the order of the matrix in this equation, and the directly obtained modal vector is complete. Moreover, the boundary conditions can be conveniently changed simply by altering the boundary spring stiffness, without the need of researching and eliminating the zero DOFs.

3. Numerical Examples and Discussions

To investigate the convergence characteristics and accuracy of the modified DQFEM in application to flexural vibrations of composite laminates, a series of numerical examples are carried out.

In the following numerical examples, the shear correction factor κ is taken as $\pi^2/12$, and a three-layered symmetric cross-ply laminate with the stacking sequence $0°/90°/0°$ is considered. The elastic constants of each single-layer are given as [34,35]

$$E_1/E_2 = v_{12}/v_{21} = 40, \; G_{12} = G_{13} = 0.6E_2, \; G_{23} = 0.5E_2, \; v_{12} = 0.25$$

To facilitate comparison with other published results, a nondimensional natural frequency parameter is defined as [34,35]

$$\Omega = (\omega b^2/\pi^2)\sqrt{\rho h/D_0}, \qquad (27)$$

in which $D_0 = E_2 h^3 / 12(1 - v_{12} v_{21})$.

3.1. Convergence Characteristics

To obtain the required number of Gauss–Lobatto nodes to ensure convergent results and the recommended values of virtual boundary spring stiffnesses that make the boundaries strictly constrained, the convergence characteristics of the present method are investigated, covering both thin ($h/b = 0.001$) and thick ($h/b = 0.2$) geometries. Additionally, the specific cases of two boundary combinations, i.e., CCCC and SSSS, are considered. For these two cases, the virtual spring stiffnesses of all the four boundaries are identical. Note that for the sake of clarity, the line spring stiffness is denoted by k_t, and the stiffnesses of torsion springs that restrict the normal rotation and tangent rotation are referred to as k_{r1} and k_{r2}, respectively.

3.1.1. Varying the Number of Gauss–Lobatto Nodes

In order to facilitate the calculation, set the same number of Gauss–Lobatto nodes along the x- and y-direction. To simulate CCCC boundary combinations, theoretically, the stiffnesses of all line and torsion springs should be assigned infinitely large values to restrict both translational and rotational DOFs of all boundaries. However, infinite values cannot be processed by numerical computations; thus, a relatively large value (i.e., 10^8) is assigned instead. Similarly, to model SSSS laminates, the stiffnesses of all line springs k_t and tangent torsion springs k_{r2} should be infinitely large to restrict the transverse deflections and tangent rotations of all edges, and a large value of 10^8 is assigned to them, while the stiffness of normal torsion springs k_{r1} should be zero to set the normal rotation free.

For square laminates ($h/b = 0.001$ and 0.2) with CCCC and SSSS boundary combinations, the variations of nondimensional natural frequencies versus the number of unilateral Gauss–Lobatto nodes are depicted in Figures 5 and 6, in which the fiducial lines indicate the convergence values of frequency parameters. Note that in the present paper, if the result with three decimal digits reaches a constant value, the calculation is seen as converged. One can find from Figures 5 and 6 that for both CCCC and SSSS plates, the minimum number of unilateral nodes required to make the nondimensional frequencies converge is 11 when $h/b = 0.001$ and 0.2.

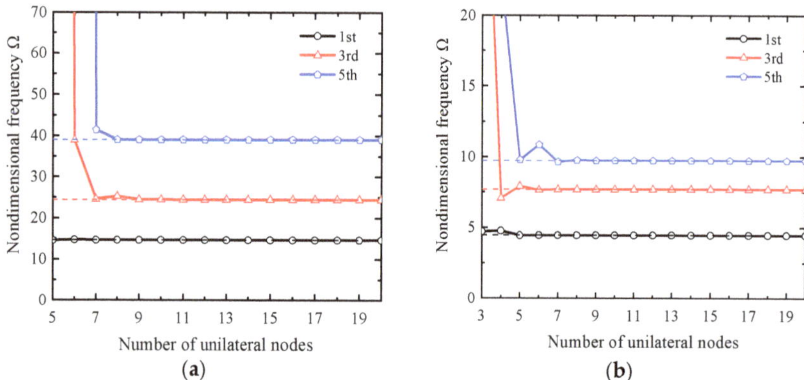

Figure 5. Variations of nondimensional frequencies Ω for CCCC laminates ($0°, 90°, 0°$) versus the number of nodes: (**a**) $h/b = 0.001$, $k_t = k_{r1} = k_{r2} = 10^8$; (**b**) $h/b = 0.2$, $k_t = k_{r1} = k_{r2} = 10^8$.

To sum up, the modified DQFEM is capable of yielding convergent results with only a few Gauss–Lobatto nodes required. In the following calculation for square laminates with thickness ratios between 0.001 and 0.2, the number of nodes per side is taken as 11 to ensure good convergence without sacrificing computational efficiency.

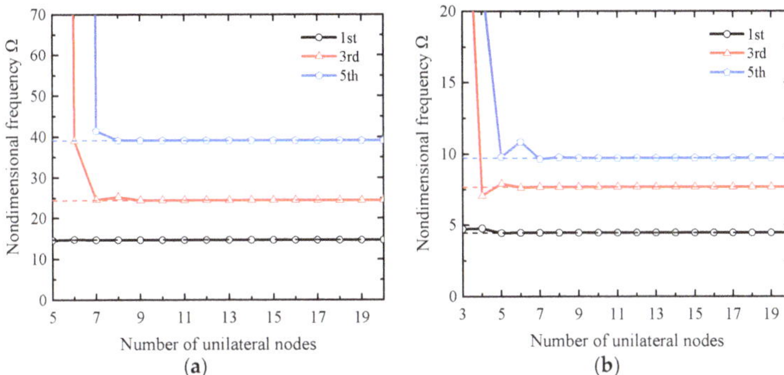

Figure 6. Variations of nondimensional frequencies Ω for SSSS laminates $(0°, 90°, 0°)$ versus the number of nodes: (**a**) $h/b = 0.001$, $k_t = k_{r2} = 10^8$, $k_{r1} = 0$; (**b**) $h/b = 0.2$, $k_t = k_{r2} = 10^8$, $k_{r1} = 0$.

3.1.2. Effect of the Boundary Spring Stiffness on Convergence

For the convenience of calculation, we assign the same value k to the stiffnesses of boundary springs corresponding to the DOFs that need to be strictly constrained. For instance, set $k_t = k_{r1} = k_{r2} = k$, when simulating a clamped edge, and $k_t = k_{r2} = k$, $k_{r1} = 0$ for a simply supported edge. During the calculation, set $M = N = 11$.

Figures 7 and 8 display the variations of nondimensional frequencies with respect to k when simulating boundary combinations of CCCC and SSSS. The straight line with an arrow points to the lower limit of k that makes the first five frequency parameters with four significant digits converge.

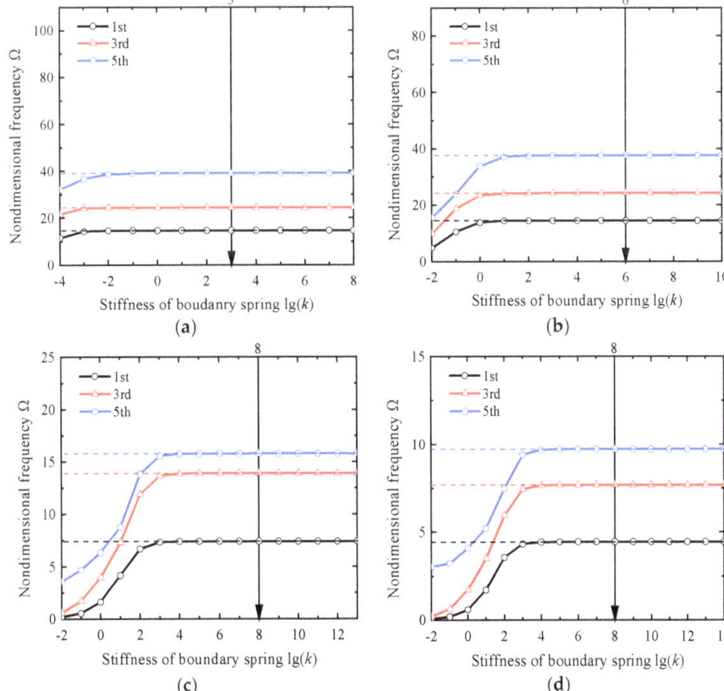

Figure 7. Variations of nondimensional frequencies Ω for CCCC laminates $(0°, 90°, 0°)$ versus the stiffness of boundary spring $(k_t = k_{r1} = k_{r2} = k)$: (**a**) $h/b = 0.001$; (**b**) $h/b = 0.01$; (**c**) $h/b = 0.1$; (**d**) $h/b = 0.2$.

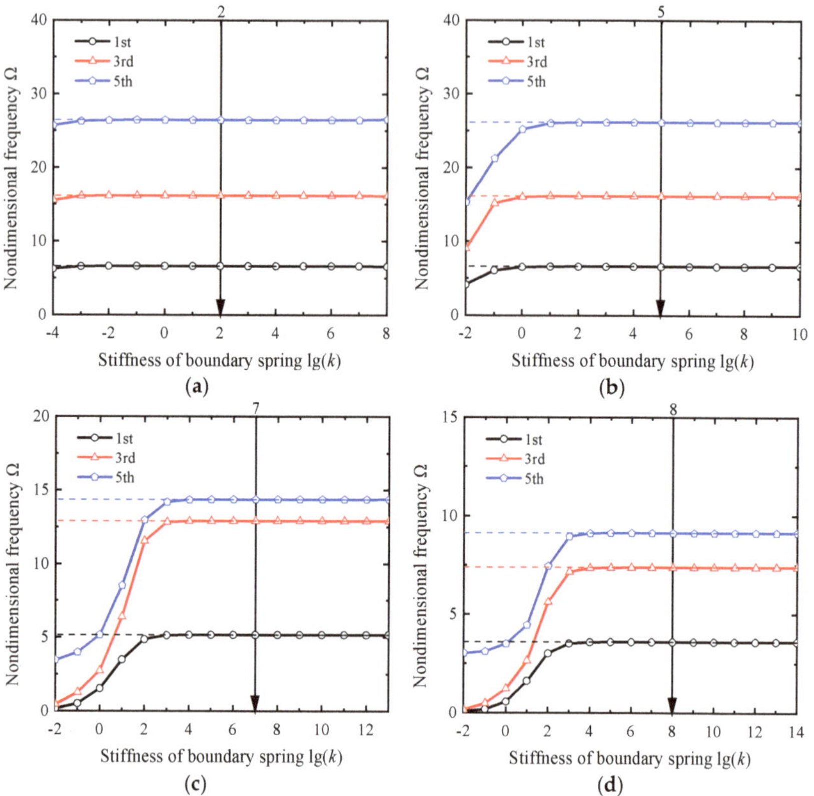

Figure 8. Variations of nondimensional frequencies Ω for SSSS laminates $(0°, 90°, 0°)$ versus the stiffness of boundary spring $(k_t = k_{r2} = k, k_{r1} = 0)$: (**a**) $h/b = 0.001$; (**b**) $h/b = 0.01$; (**c**) $h/b = 0.1$; (**d**) $h/b = 0.2$.

As is seen in Figures 7 and 8, for both CCCC and SSSS laminates with various thickness ratios ranging from 0.001 to 0.2, the frequencies experience an increase when k is relatively small so as to simulate the general elastic restraints, and the increase rate grows as the plate gets thicker. When k reaches a certain value (for instance $k = 10^2$ when $h/b = 0.01$), the frequencies will remain almost constant; this observation coincides with the basic assumption that when the spring stiffness is large enough, the corresponding DOF can be considered as strictly restricted so as to simulate the classical boundary conditions. It is noteworthy that for composite laminates with a thin geometry ($h/b = 0.001$ and 0.01), the frequency parameters are slightly influenced by the value of k as long as it is larger than 10, while the frequencies of moderately thick laminates ($h/b = 0.1$ and 0.2) are more susceptible to k.

Table 2 lists the optimal values of the boundary spring stiffnesses for clamped and simply supported boundaries. It should be emphasized that the lower limit of k is rigorously determined by the point where the first five frequency parameters accurate to four decimal places converge. This may not be shown clearly shown in Figures 7 and 8 due to the slight variation in the nondimensional frequency during the stationary part of the curve.

One can also see that the spring stiffness in Table 2 varies from 10^3 to 10^8, which is totally within the calculation ability of a personal computer. Additionally, there is an associated increase in the recommended value of k as the plate thickness increases. It should be pointed out the calculation may not converge if k is below the recommended value, and numerically ill-conditioned problems may occur if k is too large.

Table 2. Optimal values of the stiffness k for boundary springs of laminates $(0°, 90°, 0°)$.

Thickness Ratio h/b	CCCC	SSSS
0.001	10^3	10^2
0.01	10^6	10^5
0.1	10^8	10^7
0.2	10^8	10^8

3.2. Composite Laminates with Classical Boundary Conditions

3.2.1. Verification of Accuracy

To demonstrate the accuracy of Ω estimated from the modified DQFEM, a series of numerical comparisons is performed.

As is known, more accurate solutions can be generated by increasing Gauss–Lobatto nodes, but higher requirements on computing resources will be caused at the same time. Therefore, taking both accuracy and computational cost into account, and according to the discussions of the convergence characteristics above, 11×11 Gauss–Lobatto nodes are selected to discretize the square laminates in the following numerical examples. The values of boundary spring stiffnesses to simulate clamped and simply supported boundaries corresponding to various thickness ratios are listed in Table 3.

Table 3. Boundary spring stiffness for clamped and simply supported boundary.

Thickness Ratio h/b	The Stiffness k of Boundary Spring	
	Clamped Boundary (C)	Simply Supported Boundary (S)
0.001	10^3	10^2
0.01	10^6	10^5
0.05 [1]	10^7	10^6
0.1	10^8	10^7
0.15 [1]	10^8	$10^{7.5}$
0.2	10^8	10^8

[1] The corresponding stiffness are obtained by interpolation.

The first eight nondimensional natural frequencies of square laminates with various thickness ratios are calculated and listed in Tables 4–10, as well as the exact solutions by Liu [34] and numerical solutions generated by the p-Ritz method [35]. Several boundary combinations such as CCCC, SSSS, SCSC, SFSF, SSSF, SSSC, and SCSF are covered. Extensive comparisons show that the present results are highly consistent with the exact solutions for three digits. For most results, the relative errors approach zero. The non-zero relative errors exist in only a very small number of results mostly involving SFSF plates, and the maximum percentage error is less than 0.02% for the worst case. Therefore, the accuracy of the present method in free vibration of composite laminates is verified.

Table 4. The nondimensional frequency Ω for CCCC laminates.

Thickness Ratio h/b	Method	Order of Frequency							
		1st	2nd	3rd	4th	5th	6th	7th	8th
0.001	Present	14.666	17.614	24.511	35.532	39.157	40.768	44.786	50.323
	p-Ritz	14.666	17.614	24.511	35.532	39.157	40.768	44.786	50.297
0.05	Present	10.953	14.028	20.388	23.196	24.978	29.237	29.369	36.266
	p-Ritz	10.953	14.028	20.388	23.196	24.978	29.237	29.369	36.266
0.1	Present	7.411	10.393	13.913	15.429	15.806	19.572	21.489	21.620
	p-Ritz	7.411	10.393	13.913	15.429	15.806	19.572	21.489	21.620
0.15	Present	5.548	8.147	9.904	11.622	12.025	14.645	14.911	16.123
	p-Ritz	5.548	8.147	9.904	11.622	12.025	14.645	14.911	16.123
0.2	Present	4.447	6.642	7.700	9.185	9.738	11.399	11.644	12.466
	p-Ritz	4.447	6.642	7.700	9.185	9.738	11.399	11.644	12.466

Table 5. The nondimensional frequency Ω for SSSS laminates.

Thickness Ratio h/b	Method	Order of Frequency							
		1st	2nd	3rd	4th	5th	6th	7th	8th
0.001	Present	6.625	9.447	16.205	25.115	26.498	26.657	30.314	37.785
	Exact	6.625	9.447	16.205	25.115	26.498	26.657	30.314	37.785
	p-Ritz	6.625	9.447	16.205	25.115	26.498	26.657	30.314	37.785
0.05	Present	6.138	8.888	15.110	19.354	20.665	24.070	24.344	31.028
	Exact	6.138	8.888	15.110	19.354	20.665	24.070	24.344	31.028
	p-Ritz	6.138	8.888	15.110	19.354	20.665	24.070	24.344	31.028
0.1	Present	5.166	7.757	12.915	13.049	14.376	17.788	19.502	21.051
	Exact	5.166	7.757	12.915	13.049	14.376	17.788	19.502	21.051
	p-Ritz	5.166	7.757	12.915	13.049	14.376	17.788	19.502	21.051
0.15	Present	4.275	6.667	9.488	10.824	10.826	13.804	14.665	15.590
	Exact	4.275	6.667	9.488	10.824	10.826	13.804	14.665	15.590
	p-Ritz	4.275	6.667	9.488	10.824	10.826	13.804	14.665	15.590
0.2	Present	3.594	5.769	7.397	8.688	9.145	11.208	11.223	12.117
	Exact	3.594	5.769	7.397	8.688	9.145	11.208	11.223	12.117
	p-Ritz	3.594	5.769	7.397	8.688	9.145	11.208	11.223	12.117

Table 6. The nondimensional frequency Ω for SCSC laminates.

Thickness Ratio h/b	Method	Order of Frequency							
		1st	2nd	3rd	4th	5th	6th	7th	8th
0.05	Present	6.890	11.246	18.664	19.619	21.801	26.689	28.260	34.348
	Exact	6.890	11.246	18.664	19.619	21.801	26.689	28.260	34.348
	p-Ritz	6.890	11.246	18.664	19.619	21.801	26.689	28.260	34.348
0.1	Present	5.871	9.454	13.340	14.878	15.340	19.229	21.231	21.275
	Exact	5.871	9.454	13.340	14.878	15.340	19.229	21.231	21.275
	p-Ritz	5.871	9.454	13.340	14.878	15.340	19.229	21.231	21.275
0.15	Present	4.275	6.667	9.488	10.824	10.826	13.804	14.665	15.590
	Exact	4.275	6.667	9.488	10.824	10.826	13.804	14.665	15.590
	p-Ritz	4.275	6.667	9.488	10.824	10.826	13.804	14.665	15.590
0.2	Present	4.137	6.474	7.664	9.159	9.643	11.377	11.625	12.448
	Exact	4.137	6.474	7.664	9.159	9.643	11.377	11.625	12.448
	p-Ritz	4.137	6.474	7.664	9.159	9.643	11.377	11.625	12.448

Table 7. The nondimensional frequency Ω for SFSF laminates.

Thickness ratio h/b	Method	Order of Frequency							
		1st	2nd	3rd	4th	5th	6th	7th	8th
0.05	Present	5.734	5.933	7.398	11.918	19.124	19.284	19.603	20.087
	Exact	5.734	5.933	7.397	11.917	19.124	19.284	19.602	20.086
	p-Ritz	5.734	5.933	7.397	11.918	19.124	19.284	19.602	20.086
0.1	Present	4.781	4.935	6.320	10.345	12.851	12.959	13.677	16.070
	Exact	4.781	4.935	6.319	10.345	12.851	12.959	13.677	16.070
	p-Ritz	4.781	4.935	6.319	10.345	12.851	12.959	13.677	16.070
0.2	Present	3.213	3.311	4.619	7.195	7.273	7.599	8.004	10.043
	Exact	3.213	3.311	4.619	7.195	7.272	7.599	8.004	10.043
	p-Ritz	3.213	3.311	4.619	7.195	7.272	7.599	8.004	10.043

Although the above convergence study concentrates on CCCC and SSSS plates, one can see from the numerical comparisons that the conclusions regarding the required node number and recommended boundary spring stiffnesses have been successfully extended into the analysis of other boundary conditions. Additionally, it implies that a slight variation in the values of boundary spring stiffnesses within a specific interval might have an influence on the obtained results, but only to a limited extent. Therefore, it can be reasonably inferred that the present approach is numerically stable and highly accurate, regardless of boundary conditions.

Table 8. The nondimensional frequency Ω for SSSF laminates.

Thickness Ratio h/b	Method	Order of Frequency							
		1st	2nd	3rd	4th	5th	6th	7th	8th
0.05	Present	5.785	6.657	10.301	17.279	19.165	19.655	21.520	25.971
	Exact	5.785	6.657	10.301	17.279	19.165	19.655	21.519	25.970
0.1	Present	4.821	5.641	8.976	12.879	13.304	14.614	15.144	19.121
	Exact	4.821	5.641	8.976	12.879	13.304	14.614	15.144	19.121
0.2	Present	3.240	4.017	6.654	7.216	7.642	9.323	10.195	11.077
	Exact	3.240	4.017	6.654	7.216	7.642	9.323	10.195	11.077

Table 9. The nondimensional frequency Ω for SSSC laminates.

Thickness Ratio h/b	Method	Order of Frequency							
		1st	2nd	3rd	4th	5th	6th	7th	8th
0.05	Present	6.429	9.983	16.848	19.459	21.172	25.460	26.159	32.661
	Exact	6.429	9.983	16.847	19.459	21.172	25.460	26.159	32.661
0.1	Present	5.450	8.587	13.165	13.914	14.832	18.510	20.413	21.123
	Exact	5.450	8.587	13.165	13.914	14.832	18.510	20.412	21.123
0.2	Present	3.835	6.140	7.513	8.931	9.401	11.282	11.429	12.286
	Exact	3.835	6.140	7.513	8.931	9.401	11.282	11.429	12.286

Table 10. The nondimensional frequency Ω for SCSF laminates.

Thickness Ratio h/b	Method	Order of Frequency							
		1st	2nd	3rd	4th	5th	6th	7th	8th
0.05	Present	5.8293	7.1375	11.5836	19.1261	19.1837	19.8523	22.1823	27.2341
	Exact	5.8293	7.1375	11.5836	19.1261	19.1837	19.8523	22.1823	27.2341
0.1	Present	4.8650	6.0724	9.8872	12.8983	13.4994	15.6061	15.6911	19.8715
	Exact	4.8650	6.0724	9.8872	12.8983	13.4994	15.6061	15.6911	19.8715
0.2	Present	3.2877	4.3135	7.0132	7.2389	7.7982	9.5741	10.4079	11.0930
	Exact	3.2877	4.3135	7.0132	7.2389	7.7982	9.5741	10.4079	11.0930

Figure 9 presents the first three modes for SSSS, SCSC, and SSSF laminates with $h/b = 0.1$, illustrating the physical patterns of the modes.

3.2.2. Verification of Efficiency

To assess the efficiency of the present method in free vibration of composite laminates, comparisons of computation time are carried out with the classical FEM. Square laminates with thickness ratio $h/b = 0.1$ and boundary combinations of SSSS and CCCC are considered.

Varying the number of unilateral nodes, the first six nondimensional frequencies are calculated by the present method and the FEM adopting the commonly used Q4 element. It should be pointed out that all the calculations are made by running the same software program on the same computer to guarantee the effectiveness of comparisons. The variations of both the runtime and frequencies in terms of node number per edge are presented in Figures 10 and 11. Note that for clarity, only the variations of the first, third and fifth frequencies are depicted in these figures. One can see that for both CCCC and SSSS cases, when using the present method, only 11 Gauss–Lobatto nodes per edge are needed to make the first six frequency parameters with three decimal digits converge, and the calculation time is less than 0.5 s. In contrast, when using FEM, the first six modes do not converge even when the number of nodes per edge reaches 60, and the calculation time already exceeds 45 s. These results demonstrates that the present method has incomparable advantages in computation efficiency over the classical FEM, and the remarkable convergence of the current solution is also demonstrated.

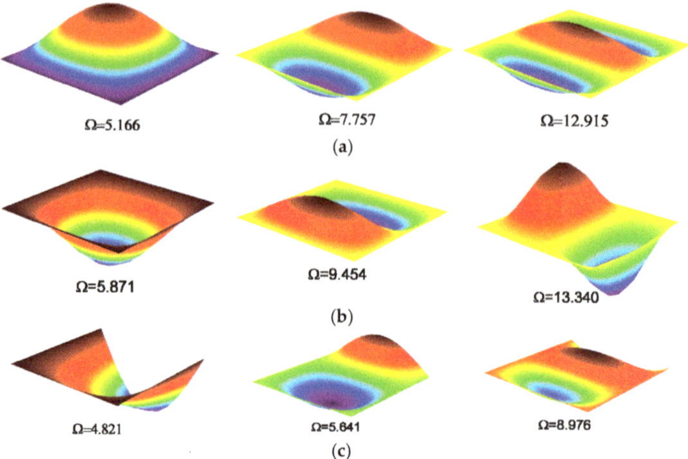

Figure 9. The first three modes for square laminates $(0°, 90°, 0°)$ with $h/b = 0.1$: (**a**) SSSS; (**b**) SCSC; (**c**) SSSF.

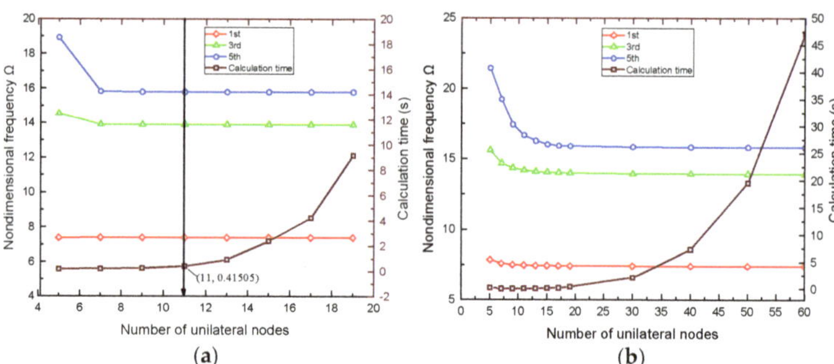

Figure 10. Variations of nondimensional frequencies and the corresponding calculation time vs. The number of unilateral nodes for CCCC laminates with $h/b = 0.1$: (**a**) the present method; (**b**) FEM using Q4 element.

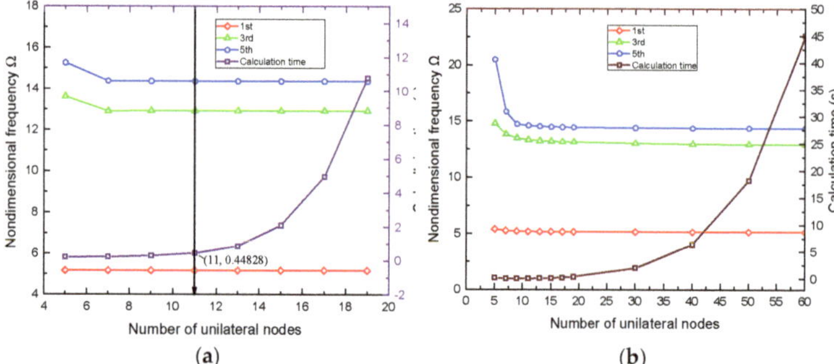

Figure 11. Variations of nondimensional frequencies and the corresponding calculation time vs. The number of unilateral nodes for SSSS laminates with $h/b=0.1$: (**a**) the present method; (**b**) FEM using Q4 element.

3.3. Composite Laminates with Elastic Boundary Conditions

The above numerical examples focus on composite laminates with classical boundaries, the free vibration characteristics of which are comprehensive in the published literatures, while those involving arbitrary elastic boundaries are relatively rare. To provide some supplementary and reference results, the following numerical examples are carried out covering three types of elastic boundaries often encountered in practical engineering. The first type referred to as E_1 makes only lateral deflection of plate boundary elastically constrained; two rotations strictly constrained, the corresponding spring stiffness for which is given as: $k_t = 10^2$, $k_{r1} = k_{r2} = 10^8$. Similarly, the second type E_2 allows two rotations elastically restrained with the boundary spring stiffness being set as $k_{r1} = k_{r2} = 10^2$ and $k_t = 10^8$, while in the third type E_3, both lateral deflection and two rotations are elastically restrained (i.e., $k_t = k_{r1} = k_{r2} = 10^2$).

The non-dimensional frequencies for composite laminates with thickness ratios of 0.01, 0.1, and 0.2 are shown in Table 11. It is shown that the natural frequencies have not changed much for composite laminates with a thin geometry ($h/b = 0.01$) regardless of boundary conditions, which coincide with the conclusions made in the previous analysis. Additionally, one can find that when only the lateral deflection of plate boundary is elastically constrained ($E_1E_1E_1E_1$), the natural frequencies decrease obviously compared to those of the fully clamped laminates, while for the case wherein only two rotations are elastically restrained ($E_2E_2E_2E_2$), there is a slight decline in the natural frequencies, which indicates that constraints on the lateral deflection rather than rotations play a more important role on the natural frequencies for composite laminates with elastic boundaries.

Table 11. The nondimensional frequency Ω for composite laminates with elastic boundary conditions.

Thickness Ratio h/b	B.C.	Order of Frequency					
		1st	2nd	3rd	4th	5th	6th
0.01	CCCC	14.4339	17.3892	24.2667	35.1818	37.7770	39.3875
	$E_1E_1E_1E_1$	14.4271	17.3823	24.2583	35.1684	37.7253	39.3352
	$E_2E_2E_2E_2$	14.4336	17.3890	24.2665	35.1817	37.7762	39.3867
	$E_3E_3E_3E_3$	14.4268	17.3820	24.2581	35.1682	37.7245	39.3344
0.1	CCCC	7.4108	10.3927	13.9129	15.4287	15.8056	19.5720
	$E_1E_1E_1E_1$	6.7022	9.5265	11.9340	13.8435	13.8624	17.2335
	$E_2E_2E_2E_2$	7.3785	10.3671	13.9005	15.4083	15.7924	19.5584
	$E_3E_3E_3E_3$	6.6796	9.5084	11.9316	13.8306	13.8579	17.2287
0.2	CCCC	4.4466	6.6419	7.6996	9.1852	9.7378	11.3991
	$E_1E_1E_1E_1$	3.5877	5.2085	5.9962	7.1819	7.5080	9.0682
	$E_2E_2E_2E_2$	4.4054	6.6109	7.6925	9.1741	9.7175	11.3933
	$E_3E_3E_3E_3$	3.5673	5.1971	5.9838	7.1720	7.5030	9.0614

Although the results presented in this section are for three types of elastic boundary combinations only, the present solution procedure can be readily applied to plates subjected to more complex boundary conditions such as point supports, partial supports, non-uniform elastic restraints, and their combinations.

4. Conclusions

This paper introduces the virtual boundary spring technique into DQFEM to deal with the flexural vibrations of composite laminates. In this new formulated method, boundary conditions are considered in the first step by including the potential energy stored in boundary springs when constructing the energy functional; thus, during the subsequent solution procedures, no special schemes are required to deal with boundary conditions, which is different from the standard DQFEM.

The most significant superiority of the present approach is that it can be universally applicable to composite laminates with any combinations of elastic boundary conditions including all the classical cases without the need of making any change to the solution procedure. Another advantage of the modified DQFEM over the standard one is that the former facilitates switches of boundary conditions, while in the latter, changing boundary conditions requires researching and eliminating the zero node displacements, which will increase computational cost.

Well-behaved convergence characteristics of the present method are demonstrated. The minimum number of unilateral Gauss–Lobatto nodes to generate convergent solutions and the recommended values of boundary spring stiffnesses are obtained as well. The nondimensional natural frequencies of square laminates under various classical boundary conditions and thickness ratios agree well with available analytical and numerical results from other analyses, which validates the high accuracy of the present method.

Some new results are presented for elastically restrained composite laminates, which can serve as reference values. Moreover, the present solution procedure can be readily extended to composite laminates with more complicated boundary conditions such as multi-point supports, partial supports, non-uniform elastic constraints, and so on.

Author Contributions: Conceptualization, W.X.; methodology, W.X.; software, X.L.; validation, X.L. and L.H.; formal analysis, W.X. and X.L.; writing—original draft preparation, W.X.; writing—review and editing, L.H.; visualization, X.L.; project administration, L.H.; funding acquisition, L.H. All authors have read and agreed to the published version of the manuscript.

Funding: This research was partially supported by the National Natural Science Foundation, China (grant number 51705436), the National Science and Technology Major Project, China (grant number 2017-I-0011-0012), and the Sichuan Province Science and Technology Support Program (grant number 2021JDRC0174).

Institutional Review Board Statement: Not applicable.

Informed Consent Statement: Not applicable.

Data Availability Statement: The data used to support the findings of this study is available from the authors upon request.

Conflicts of Interest: The authors declare no conflict of interest. The funders had no role in the design of the study; in the collection, analyses, or interpretation of data; in the writing of the manuscript; or in the decision to publish the results.

Appendix A

For completeness of the present paper, a brief review of the two-dimensional DQ rule is outlined here. The partial derivatives of $f(x, y)$ can be expressed as the following compact form:

$$\left.\frac{\partial^r f}{\partial x^r}\right|_k = \overline{A}^{(r)}\overline{f}, \quad \left.\frac{\partial^s f}{\partial y^s}\right|_k = \overline{B}^{(s)}\overline{f}, \quad \left.\frac{\partial^{r+s} f}{\partial x^r \partial y^s}\right|_k = \overline{A}^{(r)}\overline{B}^{(s)}\overline{f}, \tag{A1}$$

where

$$\overline{A}^{(r)} = \begin{bmatrix} A^{(r)} & 0 & \cdots & 0 \\ 0 & A^{(r)} & \cdots & 0 \\ \vdots & \vdots & \ddots & \vdots \\ 0 & 0 & \cdots & A^{(r)} \end{bmatrix}_{(M\times N)\times(M\times N)}, \quad A^{(r)} = \left(A_{ij}^{(r)}\right)_{M\times M}, \tag{A2}$$

$$\overline{B}^{(s)} = \begin{bmatrix} B_{11}^{(s)} & B_{12}^{(s)} & \cdots & B_{1N}^{(s)} \\ B_{21}^{(s)} & B_{22}^{(s)} & \cdots & B_{2N}^{(s)} \\ \vdots & \vdots & \ddots & \vdots \\ B_{N1}^{(s)} & B_{N2}^{(s)} & \cdots & B_{NN}^{(s)} \end{bmatrix}_{(M\times N)\times(M\times N)}, \quad B_{ij}^{(s)} = \mathrm{diag}\left(B_{ij}^{(s)}, \cdots, B_{ij}^{(s)}\right)_{M\times M}, \tag{A3}$$

$$\overline{f} = \begin{bmatrix} f_{11} & \cdots & f_{M1} & f_{12} & \cdots & f_{M2} & \cdots\cdots & f_{1N} & \cdots & f_{MN} \end{bmatrix}^{\mathrm{T}}, \tag{A4}$$

in which M and N represent the number of grid points in the x and y directions, respectively, and $k = (j-1)M + i$, $(i = 1, 2, \ldots, M; j = 1, 2, \ldots, N)$; $A_{ij}^{(r)}$ and $B_{ij}^{(s)}$ are the weighting coefficients associated with the rth-order partial derivative with respect to x and the sth-order partial derivative with respect to y.

To make the paper self-contained, an overview of the Gauss–Lobatto integration rule is also provided here. The Gauss integration of function $f(x)$ in the interval $[-1, 1]$ with a precision degree of $(2n - 3)$ is given as

$$\int_{-1}^{1} f(x)\mathrm{d}x = \sum_{j=1}^{n} C_j f(x_j), \tag{A5}$$

in which the weighting coefficients are given by

$$C_1 = C_n = \frac{2}{n(n-1)}, C_j = \frac{2}{n(n-1)[P_{n-1}(x_j)]^2}(j \neq 1, n), \tag{A6}$$

where x_j is the $(j-1)$th zero of $P'_{n-1}(x)$, the zeros of which are the eigenvalues of its companion matrix; and the Legendre polynomial $P_n(x)$ of degree n is expressed as

$$P_n(x) = \sum_{k=0}^{[n/2]} \frac{(-1)^k(2n-2k)!}{2^n k!(n-k)!(n-2k)!} x^{n-2k}. \tag{A7}$$

References

1. Sayyad, A.S.; Ghugal, Y.M. On the free vibration analysis of laminated composite and sandwich plates: A review of recent literature with some numerical results. *Compos. Struct.* **2015**, *129*, 177–201. [CrossRef]
2. Akhras, G.; Li, W. Static and free vibration analysis of composite plates using spline finite strips with higher-order shear deformation. *Compos. Part B-Eng.* **2005**, *36*, 496–503. [CrossRef]
3. Carrera, E.; Fazzolari, F.A.; Demasi, L. Vibration analysis of anisotropic simply supported plates by using variable kinematic and Rayleigh-Ritz method. *J. Vib. Acoust.-Trans. ASME* **2011**, *133*, 061017. [CrossRef]
4. Chien, R.-D.; Chen, C.-S. Nonlinear vibration of laminated plates on an elastic foundation. *Thin-Walled Struct.* **2006**, *44*, 852–860. [CrossRef]
5. Gupta, U.S.; Ansari, A.H.; Sharma, S. Buckling and vibration of polar orthotropic circular plate resting on Winkler foundation. *J. Sound Vib.* **2006**, *297*, 457–476. [CrossRef]
6. Kshirsagar, S.; Bhaskar, K. Accurate and elegant free vibration and buckling studies of orthotropic rectangular plates using untruncated infinite series. *J. Sound Vib.* **2008**, *314*, 837–850. [CrossRef]
7. Morozov, E.V.; Lopatin, A.V. Fundamental Frequency of Fully Clamped Composite Sandwich Plate. *J. Sandw. Struct. Mater.* **2010**, *12*, 591–619. [CrossRef]
8. Muthurajan, K.G.; Sankaranarayanasamy, K.; Tiwari, S.B.; Rao, B.N. Nonlinear vibration analysis of initially stressed thin laminated rectangular plates on elastic foundations. *J. Sound Vib.* **2005**, *282*, 949–969. [CrossRef]
9. Nallim, L.G.; Grossi, R.O. Natural frequencies of symmetrically laminated elliptical and circular plates. *Int. J. Mech. Sci.* **2008**, *50*, 1153–1167. [CrossRef]
10. Ren, X.; Chen, W. Free vibration analysis of laminated and sandwich plates using quadrilateral element based on an improved zig-zag theory. *J. Compos. Mater.* **2011**, *45*, 2173–2187. [CrossRef]
11. Wu, Z.; Chen, W.; Ren, X. An accurate higher-order theory and C-0 finite element for free vibration analysis of laminated composite and sandwich plates. *Compos. Struct.* **2010**, *92*, 1299–1307. [CrossRef]
12. Zhou, D.; Cheung, Y.K.; Au, F.T.K.; Lo, S.H. Three-dimensional vibration analysis of thick rectangular plates using Chebyshev polynomial and Ritz method. *Int. J. Solids Struct.* **2002**, *39*, 6339–6353. [CrossRef]
13. Dai, K.Y.; Liu, G.R.; Lim, K.M.; Chen, X.L. A mesh-free method for static and free vibration analysis of shear deformable laminated composite plates. *J. Sound Vib.* **2004**, *269*, 633–652. [CrossRef]
14. Neves, A.M.A.; Ferreira, A.J.M.; Carrera, E.; Cinefra, M.; Roque, C.M.C.; Jorge, R.M.N.; Soares, C.M.M. Static, free vibration and buckling analysis of isotropic and sandwich functionally graded plates using a quasi-3D higher-order shear deformation theory and a meshless technique. *Compos. Part B-Eng.* **2013**, *44*, 657–674. [CrossRef]
15. Ferreira, A.J.M.; Carrera, E.; Cinefra, M.; Viola, E.; Tornabene, F.; Fantuzzi, N.; Zenkour, A.M. Analysis of thick isotropic and cross-ply laminated plates by generalized differential quadrature method and a Unified Formulation. *Compos. Part B-Eng.* **2014**, *58*, 544–552. [CrossRef]
16. Liew, K.M.; Zhang, J.J.; Ng, T.Y.; Reddy, J.N. Dynamic characteristics of elastic bonding in composite laminates: A free vibration study. *J. Appl. Mech.-Trans. Asme* **2003**, *70*, 860–870. [CrossRef]

17. Xing, Y.; Liu, B. High-accuracy differential quadrature finite element method and its application to free vibrations of thin plate with curvilinear domain. *Int. J. Numer. Methods Eng.* **2009**, *80*, 1718–1742. [CrossRef]
18. Xing, Y.; Liu, B.; Liu, G. A differential quadrature finite element method. *Int. J. Appl. Mech.* **2010**, *2*, 207–227. [CrossRef]
19. Du, Y.; Pang, F.; Sun, L.; Li, H. A unified formulation for dynamic behavior analysis of spherical cap with uniform and stepped thickness distribution under different edge constraints. *Thin-Walled Struct.* **2020**, *146*, 106445. [CrossRef]
20. Kim, K.; Kim, K.; Han, C.; Jang, Y.; Han, P. A method for natural frequency calculation of the functionally graded rectangular plate with general elastic restraints. *AIP Adv.* **2020**, *10*, 085203. [CrossRef]
21. Kim, K.; Kim, S.; Sok, K.; Pak, C.; Han, K. A modeling method for vibration analysis of cracked beam with arbitrary boundary condition. *J. Ocean. Eng. Sci.* **2018**, *3*, 367–381. [CrossRef]
22. Wang, Q.; Qin, B.; Shi, D.; Liang, Q. A semi-analytical method for vibration analysis of functionally graded carbon nanotube reinforced composite doubly-curved panels and shells of revolution. *Compos. Struct.* **2017**, *174*, 87–109. [CrossRef]
23. Wang, Q.; Shao, D.; Qin, B. A simple first-order shear deformation shell theory for vibration analysis of composite laminated open cylindrical shells with general boundary conditions. *Compos. Struct.* **2018**, *184*, 211–232. [CrossRef]
24. Wang, Q.; Shi, D.; Liang, Q.; Pang, F. Free vibrations of composite laminated doubly-curved shells and panels of revolution with general elastic restraints. *Appl. Math. Modell.* **2017**, *46*, 227–262. [CrossRef]
25. Wang, Q.; Shi, D.; Liang, Q.; Pang, F. Free vibration of four-parameter functionally graded moderately thick doubly-curved panels and shells of revolution with general boundary conditions. *Appl. Math. Modell.* **2017**, *42*, 705–734. [CrossRef]
26. Wang, Q.; Shi, D.; Liang, Q.; Shi, X. A unified solution for vibration analysis of functionally graded circular, annular and sector plates with general boundary conditions. *Compos. Part B-Eng.* **2016**, *88*, 264–294. [CrossRef]
27. Wang, Q.; Shi, D.; Pang, F.; Ahad, F.e. Benchmark solution for free vibration of thick open cylindrical shells on Pasternak foundation with general boundary conditions. *Meccanica* **2017**, *52*, 457–482. [CrossRef]
28. Wang, Q.; Shi, D.; Shi, X. A modified solution for the free vibration analysis of moderately thick orthotropic rectangular plates with general boundary conditions, internal line supports and resting on elastic foundation. *Meccanica* **2016**, *51*, 1985–2017. [CrossRef]
29. Mindlin, R.D. Influence of Rotatory Inertia and Shear on Flexural Motions of Isotropic, Elastic Plates. *J. Appl. Mech.* **1951**, *18*, 31–38. [CrossRef]
30. Reissner, E. The Effect of Transverse Shear Deformation on the Bending of Elastic Plates. *J. Appl. Mech.* **1945**, *12*, A69–A77. [CrossRef]
31. Bert, C.W.; Malik, M. Free vibration analysis of thin cylindrical shells by the differential quadrature method. *J. Press. Vessel Technol.* **1996**, *118*, 1–12. [CrossRef]
32. Bert, C.W.; Malik, M. The differential quadrature method for irregular domains and application to plate vibration. *Int. J. Mech. Sci.* **1996**, *38*, 589–606. [CrossRef]
33. Malik, M.; Bert, C. Implementing multiple boundary conditions in the DQ solution of higher-order PDEs: Application to free vibration of plates. *Int. J. Numer. Methods Eng.* **1996**, *39*, 1237–1258. [CrossRef]
34. Liu, B.; Xing, Y.F.; Reddy, J.N. Exact compact characteristic equations and new results for free vibrations of orthotropic rectangular Mindlin plates. *Compos. Struct.* **2014**, *118*, 316–321. [CrossRef]
35. Liew, K.M. Solving the vibration of thick symmetric laminates by Reissner/Mindlin plate theory and the p-Ritz method. *J. Sound Vib.* **1996**, *198*, 343–360. [CrossRef]

Article

Finite Element Modeling of Interface Behavior between Normal Concrete and Ultra-High Performance Fiber-Reinforced Concrete

Xuan-Bach Luu and Seong-Kyum Kim *

Department of Civil Engineering, Kumoh National Institute of Technology, Gumi 39177, Republic of Korea; bach@kumoh.ac.kr
* Correspondence: skim@kumoh.ac.kr

Abstract: The behavior at the interface between normal strength concrete (NSC) and Ultra-High Performance Fiber-Reinforced Concrete (UHPFRC) plays a crucial role in accurately predicting the capacity of UHPFRC for strengthening and repairing concrete structures. Until now, there has been a lack of sufficient finite element (FE) models for accurately predicting the behavior at the interface between NSC and UHPFRC. This study aims to investigate the structural behavior of composite members made of NSC and UHPFRC by developing a model that accurately simulates the interface between the two materials using a linear traction-separation law. Novel parameters for the surface-based cohesive model, based on the traction-separation model, were obtained and calibrated from prior experiments using analytical methods. These parameters were then integrated into seven FE models to simulate the behavior at the interface between NSC and UHPFRC in shear, tensile, and flexural tests. The accuracy of the FE models was validated using experimental data. The findings revealed that the proposed FE models could effectively predict the structural behavior of composite NSC-UHPFRC members under various working conditions. Specifically, the maximum deviations between EXP and FEA were 6.8% in ultimate load for the shear test and 15.9% and 2.8% in ultimate displacement for the tensile and flexural tests, respectively. The model can be utilized to design the use of UHPFRC and ultra-high performance fiber-reinforced shotcrete (UHPFRS) for repairing and strengthening damaged concrete structures.

Keywords: interfacial behavior; bond strength; ultra-high-performance fiber-reinforced concrete (UHPFRC); interfacial transition zone; numerical concrete model

Citation: Luu, X.-B.; Kim, S.-K. Finite Element Modeling of Interface Behavior between Normal Concrete and Ultra-High Performance Fiber-Reinforced Concrete. *Buildings* **2023**, *13*, 950. https://doi.org/10.3390/buildings13040950

Academic Editors: Zechuan Yu and Dongming Li

Received: 6 March 2023
Revised: 27 March 2023
Accepted: 29 March 2023
Published: 3 April 2023

1. Introduction

Ultra-High Performance Fiber-Reinforced Concrete (UHPFRC) is a highly advanced material renowned for its exceptional strength, high ductility, and low permeability [1–3]. This advancement has led to increasing interest in UHPFRC as a promising solution for rehabilitating and strengthening aged and damaged concrete structures. UHPFRC's excellent bonding ability with normal strength concrete (NSC) has been particularly noteworthy, making it an attractive option for repair and strengthening applications. Therefore, the objective of this study is to develop a reliable model for predicting the bonding behavior at the interface between NSC and UHPFRC under various loading conditions.

Finite Element (FE) analysis is a vital tool for studying the behavior of composite NSC-UHPFRC members and designing them to enhance their strength and repairability. Despite its widespread use in investigating bonding behavior in composite structures, previous studies have mostly focused on specific conditions.

The cohesive zone model is a popular technique used to model the behavior between NSC and UHPFRC using a thin material layer. This approach allows for the simulation of debonding, crack initiation, and crack propagation that can occur at the interface between

NSC and UHPFRC. Tong et al. [4] developed a 2D eight node quadratic line cohesive element FE model to simulate the thin layer between NSC and UHPFRC, with the parameters calibrated from slant shear. The findings showed good agreement with experimental results in a flexural test of composite NC-UHPC members. Yu et al. [5] developed a 2D cohesive model with fixed parameters to study the bonding behavior of the NSC-UHPFRC interface. Surface roughness was modeled using representative volume elements (RVEs). The simulated traction-separation curve agreed well with experimental data, with a maximum deviation of around 27%.

Additionally, 3D models have been developed by researchers using cohesive zone models to simulate the bonding behavior between NSC and UHPFRC. For instance, Valikhani et al. [6] created a 3D FE model using ATENA software with a cohesive element calibrated from a bi-surface shear test. The simulated results matched well with the experimental results in the bi-surface test. Similarly, Hussein et al. [7] successfully used a cohesive zone model to analyze the direct tensile behavior of HSC-UHPFRC specimens, with results that matched well with the experimental data. Another study by Kadhim et al. [8] employed a cohesive zone model with contact-target elements in ANSYS software, but the results were not validated. However, while cohesive models have been introduced in some studies, their validity for accurately modeling NSC-UHPFRC composite members under different working conditions is yet to be established.

Perfect bonding using tie constraints is also a common method for modeling the NSC-UHPFRC interface, but it can lead to an overestimation of load-carrying capacity and design errors. According to Farzad et al. [9], the use of a tie constraint to simulate the bonding interface between NSC and UHPFRC in slant shear and direct shear tests led to overestimated results that were up to 150% different from the experimental data.

Several other models have been developed to simulate the behavior of the NSC-UHPFRC interface. For instance, Lampropoulos et al. [10] proposed a 3D model with a cohesion and friction coefficient of 1.5 MPa and 1.5, respectively, to simulate the behavior of a composite beam on a well-roughened substrate. However, the model's validity on the composite beam has not been tested. Farzad et al. [9] used a contact layer with a thickness of 100 μm and defined its characteristics using the CDP model to model the NSC-UHPFRC interface. They found a maximum error of 18% when comparing the results of this model with experimental data. In addition, Hor Yin et al. [11] used LS-DYNA to model the NSC-UHPC interface with equivalent beam elements, achieving good agreement between simulated and experimental results for composite beams.

As far as the authors know, no studies have investigated how different working conditions, such as the tensile, shear, and bending behavior of composite members, affect the validity of these models.

This study aims to develop a model for simulating the interfacial bonding between NSC and UHPFRC and integrate it into 3D FE models using ABAQUS software. The goal is to accurately predict the nonlinear behavior of NSC-UHPFRC composite members under various working conditions. The assumption of traction-separation law [12] is used as a starting point, as it imitates the constitutive behavior of the NSC-UHPFRC interface. Parameters for the interfacial bonding model are then determined through an analytical method based on shear and tensile tests. The interfacial model is subsequently integrated into seven ABAQUS models to verify its accuracy by comparing simulated results with experimental results for shear, tensile, and flexural tests.

The study yielded promising results, indicating that the developed parameters for the surface-based cohesive model in ABAQUS can accurately simulate the bonding behavior of NSC-UHPFRC composite members under shear, tensile, and bending loads. This innovative model holds significant potential for predicting the behavior of concrete members that have been reinforced and repaired with UHPFRC, thereby extending their service life.

This study is structured as follows: In Section 2, an analytical method is presented to determine specific parameters for the surface-based cohesive model, taking into account the diverse conditions of surface states. In Section 3, the developed model is applied to investigate the tensile, shear, and flexural behaviors of composite structures, and its accuracy is validated through comparison with experimental data. Section 4 presents the obtained results and provides a discussion of these results. Finally, Section 5 summarizes the key findings of the study and discusses their implications.

2. Interfacial Bonding Model at the NC-UHPFRSC Interface

Interfacial debonding phenomena between UHPC and NSC typically occur in three stress cases: pure shear, tensile, and mixed modes of shear and tensile or compressive stresses [13,14]. The ASSHTO LRFD Bridge Design Specifications [15] outline a shear test (Figure 1a) for a composite specimen composed of two different materials. The shear strength at the interface of different concretes cast at various times is determined using the following:

$$v_u = cA_{cv} + \mu(A_{vf}f_y + P_c) \tag{1}$$

where c is the cohesion and A_{cv} is the area between two layers of concrete. A_{vf} and f_y denote the area and the yield stress of reinforcement, respectively. μ is the friction coefficient, and P_c is the compressive force perpendicular to the shear plane.

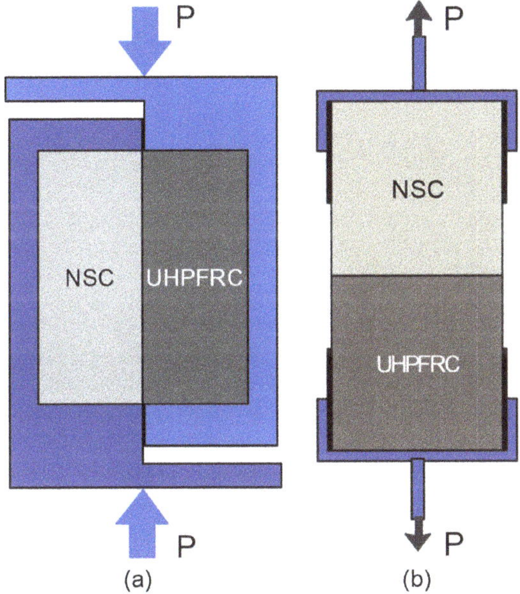

Figure 1. Types of tests: (**a**) simple shear bonding test; and (**b**) pure tensile bonding test.

The tensile strength is calculated based on the pull-off test (Figure 1b) according to ASTM C1583/C1583M [16] and is shown as follows:

$$f_t = \frac{T}{A} \tag{2}$$

where f_t is the direct tensile strength. T and A are the tensile force and the cross-sectional area of the test specimen, respectively.

The bonded interface between the NSC and UHPFRC in in situ casting can be accurately modeled using cohesive elements and surface-based cohesive behavior in ABAQUS software [17]. This modeling approach effectively represents the interface, which is usually composed of a thin layer of UHPFRC material, except in cases of abnormal surface roughness. Therefore, neglecting the surface thickness and using the surface-based cohesive behavior to model the interfacial behavior between NSC and UHPFRC is appropriate.

The surface-based cohesive behavior can be simulated using the common bi-linear traction-separation law [12], as shown in Figure 2. The parameters $K_{n(s,t)}$, $t^0_{n(s,t)}$, $\delta^0_{n(s,t)}$, and $\delta^t_{n(s,t)}$ represent the normal and two tangential stiffnesses, maximum stresses, corresponding displacements to the maximum stress, and maximum displacements at zero stress, respectively. The law of the model includes two stages. In the first stage, the stress at the interface increases linearly up to the peak $t^0_{n(s,t)}$ with the stiffness K. At the peak, damage initiation occurs and is followed by the evolution of that damage. The relationship between traction $t_{n(s,t)}$ and separation δ in the linear traction-separation law can be described by Equation (3):

$$t_{n(s,t)}(\delta) = \begin{cases} K_{n(s,t)}\delta = t^0_{n(s,t)}\dfrac{\delta}{\delta^0_{n(s,t)}}; & 0 < \delta < \delta^0_{n(s,t)} \\[2ex] t^0_{n(s,t)}\dfrac{\delta^f_{n(s,t)}-\delta}{\delta^f_{n(s,t)}-\delta^0_{n(s,t)}}; & \delta^0_{n(s,t)} < \delta < \delta^f_{n(s,t)} \\[2ex] 0; & \delta^f_{n(s,t)} < \delta \end{cases} \tag{3}$$

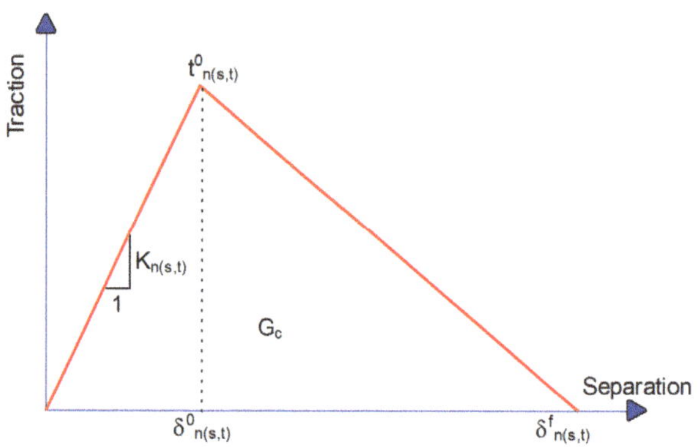

Figure 2. Linear traction-separation response.

ABAQUS offers various criteria to determine damage initiation, which is the point at which the bonding at the interface begins to degrade. In this study, the maximum stress criterion was utilized, where damage initiation occurs when the maximum contact stress ratio reaches one. This criterion is expressed in Equation (4):

$$\max\left\{\frac{\langle t_n \rangle}{t_n^0}, \frac{t_s}{t_s^0}, \frac{t_t}{t_t^0}\right\} = 1 \tag{4}$$

The Macaulay bracket notation, denoted by the symbol <>, is used in this context to represent purely compressive stress.

The damage evolution law characterizes the degradation of cohesive stiffness following damage initiation. During this stage, Equation (5) is used to describe the contact stress components at the interface that are affected by a scalar damage variable, D.

$$t_{n(s,t)}(\delta) = (1 - D)t^0_{n(s,t)} \tag{5}$$

The value of the scalar damage variable, D, is zero when there is no damage and one when the bonding has completely failed. Since the maximum stress criterion was utilized, a linear damage evolution model was adopted in this study. When the bonding experiences complete failure, D reaches a value of one, indicating complete plastic displacement. This displacement is determined by the difference between $\delta^f_{n(s,t)}$ and $\delta^0_{n(s,t)}$.

The elastic behavior is determined by Equation (6):

$$t = \left\{ \begin{array}{c} t_n \\ t_s \\ t_t \end{array} \right\} = \left[\begin{array}{ccc} K_{nn} & K_{ns} & K_{nt} \\ K_{ns} & K_{ss} & K_{ts} \\ K_{nt} & K_{st} & K_{tt} \end{array} \right] \left\{ \begin{array}{c} \delta_n \\ \delta_s \\ \delta_t \end{array} \right\} = K\delta \tag{6}$$

The nominal traction stress vector, represented by t, includes t_n for the normal traction and t_s and t_t for the two shear tractions. δ_n, δ_s, and δ_t denote corresponding separations. The uncoupled traction-separation behavior is normally utilized to model the interface's working behavior. This uncoupled behavior is applied when the pure normal separation and the pure shear separation do not affect each other [17].

For surface-based cohesive and uncoupled behavior, the thickness of the cohesive element is assumed to be equal to one [17]. Therefore, the elastic behavior can be written as follows:

$$t = \left\{ \begin{array}{c} t_n \\ t_s \\ t_t \end{array} \right\} = \left[\begin{array}{ccc} K_n & 0 & 0 \\ 0 & K_s & 0 \\ 0 & 0 & K_t \end{array} \right] \left\{ \begin{array}{c} \delta_n \\ \delta_s \\ \delta_t \end{array} \right\} \tag{7}$$

Figure 3 illustrates the typical load-displacement curve for both the shear bonding test (Figure 1a) and the pure direct tensile test (Figure 1b), shown by the blue curve. This curve is approximated as being linear over two segments, OA and AB. This relationship between load $P_{n(s,t)}$ and displacement d can be expressed using Equation (8):

$$P_{n(s,t)}(d) = \left\{ \begin{array}{c} K_{n(s,t)}d = P^0_{n(s,t)} \dfrac{d}{d^0_{n(s,t)}}; \ 0 < d < d^0_{n(s,t)} \\[2ex] P^0_{n(s,t)} \dfrac{d^f_{n(s,t)} - d}{d^f_{n(s,t)} - d^0_{n(s,t)}}; \ d^0_{n(s,t)} < d < d^f_{n(s,t)} \\[2ex] 0; \ d^f_{n(s,t)} < d \end{array} \right. \tag{8}$$

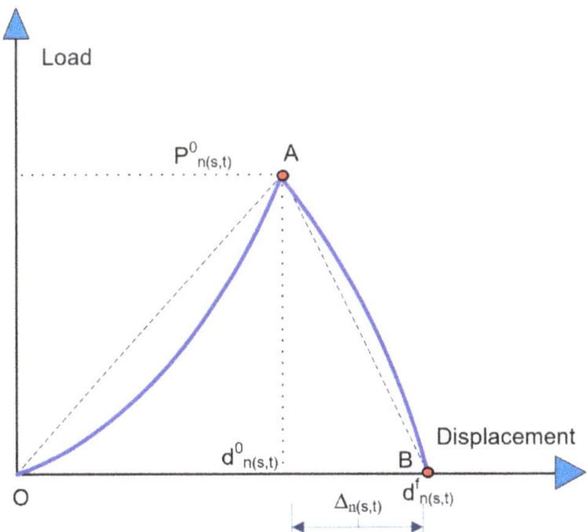

Figure 3. The typical load-displacement curve for pure shear and tensile tests.

The ultimate loads of each mode are P_n°, P_s°, and P_t°, with corresponding displacements of d_n°, d_s°, and d_t°, respectively. The highest displacement at the end of each damage evolution is represented by d_n^f, d_s^f, and d_t^f. The differences between the highest and corresponding displacements are denoted as Δ_n, Δ_s, and Δ_t.

The ultimate stresses and corresponding displacements are typically reached at the peak points of each mode's load-displacement curve. By applying these conditions to Equation (7), the corresponding Equation can be derived as follows:

$$\begin{cases} K_n\delta_n = K_n d_n^0 = t_n^0 = \dfrac{P_n^0}{A_n} \\ K_s\delta_s = K_s d_s^0 = t_s^0 = \dfrac{P_s^0}{A_s} \\ K_t\delta_t = K_t d_t^0 = t_t^0 = \dfrac{P_t^0}{A_t} \end{cases} \tag{9}$$

The interfacial areas between NSC and UHPFRC in the tensile and shear tests are denoted by A_n and $A_s(A_t)$, respectively.

Thus, the normal stiffness and two tangential stiffnesses can be determined from Equation (10), which is derived from Equation (9).

$$\begin{cases} K_n = \dfrac{P_n^0}{A_n d_n^0} \\ K_s = \dfrac{P_s^0}{A_s d_s^0} \\ K_t = \dfrac{P_t^0}{A_t d_t^0} \end{cases} \tag{10}$$

The damage of the interface is defined by using the maximum nominal stress $t_n^\circ(t_s^\circ, t_t^\circ)$ and the total/plastic displacement.

The bonding strength of an interface is primarily determined by its roughness [18,19]. In civil engineering, the roughness of a concrete surface refers to the irregularity or variation in its texture. The quantification of a concrete surface's roughness is commonly determined by its average sand-filling depth, represented as "h" in Figure 4a. Equation (11) can be used to calculate the value of "h":

$$h = \frac{V_s}{A_s} \tag{11}$$

where h represents the average depth of sand filling (mm); V_s is the volume of sand (mm^3); and A_s represents the area of the treated concrete substrate (mm^2).

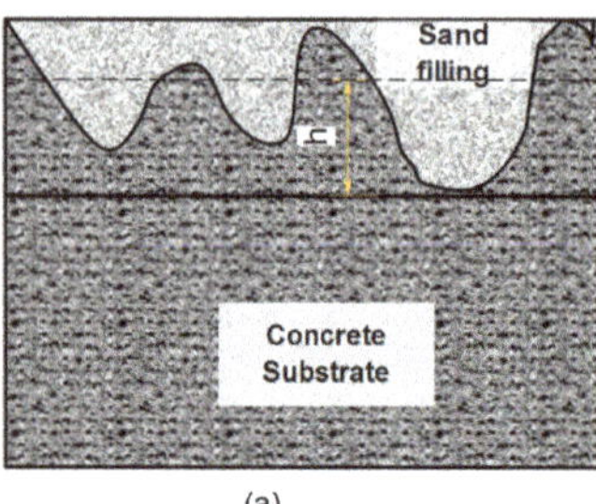

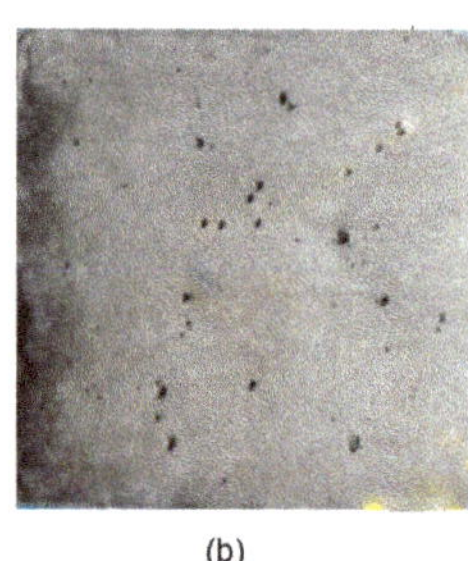
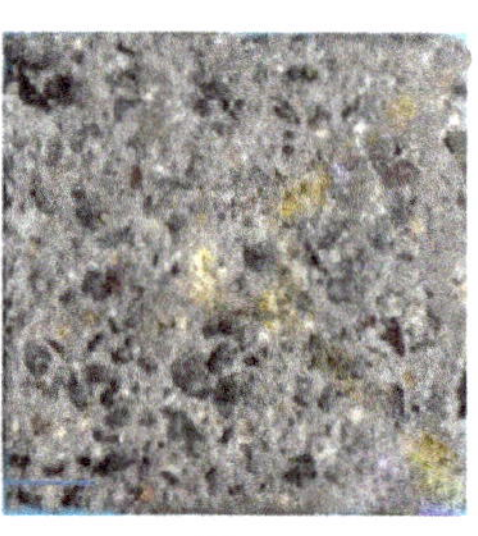

(a)   (b)   (c)   (d)

Figure 4. Surface roughness of the concrete substrate: (**a**) sand-filling method; (**b**) smooth surface (no sand filling, h = 0 mm); (**c**) mid-rough surface (sand filling depth of 0.64 mm); and (**d**) rough surface (sand filling depth of 1.29 mm).

For example, Feng et al.'s study [20] investigated the bonding strength of three types of surfaces: smooth, mid-rough, and rough (Figure 4b–d), which corresponded to sand-filling depths of 0, 0.64, and 1.29 mm, respectively.

Feng et al. [20] and Hussei et al. [21] previously investigated the shear and tensile behaviors of NSC-UHPFRC composite specimens with various surfaces. These studies tested NSC-UHPFRC specimens with smooth, mid-rough, and rough interfaces to determine load-displacement behaviors, as described in Sections 3.2 and 3.3. Table 1 summarizes the experimental results, including the ultimate loads ($P_{n(s,t)}^0$), surface areas ($A_{n(s,t)}^0$), displacements ($d_{n(s,t)}^0$), and maximum plastic displacement (Δ) for both shear and tensile tests. Additionally, the table provides the values of tractions and stiffnesses for the three types of surfaces, which were obtained using Equations (9) and (10).

The parameters for the smooth, mid-rough, and rough surfaces of the NSC-UHPFRC interface, which were derived from Table 1, are presented in Table 2 for the surface-based cohesive model used in ABAQUS. The input data required to simulate the bonding behavior between NSC and UHPFRC for different surfaces is presented in Figure 5. This includes the interaction properties and interactions defined in ABAQUS for the FE models.

Table 1. Calculation of surface-based cohesive model parameters for various surfaces.

Test	Surface	$P_{n(s,t)}{}^0$ (KN)	$A_{n(s,t)}$ (mm^2)	$d_{n(s,t)}{}^0$ (mm)	Δ (mm)	$t_{n(s,t)}$ (MPa)	$K_{n(s,t)}$ (N/mm^3)
Shear test [20]	Smooth	30.81	10,000	0.76	0.02	3.08	4.04
	Mid-rough	57.92	10,000	1.39	0.12	5.79	4.17
	Rough	65.94	10,000	1.41	0.15	6.6	4.69
Tensile test [21]	Smooth	13.35	4417.86	1.23	0.02	3.02	2.45
	Mid-rough	20.34	4417.86	2.20	0.1	4.61	2.09
	Rough	24.83	4417.86	2.56	0.24	5.63	2.2

Note: $P_{n(s,t)}{}^0$, $A_{n(s,t)}{}^0$, $d_{n(s,t)}{}^0$, and Δ represent the ultimate loads, surface areas, corresponding displacements, and maximum plastic displacement in both shear and tensile tests.

Table 2. Mechanical parameters of the surface-based cohesive model for different types of NSC-UHPFRC interfaces.

Property	Smooth Surface	Mid-Rough Surface	Rough Surface
K_n (N/mm^3)	2.45	2.09	2.20
K_s, K_t (N/mm^3)	4.04	4.17	4.69
$t_n{}^\circ$ (MPa)	3.02	4.61	5.63
$t_s{}^\circ$, $t_t{}^\circ$ (MPa)	3.08	5.79	6.60
Total/plastic displacement Δ (mm)	0.02	0.12	0.24

```
 1  **
 2  **  INTERACTION PROPERTIES
 3  **
 4  *Surface Interaction, name=surface_based cohesive model
 5  *Friction
 6   0.2,
 7  *Surface Behavior, pressure-overclosure=HARD
 8  *Surface Interaction, name=surface_based cohesive model
 9  *Cohesive Behavior
10   Kn, Ks, Kt                  %Normal and two tangential stiffnesses
11  *Damage Initiation, criterion=MAXS
12   tn0, ts0, tt0               %Normal and two tangential maximum stresses
13  *Damage Evolution, type=DISPLACEMENT
14   Δ ,                         %Plastic displacement
15  **
16  **  INTERACTIONS
17  **
18  **  Interaction: Int-1
19  *Contact, op=NEW
20  *Contact Inclusions, ALL EXTERIOR
21  *Contact Property Assignment
22   ,    , IntProp-1            %NSC surface, UHPFRC surface
23  Surf-1 , Surf-2 , surface based cohesive model
```

Figure 5. Input data for a surface-based cohesive model on various surface types. Note: The symbols * and ** respectively denote the beginning of a command and introduce a comment in the input file in ABAQUS.

3. Modeling of Composite Members of NSC and Cast In Situ UHPFRC

This section outlines the modeling process for composite members made up of NSC and cast in situ UHPFRC. The interfacial bonding between these materials is modeled using the surface-based cohesive model with the proposed parameters. The accuracy of the proposed interfacial model is validated by comparing the simulated results with experimental results obtained from shear, tensile, and flexural tests.

3.1. FE Analysis

3.1.1. Element Types

The C3D8R element (eight node linear brick with reduced integration) was selected to model NSC and UHPFRC due to its ability to capture concrete materials' tensile cracking and compressive crushing accurately [17,22,23]. The T3D2 element [17] was used to model reinforcements, which can simulate the working behavior between reinforcement and concrete. In order to model the steel pins and supports in contact simulation, a discrete rigid shell element was employed [17].

3.1.2. Material Modeling

Concrete damaged plasticity (CDP) is widely recognized as the most popular model to simulate the nonlinear response of concrete [24,25]. The CDP model can simulate cracking patterns and crack widths using d_t and d_c parameters for tension and compression (Figure 6). Therefore, in this study, the CDP model was chosen to model NSC and UHPFRC. The CDP model requires five additional parameters, including the dilation angle (ψ), the eccentricity of the plastic flow (ε), the ratio of the initial biaxial to initial uniaxial compressive strength (σ_{b0}/σ_{c0}), and the shape of the failure surface (K_c). Stress-inelastic strain relationships for compression and tension are based on models by Kent and Park [26] and Massicotte et al. [27] or experimental data. Table 3 presents the parameters of the CDP model for simulations of both NSC and UHPFRC materials.

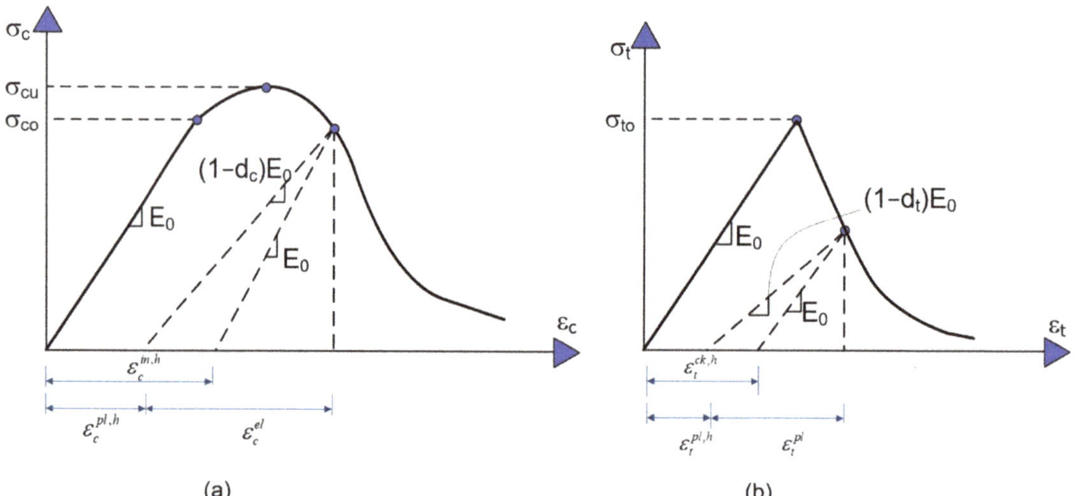

(a) (b)

Figure 6. Response of concrete to a uniaxial loading condition: (**a**) compression and (**b**) tension [23].

Table 3. Parameters of the CDP model for NSC and UHPFRC materials.

Concrete	ψ	ε	σ_{b0}/σ_{c0}	K_c	Viscosity
NC	30 [24]	0.1 [24]	1.16 [24]	0.6667 [24]	0.0001 [24]
UHPFRC	36 [28]	0.1 [28]	1.07 [29]	0.6667 [25]	0.005 [30]

3.1.3. Interfacial Interactions

The embedded constraint technique was implemented to simulate the working behavior between reinforcements and concrete, as recommended by ABAQUS [17]. The interfacial bonding behavior between NSC and UHPFRC was modeled using a surfaced-based cohesive contact model with parameters presented in Table 2.

3.1.4. Mesh Sensitivity

The accuracy of the finite element (FE) simulation is typically directly related to the mesh density used in the simulation process, as evidenced by previous research [31–34]. However, it is noted that a higher mesh density also results in greater computational costs. Therefore, a mesh sensitivity study should be conducted to select an appropriate simulation model. The subsequent sections will present the mesh sensitivity for each specific problem.

3.1.5. ABAQUS Solver

The ABAQUS/Explicit solver was selected for solving all problems in this study, as it is known to perform better in solving nonlinear and contact problems when mass scaling is not used, and the step time is one second. Analysis of all models in the study showed that the total energy remained constant and the ratio of kinetic energy to internal energy in the system did not exceed 10%. These results demonstrate that the ABAQUS/Explicit solver is suitable for solving quasistatic problems in this study [17].

3.2. Pure Shear Model

In the previous study, Feng et al. [20] investigated the effect of various factors on the shear performance at the NC-UHPFRC smooth, mid-rough, and rough interfaces. The concrete substrate and the UHPFRC have compressive strengths of 42 MPa and 120.61 MPa, respectively. Figure 7a,b show the detailed experimental setup and geometry.

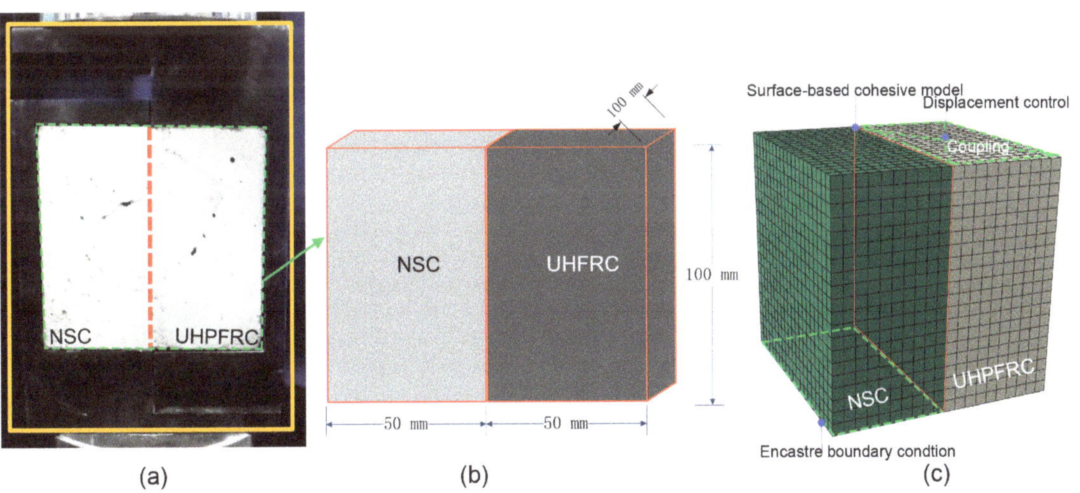

Figure 7. Pure shear test: (**a**) experimental setup [20]; (**b**) specimen dimensions; and (**c**) FE model.

The selection of a simulation model for the current problem was based on the methodology presented in Section 3. Specifically, the C3D8R element was used to model NSC and UHPFRC. The CDP model parameters for NSC and UHPFRC are provided in Tables 4 and 5, respectively. In addition, a surface-based cohesive contact model was used to model the bonding interface between NSC and UHPFRC. The model parameters are specified in Table 2, along with a penalty contact with a friction coefficient of 0.2 [35]. Figure 7c illustrates the FE model for the shear test.

Table 4. Properties of NSC along with CDP model parameters in compression and tension in the shear test.

Compressive Concrete Strength (MPa)				42	
Elastic Modulus (GPa)				30.5	
Poisson Ratio				0.18	
Strength (MPa)	d_c	ε_{in}	Strength (MPa)	d_t	ε_{ck}
16.8	0	0	3.644	0	0
37.8	0	0.001057	2.42	0.33333	0.000993
42	0	0.001227	1.366	0.625	0.001742
39.9	0.05	0.001351	0.607	0.83333	0.002958
2.94	0.93	0.003683	0.1822	0.95	0.00333
2.52	0.94	0.003772	0.14477	0.96	0.003607

Table 5. Properties of UHPFRC along with CDP model parameters in compression and tension in the shear test.

Compressive Concrete Strength (MPa)				120.61	
Elastic Modulus (GPa)				46.5	
Poisson Ratio				0.21	
Strength (MPa)	d_c	ε_{in}	Strength (MPa)	d_t	ε_{ck}
84.427	0	0	7.36241	0	0
108.549	0	0.00103	7.5	0	0.00484
120.61	0	0.00143	5	0.33333	0.00989
114.58	0.05	0.00198	2.8125	0.625	0.01244
108.549	0.1	0.00253	1.25	0.83333	0.01397
7.2366	0.94	0.01194	0.525	0.93	0.01457

In order to evaluate mesh convergence, several FE models were created with mesh sizes ranging from 2 to 20 mm. The results showed that a 5 mm mesh size provided both accurate numerical results and a reasonable computation time. Therefore, the 5 mm mesh size was chosen for the present FE model.

3.3. Pure Tensile Model

In a study by Hussein et al. [21], the bonding behavior of the interface between UHPFRC and high-strength concrete (HSC) was analyzed. The interface was tested using cylindrical specimens with a diameter and height of 75 mm. Three types of surfaces were used to investigate the bonding behavior between the materials. The experimental setup and specimen geometry are illustrated in Figure 8a,b.

The simulation model consisted of HSC, UHPFRC, and cylindrical steel nipples modeled with C3D8R elements. The CDP model was applied to HSC and UHPFRC specimens, and their material properties are presented in Tables 6 and 7. The plastic material parameter of steel was used to model the nonlinear behavior of the steel nipples. A penalty contact with a friction coefficient of 0.2 was applied at the interface of the cylindrical steel nipples and the concrete specimens. A surface-based cohesive model was used to model the interaction between HSC and UHPFRC surfaces, with the parameters shown in Table 2. The detailed simulation model of the tensile test is presented in Figure 8c. A mesh size of 5 mm was chosen based on the results of a mesh convergence study.

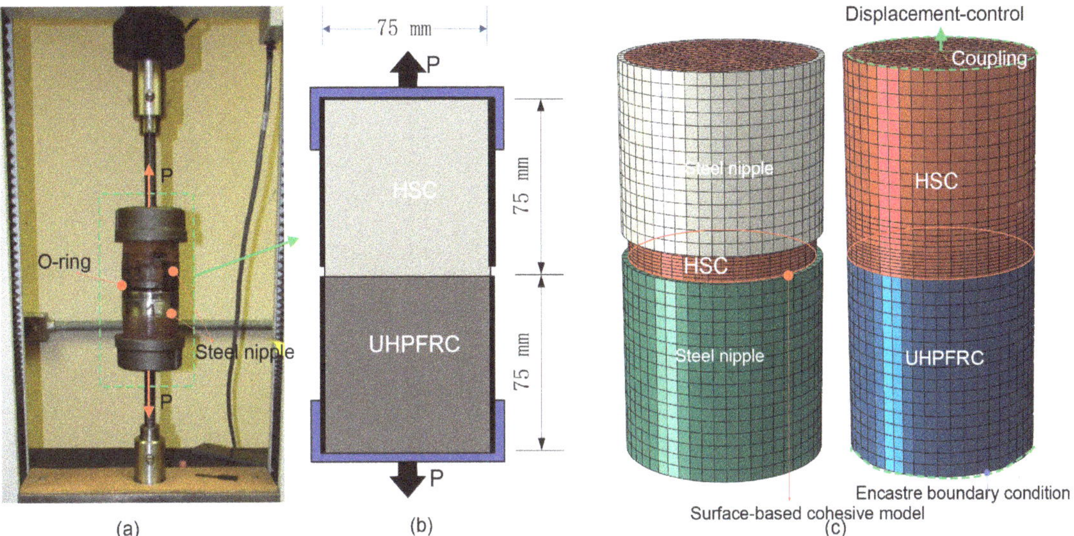

Figure 8. Pure tensile test: (**a**) experimental setup [21]; (**b**) dimensions of the specimens; and (**c**) FE model.

Table 6. Properties of HSC along with CDP model parameters in compression and tension in the tensile test.

Compressive Concrete Strength (MPa)				75	
Elastic Modulus (GPa)				41	
Poisson Ratio				0.2	
Strength (MPa)	d_c	ε_{in}	Strength (MPa)	d_t	ε_{ck}
30	0	0	5.3638	0	0
67.5	0	0.00067	3.57586	0.33333	0.00109
75	0	0.00077	2.01142	0.625	0.00191
71.25	0.05	0.00096	0.89397	0.83333	0.00325
67.5	0.1	0.00115	0.80457	0.85	0.00334
4.5	0.94	0.00464	0.21455	0.96	0.00396

Table 7. Properties of UHPFRC along with CDP model parameters in compression and tension in the tensile test.

Compressive Concrete Strength (MPa)				158.58	
Elastic Modulus (GPa)				53	
Poisson Ratio				0.21	
Strength (MPa)	d_c	ε_{in}	Strength (MPa)	d_t	ε_{in}
111.006	0	0	8.83616	0	0
142.722	0	0.00074	9	0	0.00483
158.58	0	0.00101	6	0.33333	0.00989
150.651	0.05	0.00171	3.375	0.625	0.01244
142.722	0.1	0.00241	1.5	0.83333	0.01397
9.5148	0.94	0.01447	0.63	0.93	0.01457

3.4. Model for Composite Beam Subjected to Four-Point Bending

Osta et al. [36] studied the flexural behavior of an RC beam strengthened by UHPFRC. The dimensions of the RC beam are 140 × 230 × 1600 mm, made from NSC with a compressive strength of 54 MPa. In order to enhance the structural integrity of the RC beam, a UHPFRC layer of 30 mm with a compressive strength of 121 MPa was applied at the soffit of the RC beam. Figure 9a illustrates the detailed geometry and reinforcements of the NSC-UHPFRC composite beam.

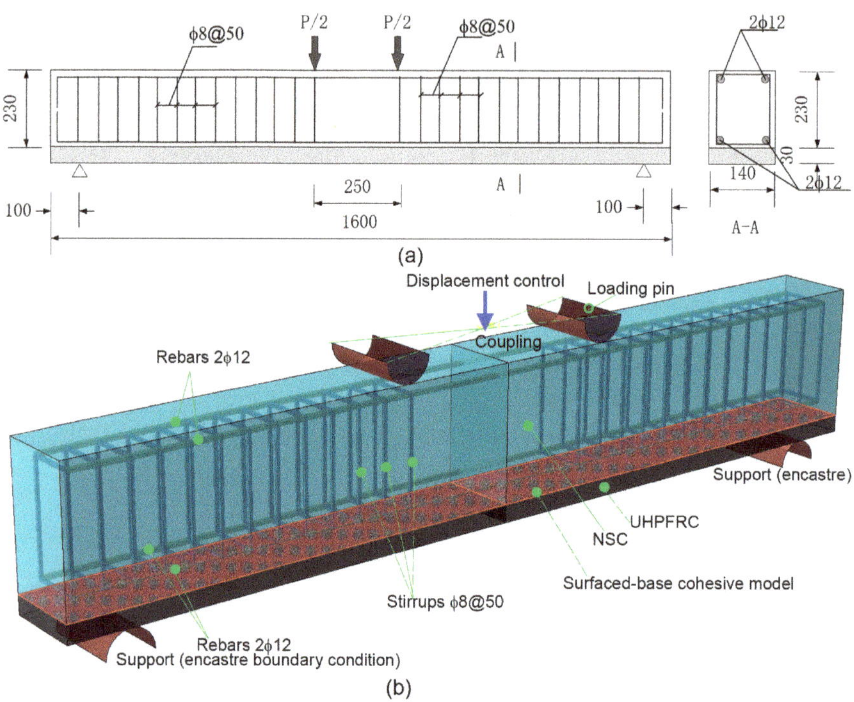

Figure 9. Flexural test: (**a**) geometry and reinforcements of the NSC-UHPFRC composite beam [36]; and (**b**) FE model.

The element types for NSC, UHPFRC, reinforcing bars, and pins were adopted as presented in Section 3.1.1. In order to model the interfacial bonding between NSC and UHPFRC, the study adopted a surface-based cohesive model with the parameters listed in Table 2. The CDP model was applied to both NSC and UHPFRC materials, while the plastic material model was selected for reinforcements. Tables 8–10 present the material properties of NSC, UHPFRC, and reinforcements. Figure 9b depicts the detailed simulation model of the NSC-UHPFRC composite beam. Following a comprehensive evaluation of a mesh sensitivity study, a mesh size of 5mm was identified as the optimal value due to its balance between computational efficiency and accuracy.

Table 8. Properties of NSC along with CDP model parameters in compression and tension in the flexural test.

Compressive Concrete Strength (MPa)				54	
Elastic Modulus (GPa)				33	
Poisson Ratio				0.19	
Strength (MPa)	d_c	ε_{in}	Strength (MPa)	d_t	ε_{ck}
21.6	0	0.0000	3.26	0.0000	0.0000
48.6	0	0.000495	2.1	0.3333	0.00022
54	0	0.000563	1.185	0.6250	0.00052
51.3	0.05	0.001716	0.52	0.8333	0.00099
5.94	0.89	0.003412	0.158	0.9500	0.0013
4.32	0.92	0.00355	0.1264	0.9600	0.0014

Table 9. Properties of UHPFRC along with CDP model parameters in compression and tension in the flexural test.

Compressive Concrete Strength (MPa)				121	
Elastic Modulus (GPa)				41.5	
Poisson Ratio				0.21	
Strength (MPa)	d_c	ε_{in}	Strength (MPa)	d_t	ε_{ck}
90.75	0	0	6.5	0	0
108.9	0	0.00098	5.005	0.23	0.0023794
121	0	0.00148	2.925	0.55	0.0079295
114.95	0.05	0.00171	0.65	0.9	0.0173843
108.9	0.1	0.00194	0.585	0.91	0.0188459
9.68	0.92	0.00574	0.325	0.95	0.0246922

Table 10. Plastic properties of reinforcements in the flexural test.

Diameter (mm)		8 and 10	
Poisson Ratio		0.3	
Elastic Modulus (GPa)		200	
	Yield Stress (MPa)	Plastic Strain	
	590	0	
	589.99	0.000668	
Plastic behavior	589.991	0.00295	
	650.7	0.23455	
	635	0.2470	

4. Results and Discussion

4.1. Results Obtained from the Analysis of Shear Tests

Figure 10 shows a strong agreement between the experimental and simulation results for three surface types in terms of peak loads and displacements. The agreement is demonstrated by small deviations between experimental and simulated ultimate loads, measuring 6.8%, 4.1%, and 5.8% for rough, mid-rough, and smooth surfaces, respectively. Corresponding small errors in displacements at ultimate loads are also observed for the same surfaces, measuring 2%, 0.9%, and 2.4%, as shown in Table 11. The highest deviation of 6.8% is found in the case of the rough surface. It can be explained that in composite members with a rough surface, the strength of the interface between UHPFRC and NSC is stronger than the NSC itself due to the majority of UHPFRC penetrating into the NSC to form the interface. Consequently, the composite member's shear resistance depends on the

NSC's strength. Therefore, any difference in the strength of the NSC between the EXP and FEM may affect the ultimate load difference between them.

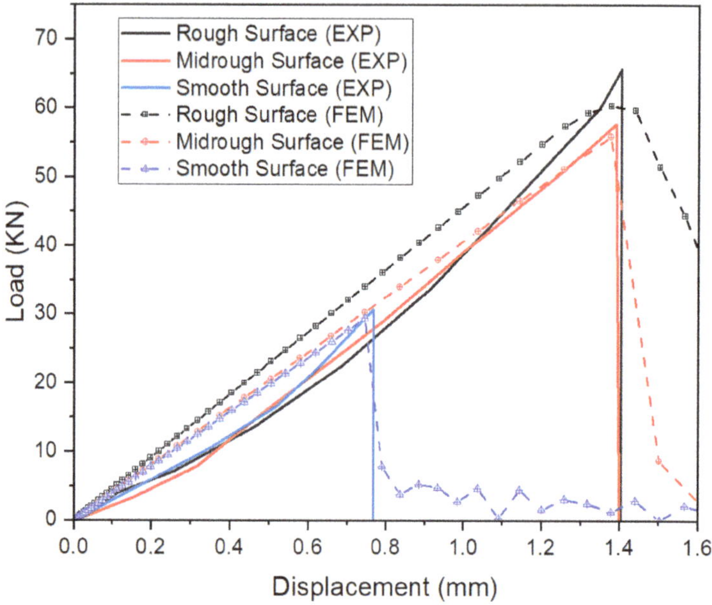

Figure 10. Comparison of the load-displacement characteristics between the EXP [20] and FEM in three distinct surface types in the shear tests.

Table 11. Summary and comparison of FE model and experimental results.

Type	Authors	Surface	P_{exp}(KN)	P_{FEM}(KN)	Δ_{exp}(mm)	Δ_{FEM}(mm)	P Error (%)	Δ Error(%)
		Rough	65.773	61.587	1.406	1.379	6.8	2.0
Shear test	Feng et al. [20]	Mid-rough	57.713	55.445	1.389	1.376	4.1	0.9
		Smooth	30.742	29.065	0.763	0.745	5.8	2.4
		Rough	24.155	21.234	2.555	2.205	13.6	15.9
Tensile test	Hussein et al. [21]	Mid-rough	19.825	20.12	2.203	2.205	1.5	0.1
		Smooth	13.180	12.954	1.233	1.205	1.7	2.3
Flexural test	Osta et al. [36]	Rough	79.302	77.223	15.460	15.041	2.7	2.8

Note: P_{exp} and P_{FEM} denote the peak load in the experimental and simulation, respectively. Δ_{exp} and Δ_{FEM} indicate the corresponding displacements at the peak load in the experimental and simulated results, respectively. P error and Δ error indicate the percent error of the peak load and corresponding displacement, respectively, between the EXP and FEM.

The study found that the load-displacement curves of the experimental and simulation results for the smooth surface were in good agreement. However, the simulation demonstrates a higher interface stiffness for the mid-rough and rough surfaces than the experimental results. This difference is due to the simultaneous participation in shear resistance of both NSC and UHPFRC. The combination of these factors causes the overall stiffness of the interface to change over time. Additionally, defects in the NSC [37] constitute a crucial factor that unsettles its stiffness under loading.

Furthermore, the separation of the mid-rough and rough specimens in the simulation was observed to occur gradually, while the experimental results indicated that the specimens fractured immediately after reaching the peak load. The discrepancy in the damage evolution phase between FEA and EXP on mid-rough and rough surfaces could be attributed to differences in the concrete material's damage behavior between the simulation and experiment. In the case of the rough and mid-rough surfaces, the composite members' damage occurred in the NSC part. However, the CDP model used to model concrete assumes that it is homogeneous and isotropic, neglecting concrete's potential variations in properties and strengths in different directions as well as the existence of microcracks. Those factors could lead to the difference in the damage evolution stage between FEA and EXP.

Figure 11 illustrates the damage patterns of composite specimens through the visualization of the tensile damage variable "DAMAGET" and the compressive damage variable "DAMAGEC" in the ABAQUS output for three distinct surface types. The simulation results accurately depicted failure modes in all three cases. In the mid-rough and rough surface cases, failure concentrated on both the NSC and interface (Figure 11a,b). In the smooth surface case, debonding occurred solely at the interface between NSC and UHPFRC (Figure 11c).

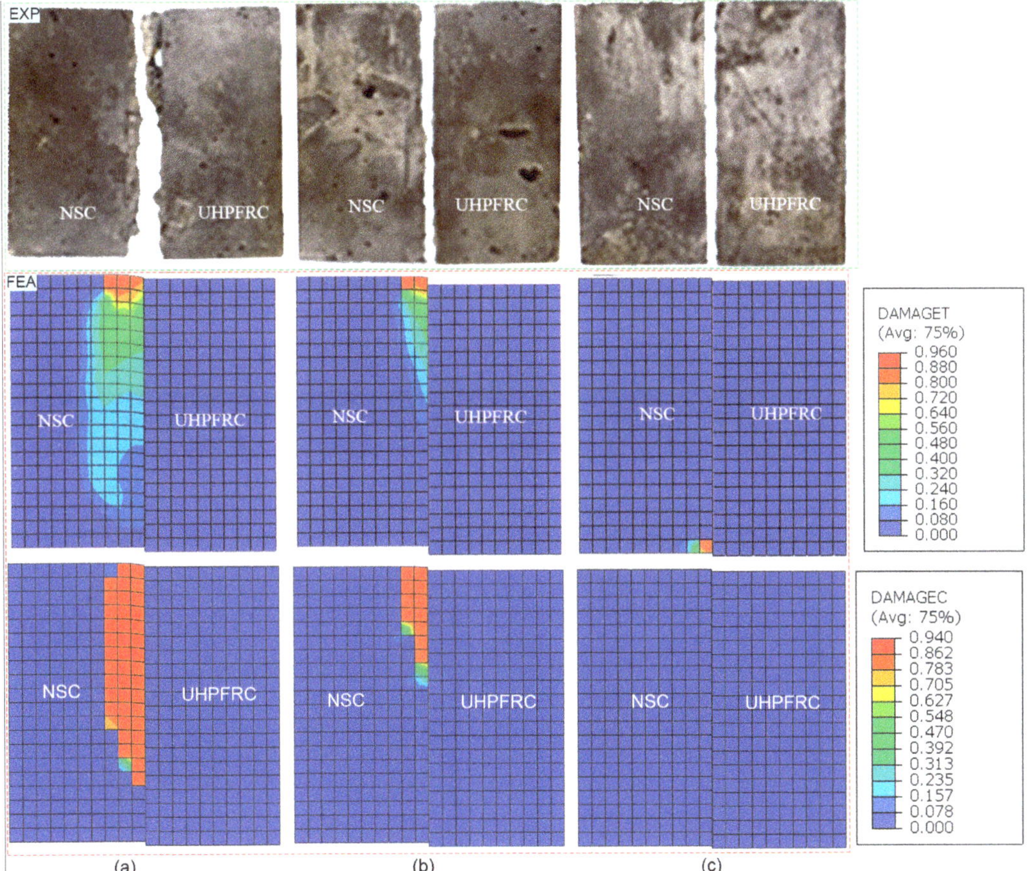

Figure 11. Comparison of the failure of the NSC-UHPFRC specimens between the EXP [20] and simulations for three distinct surface types in the shear test: (**a**) a rough surface; (**b**) a mid-rough surface; and (**c**) a smooth surface.

The failures observed between NSC and UHPFRC can be attributed to the formation of chemical bonding and mechanical interlocking. Chemical bonding occurs through the reaction between calcium oxide and silica at the NSC-UHPFRC interface, forming a calcium-silicate-hydrate gel. Mechanical interlocking is formed when UHPFRC penetrates the pores and voids of the NSC surface. On the smooth NSC surface, few pores allow for mechanical interlocking, so bonding is primarily achieved through chemical bonding. However, this bond is weaker than the shear strength of NSC, causing failure to occur at the interface of NSC and UHPFRC. Mechanical interlocking occurs between the NSC and UHPFRC surfaces on a mid-rough or rough NSC surface, creating small bridges that enhance the shear strength of the interface. A rougher NSC surface with a higher roughness degree results in a deeper average sand-fill depth, forming larger UHPFRC small bridges in the NSC. This can cause a larger failure area in the NSC specimen.

The simulation results revealed that crushing failure accounted for the majority of the area compared to cracking failure. This is because the shear force at the interface transmits the compressive load to the NSC specimen. With a higher degree of roughness in the interface between NSC and UHPFRC, more failures occurred in the top flange of the NSC. This is because, as the UHPFRC specimen moves downward, the highest node at the interface between the NSC and UHPFRC experiences a greater load than the lower nodes do. This is due to the compressive strain of each element between nodes, generating a compressive force that reduces the load at the lower node. As a result, the highest node experiences more stress than the lower nodes, which can lead to failure in the top flange of the NSC specimen. Therefore, failure in the NSC specimen occurs first in the top flange compared to other parts.

In the surface-based cohesive model used in this study, a rougher surface was assigned a greater shear strength at the interface in the parameters (Table 2). Therefore, if an element near the surface is damaged, the load is transmitted to the adjacent element, and failure develops until the corresponding compressive stress is lower than the compressive strength of NSC. In the case of a smooth surface, the compressive stress generated in the NSC by bonding with UHPFRC is lower than its compressive strength, resulting in failure at the interface instead of within the NSC.

The developed FE models are useful for predicting the shear bonding strength at the NSC and UHPFRC interfaces for different levels of roughness. This is particularly beneficial for designing effective strengthening solutions for concrete structures, such as bridge decks, that are being repaired or reinforced using UHPFRC. Accurately predicting the shear strength at the interface can ensure the durability and longevity of the structure, making it an important consideration for engineers.

4.2. Results Obtained from the Analysis of Tensile Tests

Figure 12 shows a high degree of agreement between the simulation and experimental results for the load-displacement curves overall, particularly for stiffness, peak loads, and corresponding displacements on both smooth and mid-rough surfaces. The deviation between EXP and FEA in those comparing variables is less than 3% (Table 11). Although the simulation and experimental results matched well in terms of stiffness for the rough surface, Table 11 shows a 13.6% and 15.9% discrepancy for the ultimate load and corresponding displacement, respectively. A discrepancy in the concrete's tensile strength used in the experiment and simulation could be the reason for the observed difference in ultimate load and corresponding displacement. This is because the rough surface has a higher tensile strength (5.63 MPa) than the NSC (5.36 MPa), as shown in Table 2. Consequently, the tensile strength of the NSC-UHPFRC composite specimen during tensile testing is primarily determined by the strength of the NSC.

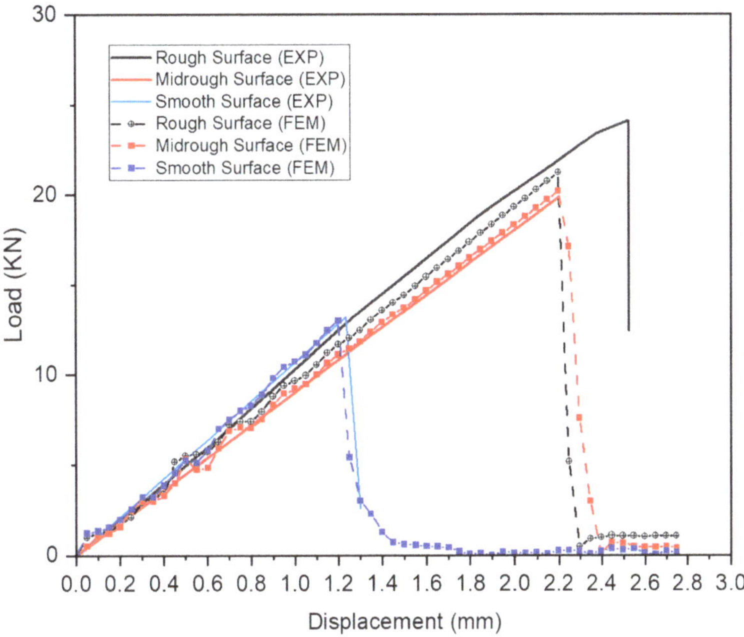

Figure 12. Comparison of the load-displacement characteristics between the EXP [21] and FEM for three distinct surface types in the tensile test.

Regarding damage patterns, Figure 13 demonstrates that the simulation results are consistent with the experimental results. The simulation results show greater damage in the NSC for the mid-rough and rough surfaces. Additionally, on the rough surface, the failure of NSC occurred further away from the interface than on the mid-rough surface. Furthermore, debonding mostly occurred at the NSC and UHPFRC interfaces for the smooth surface.

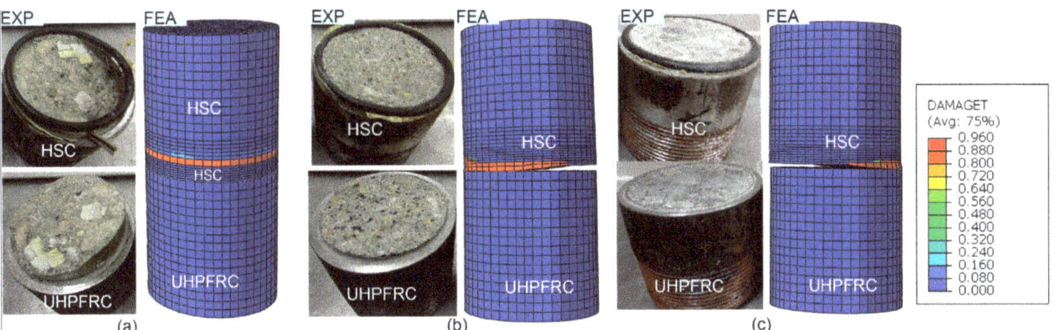

Figure 13. Comparison of the failure of the HSC-UHPFRC specimens between the EXP [21] and FEM simulations in three distinct surface types in the tensile test: (**a**) rough interface; (**b**) mid-rough interface; and (**c**) smooth interface.

The reasons for these results can be explained as follows: The rough surface has a larger tensile strength at the interface, causing the tensile strain of elements near the interface to be lower than those farther away during tensile loading. Due to this behavior, the concrete elements farthest from the interface, particularly those located at the position of the steel nipple in the NSC, are more prone to failure. In contrast, the mid-surface and smooth surface have a lower tensile strength at the interface than NSC, leading to a failure tendency at the interface. During failure, the NSC-UHPFRC interface undergoes a reduction in area, which results in an increase in its tensile strength at that specific location. Consequently, some elements of the NSC fail because their tensile strength exceeds the ultimate tensile strength of the NSC. The extent of concrete damage depends on the interface's tensile strength. As the interface's tensile strength increases, the concrete's damaged area also increases. This finding explains why the damaged area of NSC in the mid-rough surface was greater than that in the smooth surface, as shown in Figure 13b,c.

By accurately predicting the tensile behavior at the HSC-UHPFRC interface, this FE model can be used to optimize interface design and maximize the tensile bonding strength between HSC and UHPC. This optimization can ultimately improve the performance and durability of concrete structures. Additionally, the model can be utilized to simulate different scenarios and loading conditions, enabling engineers to evaluate the structural behavior of the interface and make informed decisions about design modifications and repairs.

Based on the two analyses mentioned above, when the interface between HSC and UHPFRC has a higher roughness level, the concrete substrate experiences either shear or tensile failure, and the composite member's shear or tensile strength capacity depends on the corresponding strength of the concrete substrate. Therefore, it is recommended that substrate surfaces be prepared with sufficient roughness to ensure that failure occurs in the substrate during the repair and strengthening of concrete members.

4.3. Results Obtained from Analysis of Flexural Test

Figure 14 demonstrates that the load-displacement curves of the simulated results are aligned with the experimental results for both the reference beam and composite beam. This high degree of concordance is evident from the small difference of around 3% between the simulation and experimental results for both peak load and corresponding displacement (Table 11). Additionally, the FE model captures the ductile behavior of the composite beam as observed in the experimental results, with the initiation of the first crack observed at approximately 22 KN and the first yielding of the reinforcements estimated to occur at approximately 70 KN. However, the simulation results reveal a higher initial stiffness compared to the experimental results. The disparity in the initial stiffness of an RC beam between the experimental and simulated results can be attributed to several factors. One such factor is the presence of voids and defects inside the experimental beam, which can reduce its stiffness [37]. Additionally, the modeling assumptions made in the simulation, such as the use of the embedded constraint technique, may not accurately reflect the actual response between the reinforcement and concrete, potentially leading to an overestimation of the beam's stiffness in the simulated model.

Figure 15 demonstrates a strong correlation in damage patterns between the simulation and experimental results. The results indicate that the debonding between the NSC and UHPFRC primarily occurred at the interface, highlighting the critical role of the bond between the two materials in the overall performance of composite beams. Furthermore, the experimental results revealed the presence of crushing damage at the location of applied loads. This damage may be attributed to the high concentration of stresses in this region, which can exceed the material's load-bearing capacity. These findings underscore the importance of considering both the bond between the NSC and UHPFRC and the distribution of loads and stresses when designing reinforced and retrofitted concrete structures.

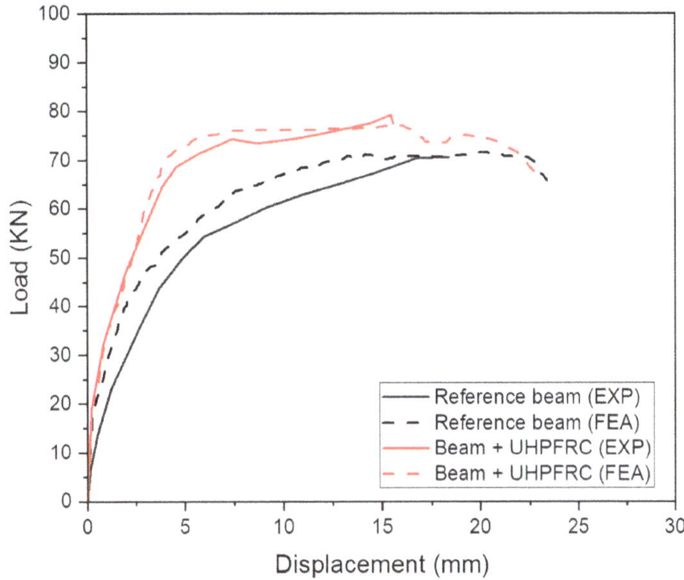

Figure 14. Comparison of load-displacement characteristics between the EXP [36] and FEM simulations in the flexural test for rough surfaces.

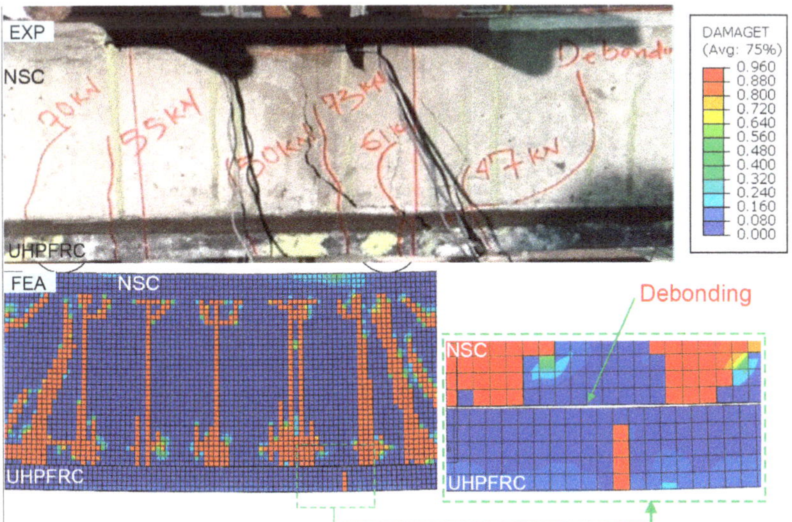

Figure 15. Comparison of the failure of the NSC-UHPFRC specimens between the EXP [36] and FEM simulations for the rough surface in the flexural test.

5. Conclusions

This study utilized the ABAQUS software to investigate the response of composite members made of NSC (or HSC) and UHPFRC. The bonding between the NSC and UHPFRC was modeled using a surface-based cohesive model derived and calibrated based on previous experimental results. This cohesive model was then integrated into seven FE models to simulate and validate the behavior of NSC-UHPFRC composite members under various loading conditions. The key conclusions derived from this study are as follows:

(1) The developed FE models show good agreement with experimental data for the overall response of NSC (HSC)-UHPFRC composite members under different working conditions. In the rough surface tests, the shear test shows a maximum deviation of 6.8% in the ultimate load, while the tensile test shows a 15.9% deviation in the ultimate displacement;

(2) A linear traction-separation model based on experimental data from pure shear and tensile tests is used to develop an analytical method for determining the parameters of a surface-based cohesive model in ABAQUS;

(3) Novel parameters for the surface-based cohesive model are proposed to simulate the bonding between NSC (HSC) and cast in situ UHPFRC, as presented in Table 2.

The present study provides valuable insights into the potential use of UHPFRC and UHPFRS for reinforcing and retrofitting damaged concrete structures. While the findings indicate a high potential for UHPFRC in these applications, it is essential to acknowledge that the uniform surface roughness observed in this study may not fully reflect the complex conditions encountered in real-world scenarios, where the bonding surfaces of NSC and UHPFRC can exhibit non-uniform roughness. Therefore, future research should consider the impact of surface roughness variations on the performance of UHPFRC in practical settings to ensure its optimal use in reinforcing and retrofitting structural members.

Author Contributions: Supervision, S.-K.K.; methodology, X.-B.L.; software, X.-B.L.; validation, X.-B.L.; writing—review and editing, S.-K.K. and X.-B.L.; funding acquisition, S.-K.K. All authors have read and agreed to the published version of the manuscript.

Funding: This research was funded by a National Research Foundation of Korea (NRF) grant funded by the Korean government (MIST) (No.2021R1C1C1013130).

Data Availability Statement: Not applicable.

Acknowledgments: This work was supported by a National Research Foundation of Korea (NRF) grant funded by the Korean government (MIST) (No.2021R1C1C1013130).

Conflicts of Interest: The authors declare no conflict of interest.

References

1. Oesterlee, C. Structural Response of Reinforced UHPFRC and RC Composite Members. Ph.D. Thesis, Swiss Federal Institute of Technology Lausanne, Lausanne, Switzerland, 2010.
2. Makita, T.; Brühwiler, E. Tensile Fatigue Behaviour of Ultra-High Performance Fibre Reinforced Concrete Combined with Steel Rebars (R-UHPFRC). *Int. J. Fatigue* **2014**, *59*, 145–152. [CrossRef]
3. Yoo, D.Y.; Banthia, N. Mechanical Properties of Ultra-High-Performance Fiber-Reinforced Concrete: A Review. *Cem. Concr. Compos.* **2016**, *73*, 267–280. [CrossRef]
4. Tong, T.; Yuan, S.; Wang, J.; Liu, Z. The Role of Bond Strength in Structural Behaviors of UHPC-NC Composite Beams: Experimental Investigation and Finite Element Modeling. *Compos. Struct.* **2021**, *255*, 112914. [CrossRef]
5. Yu, J.; Zhang, B.; Chen, W.; Liu, H.; Li, H. Multi-Scale Study on Interfacial Bond Failure between Normal Concrete (NC) and Ultra-High Performance Concrete (UHPC). *J. Build. Eng.* **2022**, *57*, 104808. [CrossRef]
6. Valikhani, A.; Jahromi, A.J.; Mantawy, I.M.; Azizinamini, A. Numerical Modelling of Concrete-to-UHPC Bond Strength. *Materials* **2020**, *13*, 1379. [CrossRef]
7. Hussein, H.H.; Walsh, K.K.; Sargand, S.M.; Al Rikabi, F.T.; Steinberg, E.P. Modeling the Shear Connection in Adjacent Box-Beam Bridges with Ultrahigh-Performance Concrete Joints. I: Model Calibration and Validation. *J. Bridg. Eng.* **2017**, *22*, 1–14. [CrossRef]
8. Kadhim, M.M.A.; Jawdhari, A.; Peiris, A. Development of Hybrid UHPC-NC Beams: A Numerical Study. *Eng. Struct.* **2021**, *233*, 111893. [CrossRef]
9. Farzad, M.; Shafieifar, M.; Azizinamini, A. Experimental and Numerical Study on Bond Strength between Conventional Concrete and Ultra High-Performance Concrete (UHPC). *Eng. Struct.* **2019**, *186*, 297–305. [CrossRef]
10. Lampropoulos, A.P.; Paschalis, S.A.; Tsioulou, O.T.; Dritsos, S.E. Strengthening of Reinforced Concrete Beams Using Ultra High Performance Fibre Reinforced Concrete (UHPFRC). *Eng. Struct.* **2016**, *106*, 370–384. [CrossRef]
11. Yin, H.; Shirai, K.; Teo, W. Numerical Model for Predicting the Structural Response of Composite UHPC–Concrete Members Considering the Bond Strength at the Interface. *Compos. Struct.* **2019**, *215*, 185–197. [CrossRef]
12. Abdel Wahab, MM Simulating Mode I Fatigue Crack Propagation in Adhesively-Bonded Composite Joints. *Fatigue and Fracture of Adhesively-Bonded Composite Joints*; Vassilopoulos, A.P., Ed.; Woodhead Publishing: Sawston, UK, 2015; pp. 323–344. ISBN 978-0-85709-806-1.

13. Chabot, A.; Hun, M.; Hammoum, F. Mechanical Analysis of a Mixed Mode Debonding Test for " Composite" Pavements. *Constr. Build. Mater.* **2013**, *40*, 1076–1087. [CrossRef]
14. Bissonnette, B.; Courard, L.; Fowler, D.W.; Granju, J.-L. *Bonded Cement-Based Material Overlays for the Repair, the Lining or the Strengthening of Slabs or Pavements: State-of-the-Art Report of the RILEM Technical Committee 193-RLS*; Springer Science & Business Media: Berlin, Germany, 2011; Volume 3.
15. AASHTO. *AASHTO LRFD Bridge Design Specifications*; American Association of State Highway and Transportation Officials: Washington, DC, USA, 2017; ISBN 9781560516545.
16. ASTM International. *1583: Standard Test Method for Tensile Strength of Concrete Surface and the Bond Strength or Tensile Strength of Concrete Repair and Overlay Materials by Direct Tension (Pull-off Method)*; ASTM International: West Conshohocken, PA, USA, 2004.
17. Hibbitt, K.; Karlsson, B.I. *ABAQUS: User's Manual*; Hibbitt, Karlsson & Sorensen: Birmingham, AL, USA, 2013.
18. Gadri, K.; Guettala, A. Evaluation of Bond Strength between Sand Concrete as New Repair Material and Ordinary Concrete Substrate (The Surface Roughness Effect). *Constr. Build. Mater.* **2017**, *157*, 1133–1144. [CrossRef]
19. Júlio, E.N.B.S.; Branco, F.A.B.; Silva, V.D. Concrete-to-Concrete Bond Strength. Influence of the Roughness of the Substrate Surface. *Constr. Build. Mater.* **2004**, *18*, 675–681. [CrossRef]
20. Feng, S.; Xiao, H.; Liu, M.; Zhang, F.; Lu, M. Shear Behaviour of Interface between Normal-Strength Concrete and UHPC: Experiment and Predictive Model. *Constr. Build. Mater.* **2022**, *342*, 127919. [CrossRef]
21. Hussein, H.H.; Walsh, K.K.; Sargand, S.M.; Steinberg, E.P. Interfacial Properties of Ultrahigh-Performance Concrete and High-Strength Concrete Bridge Connections. *J. Mater. Civ. Eng.* **2016**, *28*, 04015208. [CrossRef]
22. Willam, K.J. Constitutive Model for the Triaxial Behaviour of Concrete. *Proc. Intl. Assoc. Bridg. Structl. Engrs* **1975**, *19*, 1–30.
23. Lapidus, L.; Pinder, G.F. *Numerical Solution of Partial Differential Equations in Science and Engineering*; John Wiley& Sons: Hoboken, NJ, USA, 2011.
24. Johnson, S. *Comparison of Nonlinear Finite Element Modeling Tools for Structural Concrete*; University of Illinois: Urbana, IL, USA, 2006.
25. Kmiecik, P.; Kamiński, M. Modelling of Reinforced Concrete Structures and Composite Structures with Concrete Strength Degradation Taken into Consideration. *Arch. Civ. Mech. Eng.* **2011**, *11*, 623–636. [CrossRef]
26. Kent, D.C.; Park, R. Flexural Members With Confined Concrete. *J. Struct. Div.* **1971**, *97*, 1969–1990. [CrossRef]
27. Massicotte, B.; Elwi, A.E.; MacGregor, J.G. TensionStiffening Model for Planar Reinforced Concrete Members. *J. Struct. Eng.* **1990**, *116*, 3039–3058. [CrossRef]
28. Bahij, S.; Adekunle, S.K.; Al-Osta, M.; Ahmad, S.; Al-Dulaijan, S.U.; Rahman, M.K. Numerical Investigation of the Shear Behavior of Reinforced Ultra-High-Performance Concrete Beams. *Struct. Concr.* **2018**, *19*, 305–317. [CrossRef]
29. Curbach, M.; Speck, K. Ultra High Performance Concrete under Biaxial Compression. In Proceedings of the Second International Symposium on Ultra High Performance Concrete, Kassel, Germany, 5–7 March 2008; Kassel University Press GmbH: Kassel, Germany, 2008; pp. 477–484.
30. Da Silva, V.D. A Simple Model for Viscous Regularization of Elasto-Plastic Constitutive Laws with Softening. *Commun. Numer. Methods Eng.* **2004**, *20*, 547–568. [CrossRef]
31. Wang, Z.M.; Huang, Y.J.; Yang, Z.J.; Liu, G.H.; Wang, F. Efficient Meso-Scale Homogenisation and Statistical Size Effect Analysis of Concrete Modelled by Scaled Boundary Finite Element Polygons. *Constr. Build. Mater.* **2017**, *151*, 449–463. [CrossRef]
32. Chen, J.H.; Xu, W.F.; Xie, R.Z.; Zhang, F.J.; Hu, W.J.; Huang, X.C.; Chen, G. Sample Size Effect on the Dynamic Torsional Behaviour of the 2A12 Aluminium Alloy. *Theor. Appl. Mech. Lett.* **2017**, *7*, 317–324. [CrossRef]
33. Jumaa, G.B.; Yousif, A.R. Numerical Modeling of Size Effect in Shear Strength of FRP-Reinforced Concrete Beams. In *Structures*; Elsevier: Amsterdam, The Netherlands, 2019; Volume 20, pp. 237–254.
34. Sun, Q.; Zhu, H.; Wang, G.; Fan, J. Effects of Mesh Resolution on Hypersonic Heating Prediction. *Theor. Appl. Mech. Lett.* **2011**, *1*, 22001. [CrossRef]
35. Européen, C. *Eurocode 2: Design of Concrete Structures—Part 1-1: General Rules and Rules for Buildings*; British Standard Institution: London, UK, 2004; p. 66.
36. Al-Osta, M.A.; Isa, M.N.; Baluch, M.H.; Rahman, M.K. Flexural Behavior of Reinforced Concrete Beams Strengthened with Ultra-High Performance Fiber Reinforced Concrete. *Constr. Build. Mater.* **2017**, *134*, 279–296. [CrossRef]
37. Hainsworth, S.V.; Chandler, H.W.; Page, T.F. Analysis of Nanoindentation Load-Displacement Loading Curves. *J. Mater. Res.* **1996**, *11*, 1987–1995. [CrossRef]

 *buildings*

Article

Prediction of Unstable Hydrodynamic Forces on Submerged Structures under the Water Surface Using a Data-Driven Modeling Approach

Van My Nguyen [1], Hoang Nam Phan [1,*] and Thanh Hoang Phan [2]

[1] Faculty of Road and Bridge Engineering, The University of Danang—University of Science and Technology, Da Nang 550000, Vietnam
[2] School of Mechanical Engineering, Pusan National University, Busan 46241, Korea
* Correspondence: phnam@dut.udn.vn

Citation: Nguyen, V.M.; Phan, H.N.; Phan, T.H. Prediction of Unstable Hydrodynamic Forces on Submerged Structures under the Water Surface Using a Data-Driven Modeling Approach. *Buildings* **2022**, *12*, 1683. https://doi.org/10.3390/ buildings12101683

Academic Editors: Dongming Li and Zechuan Yu

Received: 14 September 2022
Accepted: 8 October 2022
Published: 13 October 2022

Publisher's Note: MDPI stays neutral with regard to jurisdictional claims in published maps and institutional affil- iations.

Abstract: Catastrophic failures of partially or fully submerged structures, e.g., offshore platforms, hydrokinetic turbine blades, bridge decks, etc., due to the dynamic impact of free surface flows such as waves or floods have revealed the need to evaluate their reliability. In this respect, an accurate estimation of hydrodynamic forces and their relationship to instability in structures is required. The computational fluid dynamics (CFD) solver is known as a powerful tool to identify dynamic characteristics of flow; however, it commonly consumes a huge computational cost, especially in cases of re-simulations needed. In this paper, an efficient surrogate model based on the Gaussian process is developed to rapidly predict the nonlinear hydrodynamic pressure coefficients on submerged bodies near the water surface. For this purpose, a CFD model is first developed, which is based on a two-dimensional incompressible Navier–Stokes solver incorporating free surface treatment and turbulent flow models. Then, an experimental design is adopted to generate initial training samples considering the effect of the submerged body shape ratio and flow Re number. Surrogate models of hydrodynamic pressure coefficients and their instability based on Gaussian process modeling are established using the outcome from the CFD simulations, where optimal trend and correlation functions are also investigated. Once surrogate models are obtained, the mean and oscillation amplitudes of hydrodynamic pressure coefficients on a submerged rectangular body, which represents the shape of most civil structures, can be rapidly predicted without the attempt at re-simulation. The findings can be practically applied in rapidly assessing hydrodynamic forces and their instability of existing submerged civil structures or in designing new structures, where a suitable shape ratio should be adopted to avoid flow-induced instability of hydrodynamic forces.

Keywords: submerged structure; free surface flow; hydrodynamic force; computational fluid dynamics; surrogate model; Gaussian process; data-driven approach

1. Introduction

Past failures of partially or fully submerged civil structures, e.g., offshore platforms, hydrokinetic turbine blades, bridge decks, etc., due to the dynamic impact caused by floods, waves or tsunamis have revealed the significant need for evaluating the reliability of these structures under hydrodynamic forces [1–3]. Regarding this issue, one of the challenges is the accurate evaluation of multiphase flows and their impact on submerged structures. This is essential not only for optimum designs of the newly designed structures but also for the estimation of the degree of risk for existing ones.

Large-scale structures under free surface flows commonly require time-consuming experimental analyses. In addition, flow characteristics, e.g., flow patterns, velocities and pressure fields, are uncertain and nonlinear; hence, resulting in a very high cost and time re- quired for many experimental setups in order to observe accurate flow characteristics [4,5].

While analytical models for hydrodynamic forces on submerged bodies have been developed and incorporated into design codes, they are often overestimated and with a degree of error [6–8]. An alternative is the development of numerical models for free surface flows and their impact on structures; this can reduce much effort in terms of cost and time as compared with experimental models [9]. The numerical models are mainly based on computational fluid dynamic (CFD) approaches that solve Navier-Stokes equations along with the treatment of the free surface. However, due to the uncertain flow characteristics, varying with each particular case, the design of these numerical simulations becomes complicated and requires numerous computing resources, especially in cases of the large number of samples considered in a reliability analysis [10,11].

With the development of computer science, besides various computer models which have been adopted in many fields of CFD [12], machine learning techniques have been developed and widely used to predict multiphase flow characteristics [13–15], in which, data-driven techniques, e.g., neural network [13], support vector machine [14] and Gaussian process [15], have been implemented to build surrogate models for the prediction of the multiphase flow pattern, as well as the hydrodynamic pressure distribution. These surrogate models, which are used when an outcome of interest cannot be easily measured or computed, can efficiently reduce the computational effort and rapidly estimate flow characteristics in the context of uncertainty treatments and reliability analyses. The first two methods (i.e., neural network and support vector machine) commonly require an adequate dataset for a reliable prediction, which depends on the number of input and output parameters, while the Gaussian process regression makes it possible to predict the model response with little observed data. In addition, the Gaussian process offers a flexible kernel method for regression due to various available trends and correlation functions. Therefore, in many complex problems, this technique is more suitable and efficient in terms of reducing computational costs [11,16].

The objective of this paper is to develop surrogate models, which are based on Gaussian process modeling, to rapidly predict nonlinear hydrodynamic pressure coefficients and their instability effect on submerged bodies. As a case study, rectangular submerged bodies near the water surface are considered. This type of shape is standard and represents the shape of many engineering applications such as bridge decks, offshore platforms and hydrokinetic turbine blades. Firstly, a modeling approach of the flow passing a submerged body is presented based on a two-dimensional incompressible Navier–Stokes solver. The free surface is treated using the volume of fluid method and the effect of the turbulent flow is also considered by using the shear stress transport turbulence model. Then, an appropriate experimental design is used to generate initial training samples considering the effect of the aspect ratio of the submerged body and the Re number of the flow. Surrogate models of hydrodynamic pressure coefficients based on the Gaussian process modeling are established using the outcome from the CFD simulations, where optimal trend and correlation functions are also investigated. Once surrogate models are obtained, the mean and oscillation amplitudes of hydrodynamic pressure coefficients of the free surface flow on a submerged cylinder with an arbitrary aspect ratio can be accurately and rapidly predicted without the need for attempt at re-simulation. The findings from the work can be practically applied in rapidly assessing hydrodynamic force and its instability effect on existing submerged civil structures, or in designing new structures, where a suitable shape ratio range is recommended to avoid the detrimental effects of flow-induced instability from hydrodynamic forces.

2. Numerical Model of Free Surface Flow

In this study, a two-dimensional (2D) incompressible Reynolds-averaged Navier-Stokes (RANS) homogeneous two-phase mixture model is adopted to simulate the nonlinear interactions of a submerged cylinder beneath the free surface [9,17]. The model solves the mixture continuity and momentum equations to obtain the mean flow velocity and pressure fields. The RANS model is closed by including a turbulence model to predict fluctuating velocity components; thus, the shear stress transport $k - \omega$ model is adopted. In addition, to capture complex free surface behaviors and non-linear hydrodynamic pressure coefficients, the interface between the water and air phases is numerically treated using the volume-of-fluid (VOF) method [18,19]. These equations can be described in the Cartesian,

$$\frac{\partial u_j}{\partial x_j} = 0 \tag{1}$$

$$\frac{\partial u_i}{\partial t} + u_j \frac{\partial u_i}{\partial x_j} = -\frac{1}{\rho_m} \frac{\partial p}{\partial x_i} + g_i + \mu_m \frac{\partial^2 u_i}{\partial x_j} - \frac{\partial \overline{u_i' u_j'}}{\partial x_j}, \tag{2}$$

$$\frac{\partial \alpha_g}{\partial t} + \frac{\partial (\alpha_g u_j)}{\partial x_j} = 0, \tag{3}$$

Here, the subscripts i, $j = 1$, 2 represent two directions of x and y in the computational domain, respectively; t is the computation time; p denotes the pressure, u and u' are the mean and the fluctuating velocities; g is the gravitational term; and ρ_m and μ_m are the mixture density and viscosity, respectively.

The interface position is numerically treated via the phasic volume fraction of the gas phase, α_g. Here, the pure gas phase is obtained in the case of α_g equals 1.0, while the pure water phase is obtained in the case of α_g equals 0.0. The interfaces with a limited thickness between two phases are identified by the values in the range from 0.0 to 1.0. The properties of the mixture phase at the interface are predicted by a function of the volume of fraction of the individual phase, an example, for the calculation of the mixture values for density and mixture valuables

$$\rho_m = \alpha_g \rho_g + (1 - \alpha_g)\rho_w, \tag{4}$$

$$\mu_m = \alpha_g \mu_g + (1 - \alpha_g)\mu_w, \tag{5}$$

where ρ_w, μ_w and ρ_g, μ_g are the density and viscosity of the individual water and gas phases, respectively.

The numerical discretization of the equation system in the generally structured grid is based on the finite volume method with a pressure-based solver. For time discretization, the first-order implicit method is applied. The first-order upwind scheme is adopted for both the convective and viscous terms and the advection equation is approximated using the implicit compressive scheme. The main reason behind using first-order and implicit schemes is to obtain better convergence than high-order and explicit schemes, especially for strongly deformable free surfaces with breaking wave phenomena. All simulation results are performed using the Ansys Fluent software [20].

3. Surrogate Model of Hydrodynamic Pressure Coefficients

3.1. Basic Formulations

Gaussian process regression uses a set of observed training data to predict spatially correlated data, which postulates a combination of a functional basis and departure in the following form [21],

$$Y_i = \sum_{j=1}^{p} \beta_j f_j(x^{(i)}) + Z(x^{(i)}) \text{ with } i = 0, \dots, m, \tag{6}$$

where the first term is the unknown multivariate polynomial function, $f = \left\{ f_j(x^{(i)}) \right\}$ with $j = 1, \dots, p$, called the trend, and $Z(x)$ is the realization of the Gaussian process having zero mean and variance σ^2; $Z(x)$ is expressed as

$$Cov[Z(x), Z(x')] = \sigma^2 R(x - x', \ell), \tag{7}$$

where σ^2 is the variance of the Gaussian process, whereas R is the correlation function which is the function of the difference $x - x'$ and scale parameters ℓ ($\ell_i > 0, i = 1, \dots, n$).

Several correlation functions are proposed, e.g., the exponential (Equation (8)), Gaussian (Equation (9)), and Matérn-3/2 (Equation (10)),

$$R\left(x - x', \ell\right) = \sum_{i=1}^{n} e^{-1/\ell_i|x_i - x_i'|},\tag{8}$$

$$R\left(x - x', \ell\right) = \sum_{i=1}^{n} e^{-\frac{1}{2}\left(\frac{|x_i - x_i'|}{\ell_i}\right)^2},\tag{9}$$

$$R\left(x - x', \ell\right) = \sum_{i=1}^{n} \left(1 + \frac{\sqrt{3}}{\ell_i}|x_i - x_i'|\right) e^{-\sqrt{3}/\ell_i|x_i - x_i'|}\tag{10}$$

The vector of the prediction $\hat{Y}_0$ and the true response $Y = \{Y_i\}$ with $i = 1, \ldots, m$ is normally distributed,

$$\begin{Bmatrix} \hat{Y}_0 \\ Y \end{Bmatrix} \sim N_{1+m}\begin{Bmatrix} f_0^T \beta \\ F\beta \end{Bmatrix}, \sigma^2 \begin{bmatrix} 1 & r_0^T \\ r_0 & R \end{bmatrix},\tag{11}$$

where f_0 is the vector of regression models evaluated at $x^{(0)}$, F is the regression matrix, r_0 is the vector of cross-correlations between the point $x^{(0)}$, $r_{0i} = R\left(x^{(0)} - x^{(i)}, \ell\right)$, R is the correlation matrix of the true response, $R_{ij} = \left(x^{(i)} - x^{(j)}, \ell\right)$.

The best linear unbiased predictor of the unknown quantity of interest y_0 is the Gaussian random variate $\hat{Y}_0$ with mean and variance,

$$\mu_{\hat{Y}_0} = f_0^T \hat{\beta} + r_0^T R^{-1}(y - F\hat{\beta}),\tag{12}$$

$$\sigma_{\hat{Y}_0}^2 = \sigma^2 \left(1 - r_0^T R^{-1} r_0 + \left(F^T R^{-1} r_0 - f_0\right)^T \left(F^T R^{-1} F\right)^{-1} \left(F^T R^{-1} r_0 - f_0\right)\right),\tag{13}$$

where $\hat{\beta} = \left(F^T R^{-1} F\right)^{-1} F^T R^{-1} y$.

The maximum likelihood estimation technique is better suited for deriving estimators. Here, the likelihood of the observations y is defined concerning its multivariate normal distribution, which depends on β, σ^2, and ℓ,

$$L\left(y|\beta, \sigma^2, \ell\right) = \frac{1}{\left((2\pi\sigma^2)^m [\det R(\ell)]\right)^{1/2}} \exp\left[-\frac{1}{2\sigma^2}\left(y - F\beta^T\right) R(\ell)^{-1}(y - F\beta)\right].\tag{14}$$

By maximizing the quantity described in Equation (14), the following analytical estimates of β and σ^2 that are functions of ℓ are obtained as

$$\hat{\beta}(\ell) = \left(F^T R(\ell)^{-1} F\right)^{-1} F^T R(\ell)^{-1} y,\tag{15}$$

$$\hat{\sigma}^2(\ell) = \frac{1}{m}\left(y - F\hat{\beta}^T\right) R(\ell)^{-1}(y - F\hat{\beta}).\tag{16}$$

By substituting these two solutions into Equation (14), its corresponding opposite log-likelihood reads

$$-\log L\left(y|\beta, \sigma^2, \ell\right) = \frac{m}{2}\log(\psi(\ell)) + \frac{m}{2}(\log(2\pi) + 1) \text{ with } \psi(\ell) = \hat{\sigma}^2(\ell)[\det R(\ell)]^{\frac{1}{2}},\tag{17}$$

and thus, the maximum likelihood estimate of ℓ is given as

$$\hat{\ell} = \underset{\ell}{\mathrm{argmin}}\,\psi(\ell).\tag{18}$$

3.2. Procedure of Surrogate Model-Based Hydrodynamic Pressure Coefficient Prediction

The overall procedure for the development of a Gaussian process based-surrogate model is as follows:

(i)   In the first step, the most important variables and their distribution functions should be identified. An appropriate DOE is then conducted within the range of interest variables. As a result, several initial training samples are generated and corresponding CFD models are then built.

(ii)   CFD simulations of the flow field are conducted for each combination of training conditions. The flow field characteristics, as well as hydrodynamic pressure coefficients, are obtained at each simulation. In this study, the mean and oscillation values of hydrodynamic pressure coefficients, i.e., drag, lift and moment, are considered as the model responses.

(iii)   Once the training dataset has been established based on the DOE and the corresponding model responses, a surrogate model of the model response is built using the Gaussian process modeling incorporated in a Matlab-based software, Uqlab [22]. In this step, different trend and correlation functions that compose the Gaussian process are tested. An optimal surrogate model is finally obtained based on the error estimation from the cross-validation.

4. Case study of Submerged Bodies beneath the Water Surface

4.1. CFD Simulation and Design of Experiments

In this study, a rectangular shape body fully submerged beneath a free surface is selected, which represents the shape of most bridge deck or other civil structure components under the free surface flow. The computational domain and simulation conditions are adopted following the experimental work by Chu et al. [23], in which the problem of an open channel with a particular size is shown schematically in Figure 1. The rectangular body submerges in the water at a depth of h and a distance between the channel bed and the cylinder S. To reduce the computational cost, a planar symmetric numerical model is used under the main assumption that there are no effects in the spanwise direction. This assumption was also adopted for numerical computations of free surface flows over a submerged body in many studies [9]. The boundary conditions are applied as follows: (i) the fixed uniform velocity is specified at the inlet condition, (ii) at the outlet condition, the extrapolation values are applied for the pressure and velocity fields, (iii) at the top boundary condition, open conditions are applied and (iv) at the bottom line and cylinder, non-slip wall conditions are used.

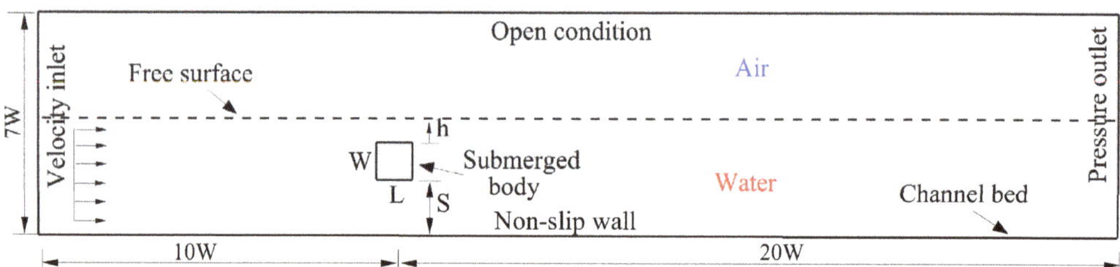

Figure 1. Computational model of the free surface flow around a submerged body.

The mesh distribution for the whole computational domain and zoomed regions near the submerged body are presented in Figure 2. Here, the meshing strategy with a high-resolution value close to the body and free surface is used to obtain high accuracy predictions for the pressure and velocity fields, particularly for the free surface shape motion. The grid and time step sensitivity tests were performed through convergency analyses in the previous work [9]; thus the fine grid with a total number of nodes of 112,649, the $y+$ value at the body surface of 1.1, and the time step of 0.002 (s) are used in the present simulations.

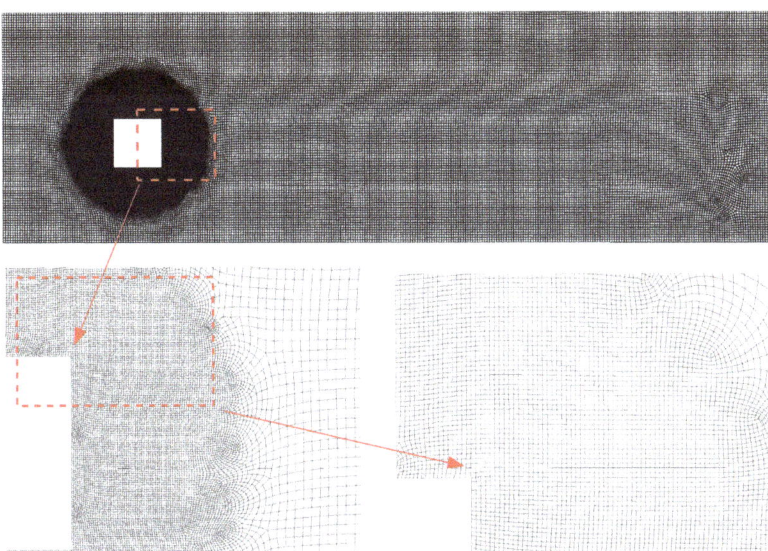

Figure 2. The mesh domain and mesh distribution around the submerged body.

The complex simulation of the above-mentioned CFD model reveals the need for developing a more efficient surrogate model to possibly and rapidly identify nonlinear hydrodynamic pressure coefficients on a submerged body under free surface flows. The development of a surrogate model or metamodel first requires the generated samples that involve modeling parameters. In this study, Latin Hypercube Sampling (LHS) [24] is utilized to generate training samples; this technique is widely used and has demonstrated its efficiency in the construction of surrogate models, especially ones based on the Gaussian process [25,26].

Many studies have demonstrated that the shape ratio (AR) (i.e., the length and depth ratios of the rectangular body, $AR = L/W$) and the Re number are the most significant parameters that affect the flow characteristic and impact hydrodynamic pressure [8,10]. Therefore, these two parameters are chosen as input random variables in the study with the ranges selected and presented in Table 1, which are assumed to be a uniform distribution. The other geometry parameters, such as the depth h and the clearance distance S, are deterministic.

Table 1. Modeling parameters for the design of experiments.

Parameter	Distribution Function	Lower Value	Upper Value
Shape ratio AR	Uniform	0.2	4
Re number	Uniform	8000	16,000

By using the LHS on two modeling parameters, a total of 40 samples are generated, as distributed in Figure 3. It should be noticed that there is no specific standard for the number of initial training samples, depending on the number of input variables, particular problem and training method. Since the Gaussian process has not needed the pre-assuming of a specified model and just requires a small number of initial training samples, an optimized design of 40 samples is chosen, as proposed by [26].

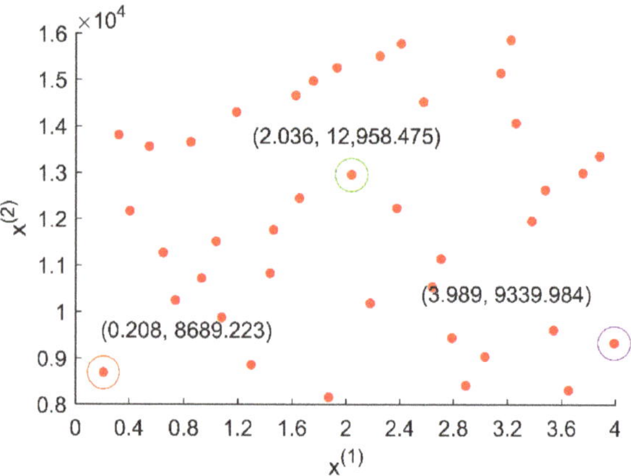

Figure 3. Design of experiments using LHS.

As examples of the flow characteristic, Figure 4 shows the velocity field and pressure contour for three cases of *AR* and *Re* values, marked by large circles in Figure 3. The free surface shape is plotted by the red solid line using the air volume fraction value of 0.5. Under the presence of the submerged body, significantly increasing free surface flow and reduction in the depth at the upward and the downward regions is observed. Therefore, a high-velocity water flow in the downstream region is formed. In addition, submerged wake vortices behind the body are generated under the effects of the inclination of the free surface, as shown by streamline fields on the left side of Figure 4. This nonlinear evolution behavior is found to be a major mechanism that increases the hydrodynamic force coefficients in comparison with the unbounded free surface flow. The pressure distributions around the submerged body are also shown on the right side of Figure 4. In the presence of the free surface, asymmetric low-pressure regions at the top and bottom of the submerged body are observed.

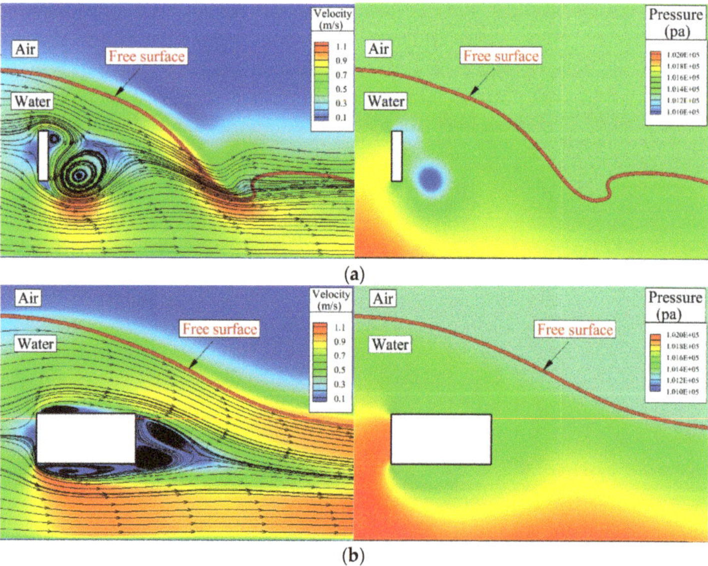

Figure 4. *Cont.*

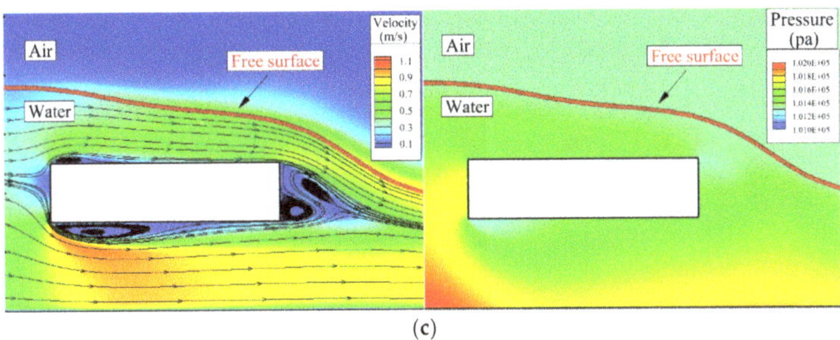

(c)

Figure 4. Velocity and pressure fields around the submerged body with the free surface interactions: (**a**) $AR = 0.208$, $Re = 8689.223$, (**b**) $AR = 2.036$, $Re = 12{,}958.475$, (**c**) $AR = 3.989$, $Re = 9339.984$.

For each sample, hydrodynamic pressure coefficients (i.e., drag, lift and moment coefficients) acting on submerged bodies are obtained from CFD analyses. Figure 5 shows an example of the time evolution of hydrodynamic pressure coefficients for $AR = 0.208$, $Re = 8689.223$ (Figure 5a), $AR = 2.036$, $Re = 12{,}958.475$ (Figure 5b), and $AR = 3.989$, $Re = 9339.984$ (Figure 5c). The numerical results show that the hydrodynamic force coefficients are significantly varied under the nonlinear interaction with the free surface. In the cases of lower AR values ($AR = 0.208$), the force coefficients are in periodic evolutions with time and are characterized by a mean value and oscillated magnitude. These predicted values are determined by an averaged method over five oscillation cycles. In the case of higher values ($AR = 2.036$ and 3.989), stable behaviors of the force coefficients without oscillation features are observed.

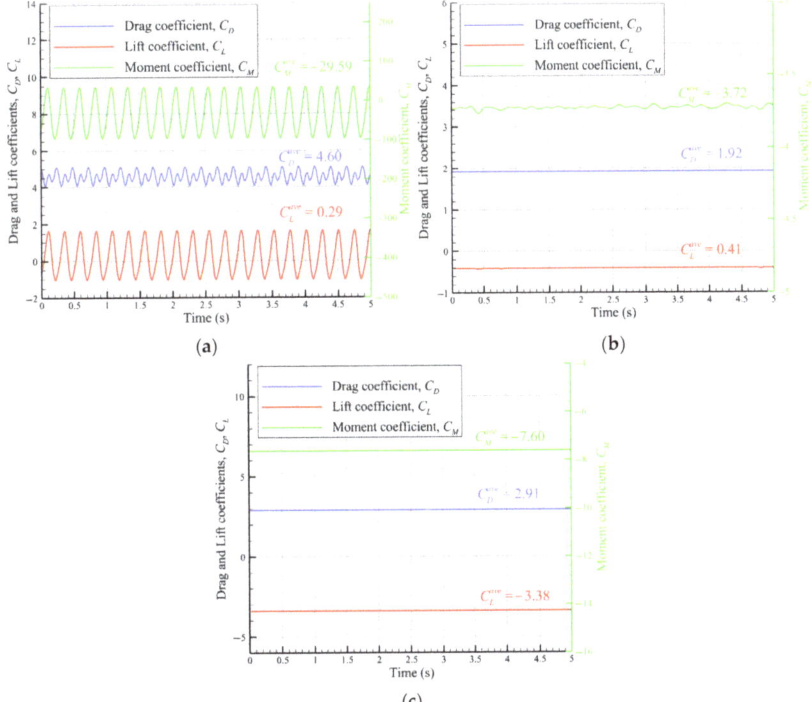

Figure 5. The time evolution of hydrodynamic force coefficients: (**a**) $AR = 0.208$, $Re = 8689.223$, (**b**) $AR = 2.036$, $Re = 12{,}958.475$, (**c**) $AR = 3.989$, $Re = 9339.984$.

Similarly, the outcomes of interest including six above-mentioned quantities (the mean values of C_L, C_D, C_M and their oscillations) are obtained and summarized in Table A1 (Appendix A), resulting in a total of 40 examples of training data in the dataset. The observed responses from the dataset are later used to train surrogate models of the outcomes.

4.2. Surrogate Model Development

Based on the above model output, the surrogate model is then developed using the above-mentioned Gaussian process modeling. The major advantage of this modeling approach is that it requires less observed data for the regression as compared to other data-driven techniques, such as support vector machine or artificial neural network. To optimize the surrogate model for the prediction, several trend and correlation functions are tested in this study. Due to the nonlinearity of the hydrodynamic pressure coefficients, nonlinear regression and correlation models are selected. In particular, for the trend, polynomial (one, two, and three degrees) functions are employed. On the other term, Matérn 3/2, exponential and Gaussian (in Equations (8)–(10)) correlation functions are examined.

To estimate the accuracy of each tested surrogate model, the leave-one-out (LOO) cross-validation is adopted, where one point is randomly ignored for the cross-validation and the other points are for training the surrogate model. This procedure is repeated until all the points are used. Therefore, to perform the LOO, one point $x^{(i)}$ from the initial DOE is subsequently removed and the surrogate model $\hat{Y}_{0,(-i)}\left(x^{(i)}\right)$ is built from the remaining points of the design. The LOO cross-validation error is calculated on the true response design and its corresponding predicted responses as

$$\epsilon_{LOO} = \frac{1}{n} \frac{\sum_{i=1}^{n}\left[Y(x_i) - \hat{Y}_{0,(-i)}(x_i)\right]^2}{\mathrm{Var}\ (Y)}, \tag{19}$$

where $\mathrm{Var}(Y)$ defines the estimated variance of the output variable.

Table 2 shows the cross-validation error of the tested surrogate models as combinations of different trend and observation functions. It can be observed that the estimated LOO errors vary with different combinations and outcomes of interest. In most of the cases, the combination of 3rd degree polynomial function and Matérn 3/2 correlation function results in the best performance with a minimum mean error of the prediction for the outcomes (highlighted in bold in Table 2). Also of note is that combinations of 2nd degree polynomial-Matérn 3/2 and 3rd degree polynomial-Gaussian also exhibit a good prediction.

Table 2. Cross-validation error estimation of the tested surrogate models.

Trend Function	Correlation Function	LOO Error						Mean Error
		C_D (Mean)	C_D (Oscil.)	C_L (Mean)	C_L (Oscil.)	C_M (Mean)	C_M (Oscil.)	
1st degree polynomial		0.0222	0.0244	0.0010	0.0631	0.0956	0.0350	0.0402
2nd degree polynomial	**Matérn 3/2**	0.0102	0.0250	0.0023	0.0620	0.1004	0.0371	0.0395
3rd degree polynomial		**0.0074**	**0.0270**	**0.0045**	**0.0610**	**0.0745**	**0.0259**	**0.0334**
1st degree polynomial		0.0264	0.0449	0.0100	0.0728	0.1306	0.0672	0.0586
2nd degree polynomial	Exponential	0.0079	0.0452	0.0068	0.0787	0.1131	0.0486	0.0500
3rd degree polynomial		0.0072	0.0616	0.0060	0.1173	0.0906	0.0367	0.0532
1st degree polynomial		0.0314	0.2446	0.0026	0.0590	0.1250	0.0441	0.0845
2nd degree polynomial	Gaussian	0.0155	0.2637	0.0042	0.3129	0.1111	0.0327	0.1234
3rd degree polynomial		0.0105	0.0316	0.0046	0.0677	0.0841	0.0216	0.0367

Examples of optimal surrogate models for the drag coefficient outcomes, i.e., mean and oscillation amplitude quantities, are shown in Figure 6, where the red dots represent the DOE. The estimated parameters of all six models for six quantities of interest are summarized in Table A2 (Appendix B). Once a surrogate model is built with its estimated parameters, the hydrodynamic pressure coefficients of an arbitrary design of AR and Re can be rapidly predicted without the need for an attempt at re-simulation.

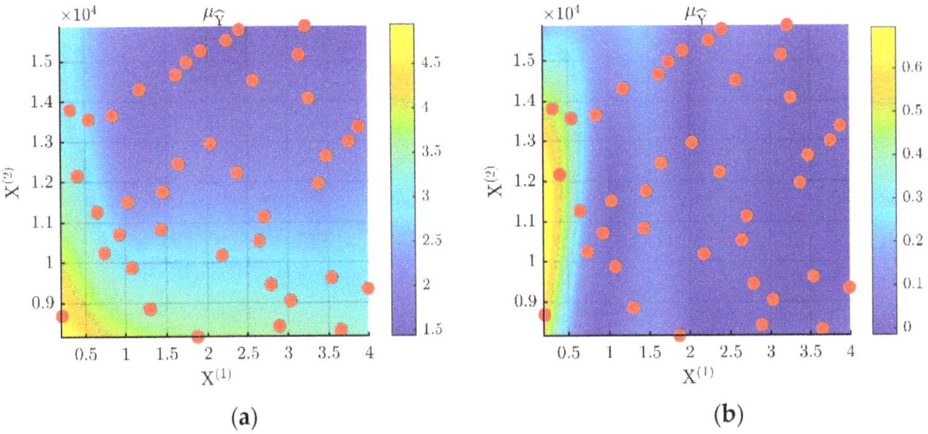

(a)

(b)

Figure 6. Examples of surrogate models of the drag coefficient in terms of $\mu_{\hat{Y}}$: (**a**) Mean and (**b**) Oscillation amplitude.

4.3. Validation of Surrogate Models with a Test Set

A test set of the hydrodynamic pressure coefficients is obtained from the previous work to validate the developed surrogate models. In particular, five different designs of the submerged body under free surface flows were numerically performed, which were uniformly composed by $Re = 11{,}850$ and five body shape ratios, $AR = 0.25, 0.5, 1.0, 2.0,$ and 4.0. As a comparison, the plots of observed mean and oscillation magnitude together with those predicted for three hydrodynamic pressure coefficients are shown in Figure 7. Careful readers can see a good fit between the observed and predicted values both for the mean and oscillation quantities. To quantify the goodness of fit between the observed and predicted data, the mean square error (MSE) and coefficient of determination (R^2) are calculated and presented in Table 3. It can be re-confirmed that the surrogate models predict the mean and oscillation amplitude of the hydrodynamic pressure coefficients with a high degree of accuracy. Hence, the developed models are reliable in prediction and efficient in terms of computational effort.

Table 3. Error estimations of the observed and predicted hydrodynamic pressure coefficients in the case of $Re = 11{,}850$.

Quantity of Interest	MSE	R^2
C_D (Mean)	0.0033	0.9776
C_D (Oscil.)	0.0009	0.9868
C_L (Mean)	0.0060	0.9866
C_L (Oscil.)	0.0150	0.9955
C_M (Mean)	4.4359	0.9727
C_M (Oscil.)	8.8146	0.9030

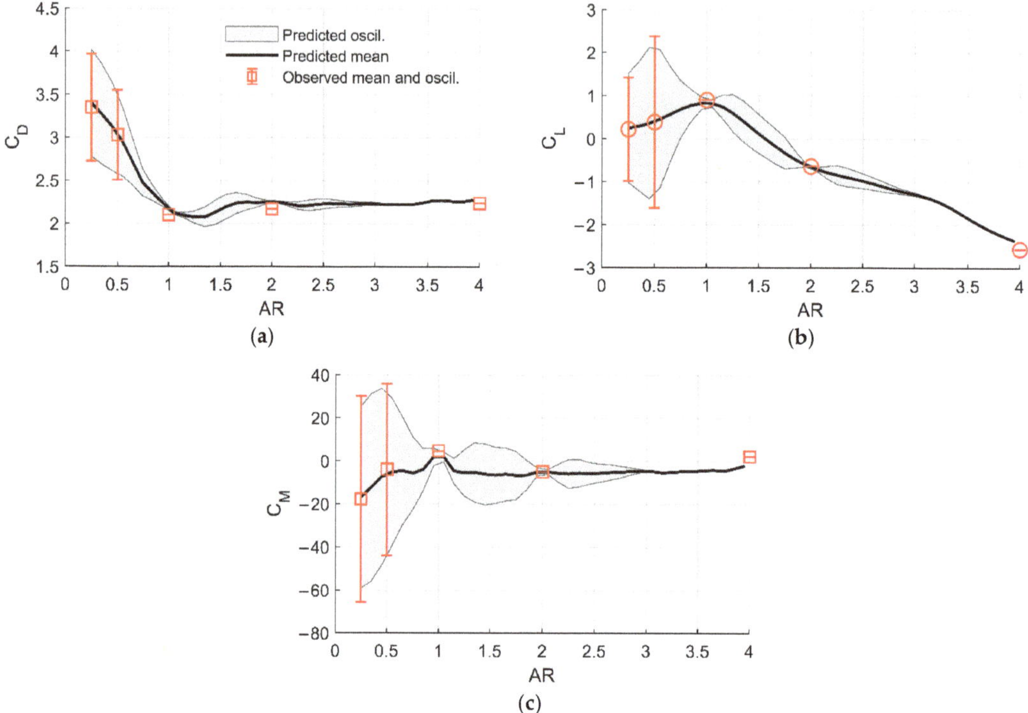

Figure 7. Comparisons of predicted hydrodynamic pressure coefficients with observed values from Nguyen et al. [9]: (**a**) Drag, (**b**) Lift, and (**c**) Moment coefficients.

In the practice design of potentially submerged civil structures, such as bridge decks, offshore platforms and hydrokinetic turbine blades, it is important to avoid unstable regions caused by the oscillation of the hydrodynamic forces. By considering a wide range of Re number and shape aspect ratio (Re = 6000–20,000, AR = 0.1–10), the unstable or oscillation regions are plotted based on the developed surrogate models for the three coefficients, as shown in Figure 8, in which the bar color represents the oscillation amplitude of the examined coefficients. It can be observed that the most unstable region appears with small aspect ratios of the submerged body. The amplitude of the oscillation mostly decreases with the increase of both Re and AR. In the cases of $AR > 2$, the oscillation amplitude significantly drops and almost equals zero. These observations are criteria for the dynamic instability assessment of existing submerged civil structures or for practice design of new ones under the free surface flow to avoid adverse effects of the dynamic impact.

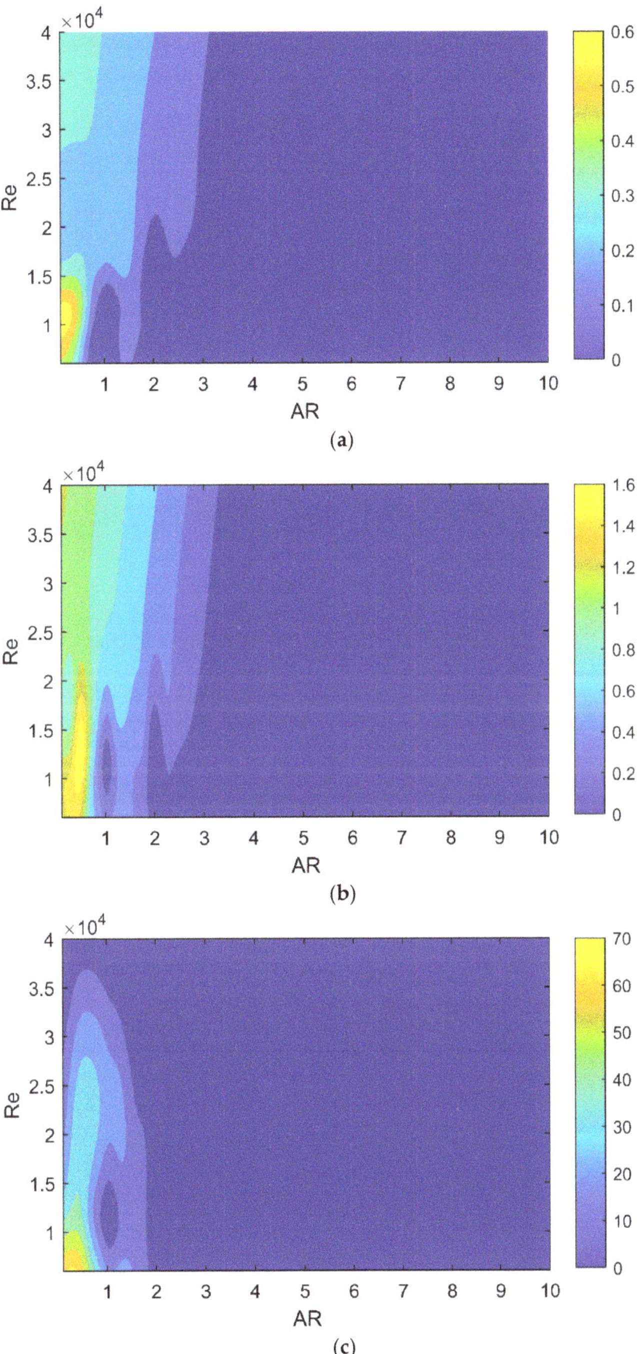

Figure 8. Predictions of unstable zones of hydrodynamic pressure coefficients: (**a**) Drag, (**b**) Lift, and (**c**) Moment coefficients.

5. Conclusions

This study aimed to develop a computationally efficient and accurate surrogate model to estimate hydrodynamic pressure coefficients on submerged bodies beneath the water surface. Using the LHS sampling method, several computational fluid dynamics analyses, based on a Navier-Stokes solver implemented with the shear stress transport turbulence model and the volume of fluid method, were performed to extract hydrodynamic pressure coefficients and their instability.

From the outcomes of the CFD analyses, a Gaussian process modeling-based surrogate model was trained to predict the hydrodynamic pressure coefficients around submerged bodies with a rectangular shape considering a range of shape ratio and Re number values.

As cross-validation for several testing surrogate models, the optimized model was found to be a combination of the 3rd degree polynomial and Matérn 3/2 correlation functions.

Since the surrogate models were developed, the hydrodynamic pressure coefficients were then predicted for a wide range of input parameters. The finding from the study highlighted the efficiency of the surrogate model in rapidly estimating the hydrodynamic pressure coefficients in place of complex and expensive CFD analyses.

By plotting unstable regions of the hydrodynamic pressure coefficients within the ranges of the shape ratio and Re number, it is concluded that the most unstable region appeared at small aspect ratios of the submerged body. In most of the cases, the oscillation amplitude significantly dropped with the increase of both AR and Re and reached almost zero with $AR > 2$.

The surrogate model in this study can be practically applied in rapidly assessing the hydrodynamic force and its instability effect on existing submerged civil structures, or in designing new structures, where a suitable shape ratio should be adopted to avoid flow-induced instability of hydrodynamic forces.

The present study can also be enabled and facilitate future sensitivity, fragility and reliability studies across a broad range of submerged bodies and flow conditions that are involved in civil structures under flood and wave flows.

Author Contributions: Conceptualization, V.M.N. and H.N.P.; methodology, V.M.N. and H.N.P.; software, H.N.P. and T.H.P.; validation, V.M.N., H.N.P. and T.H.P.; formal analysis, H.N.P. and T.H.P.; investigation, H.N.P. and T.H.P.; writing—original draft preparation, H.N.P. and T.H.P.; writing—review and editing, V.M.N., H.N.P. and T.H.P. All authors have read and agreed to the published version of the manuscript.

Funding: This research is funded by Funds for Science and Technology Development of the University of Danang under project number B2020-DN02-80.

Data Availability Statement: Not applicable.

Conflicts of Interest: The authors declare no conflict of interest.

Appendix A

Table A1. Observed response from CFD simulations for the initial training samples.

Input Variable		Output Hydrodynamic Pressure Coefficients					
AR	Re	C_D		C_L		C_M	
		Mean	Oscillation Amplitude	Mean	Oscillation Amplitude	Mean	Oscillation Amplitude
1.623	14,665.087	1.653	0.146	0.405	0.492	−3.511	12.653
3.476	12,640.045	2.087	0	−1.465	0	−3.957	0
2.403	15,783.894	1.451	0.087	0.260	0.305	−1.895	5.741

Table A1. *Cont.*

Input Variable		Output Hydrodynamic Pressure Coefficients					
AR	Re	C_D		C_L		C_M	
		Mean	Oscillation Amplitude	Mean	Oscillation Amplitude	Mean	Oscillation Amplitude
3.144	15,153.660	1.632	0.018	−0.273	0.084	−1.909	0
2.036	12,958.475	1.920	0.007	0.410	0.042	−3.720	1.029
2.890	8416.533	3.441	0	−2.661	0	−8.266	1.504
1.297	8857.737	3.429	0.102	−0.361	0.433	−9.072	15.586
0.208	8689.223	4.600	0.578	0.290	1.433	−29.590	57.009
2.704	11,139.096	2.458	0.030	−1.416	0.103	−5.916	3.124
1.435	10,833.029	2.569	0.139	−0.065	0.446	−7.160	14.355
2.244	15,512.939	1.487	0.055	0.283	0.333	−2.195	6.928
3.538	9615.309	2.959	0	−2.973	0	−6.677	0
1.926	15,258.790	1.521	0.029	0.382	0.145	−0.823	3.406
1.183	14,304.394	1.737	0.128	0.600	0.465	−1.907	11.747
1.751	14,974.758	1.589	0.098	0.396	0.464	−3.120	11.481
3.755	13,004.791	2.040	0	−1.717	0	−3.330	0
3.879	13,368.033	1.987	0	−1.747	0	−1.323	0
3.377	11,959.480	2.199	0	−1.608	0	−4.654	0
3.651	8313.764	3.372	0	−3.584	0	−7.782	0
0.927	10,719.160	2.701	0.040	0.650	0.190	−1.448	6.000
0.404	12,170.650	3.099	0.541	0.303	1.663	−9.008	42.218
0.846	13,654.136	2.122	0.142	0.597	0.472	−1.460	12.568
1.459	11,762.251	2.170	0.143	0.171	0.463	−6.291	14.129
0.647	11,271.221	2.905	0.262	0.575	1.080	−5.339	26.001
2.786	9449.139	3.077	0.012	−2.200	0.037	−7.400	2.408
1.869	8157.854	3.678	0.036	−1.774	0.158	−11.104	6.056
3.258	14,079.809	1.828	0.001	−0.660	0.025	−2.805	0
0.736	10,250.214	3.081	0.128	0.617	0.650	−7.770	20.596
3.220	15,872.356	1.509	0.016	−0.151	0.081	−1.194	0
1.077	9880.575	2.986	0.028	0.336	0.153	−3.508	4.934
2.641	10,540.904	2.681	0.031	−1.625	0.105	−6.511	3.687
1.650	12,453.679	2.040	0.122	0.010	0.427	−5.513	12.430
0.319	13,810.938	2.893	0.483	0.162	1.227	−11.899	39.283
1.034	11,514.888	2.253	0.013	0.775	0.071	2.268	2.172
3.030	9043.564	3.200	0	−2.554	0	−7.586	0.260
2.176	10,188.772	2.835	0.024	−1.431	0.146	−7.570	5.035
2.371	12,232.946	2.111	0.064	−0.759	0.223	−5.168	6.053
0.544	13,561.550	2.693	0.382	0.291	1.676	−3.352	34.293
2.572	14,520.981	1.699	0.063	−0.116	0.225	−2.922	4.254
3.989	9339.984	2.910	0	−3.380	0	−7.600	0

Appendix B

Table A2. Estimated parameters of the Gaussian process-based surrogate model.

Surrogate Model	β	σ^2	ℓ
C_D (Mean)	$(2.120; -0.665; 0.150; 0.235; -0.003; 0.002; -0.138; -0.0690; 0.037)$	0.009	(0.220; 0.9696)
C_D (Oscil.)	$(0.445; 0.0807; -0.402; -0.027; -0.067; 0.0667; 0.003; 0.052; 0.034; -0.063)$	3.661	(1.886; 8.554)
C_L (Mean)	$(-1.531; 0.743; -0.521; -0.037; 0.152; 0.193; -0.061; -0.060; 0.031; -0.094)$	58.317	(2.010; 9.945)
C_L (Oscil.)	$(0.140; 0.070; -0.181; 0.014; 0.198; 0.010; -0.008; -0.101; 0.004; -0.035)$	0.036	(0.137; 9.991)
C_M (Mean)	$(-3.577; 2.452; -3.368; -0.934; -1.127; 0.049; 0.03; 2.114; 1.166; 0.033)$	2.186	(0.148; 1.225)
C_M (Oscil.)	$(3.223; 0.982; -3.512; 0.696; 6.151; 0.755; -0.183; -3.423; -0.919; -1.282)$	9.040	(0.178; 1.785)

References

1. Palermo, D.; Nistor, I.; Saatcioglu, M.; Ghobarah, A. Impact and damage to structures during the 27 February 2010 Chile tsunami Canadian. *J. Civ. Eng.* **2013**, *40*, 750–758. [CrossRef]
2. Kim, H.; Sim, S.H.; Lee, J.; Lee, Y.J.; Kim, J.M. Flood fragility analysis for bridges with multiple failure modes. *Adv. Mech. Eng.* **2017**, *9*. [CrossRef]
3. Balomenos, G.P.; Padgett, J.E. Fragility analysis of pile-supported wharves and piers exposed to storm surge and waves. *J. Waterw. Port Coast. Ocean Eng.* **2018**, *144*, 04017046. [CrossRef]
4. Liu, T.L.; Guo, Z.M. Analysis of wave spectrum for submerged bodies moving near the free surface. *Ocean Eng.* **2013**, *58*, 239–251. [CrossRef]
5. Ren, H.; Fu, S.; Liu, C.; Zhang, M.; Xu, Y.; Deng, S. Hydrodynamic Forces of a Semi-Submerged Cylinder in an Oscillatory Flow. *Appl. Sci.* **2020**, *10*, 6404. [CrossRef]
6. Olaya, S.; Bourgeot, J.; Benbouzid, M.E.H. Hydrodynamic Coefficient Computation for a Partially Submerged Wave Energy Converter. *IEEE J. Ocean. Eng.* **2015**, *40*, 522–535. [CrossRef]
7. Lagrange, R.; Delaune, X.; Piteau, P.; Borsoi, L.; Antunes, J. A new analytical approach for modeling the added mass and hydrodynamic interaction of two cylinders subjected to large motions in a potential stagnant fluid. *J. Fluids Struct.* **2018**, *77*, 102–114. [CrossRef]
8. Chen, X.; Liang, H. Wavy properties and analytical modeling of free-surface flows in the development of the multi-domain method. *J. Hydrodyn. Ser. B* **2018**, *28*, 971–976. [CrossRef]
9. Nguyen, V.M.; Le, A.T.; Phan, H.N.; Nguyen, Q.K.; Chau, V.T.; Phan, T.H. On the behavior of nonlinear hydrodynamic pressure coefficients of a submerged cylinder beneath the water surface. *Vietnam. J. Mech.* **2021**, *43*, 371–387. [CrossRef]
10. Hoang, P.H.; Phan, H.N.; Nguyen, D.T.; Paolacci, F. Kriging Metamodel-Based Seismic Fragility Analysis of Single-Bent Reinforced Concrete Highway Bridges. *Buildings* **2021**, *11*, 238. [CrossRef]
11. Phan, H.N.; Paolacci, F.; Di Filippo, R.; Bursi, O.S. Seismic vulnerability of above-ground storage tanks with unanchored support conditions for Na-tech risks based on Gaussian process regression. *Bull. Earthq. Eng.* **2020**, *18*, 6883–6906. [CrossRef]
12. Bhatti, M.M.; Marin, M.; Zeeshan, A.; Abdelsalam, S.I. Recent trends in computational fluid dynamics. *Front. Phys* **2020**, *8*, 593111. [CrossRef]
13. Ahmadi, M.A.; Chen, Z. Machine learning models to predict bottom hole pressure in multi-phase flow in vertical oil production wells. *Can. J. Chem. Eng.* **2019**, *97*, 2928–2940. [CrossRef]
14. Guillén-Rondon, P.; Robinson, M.D.; Torres, C.; Pereya, E. Support vector machine application for multiphase flow pattern prediction. In Proceedings of the Workshop on Data Mining for Geophysics and Geology (DMG2), San Diego, CA, USA, 5 May 2018.
15. Ganti, H.; Khare, P. Data-driven surrogate modeling of multiphase flows using machine learning techniques. *Comput. Fluids* **2020**, *211*, 104626. [CrossRef]
16. Quinonero-Candela, J.; Rasmussen, C.E.; Williams, C.K. Approximation Methods for Gaussian Process Regression. In *Large-Scale Kernel Machines*; MIT Press: Cambridge, MA, USA, 2007.
17. Duy, T.N.; Nguyen, V.T.; Phan, T.H.; Park, W.G. An enhancement of coupling method for interface computations in incompressible two-phase flows. *Comput. Fluids* **2021**, *214*, 104763. [CrossRef]
18. Phan, H.N.; Lee, J.H. Flood Impact Pressure Analysis of Vertical Wall Structures using PLICVOF Method with Lagrangian Advection Algorithm. *J. Comput. Struct. Eng. Inst. Korea* **2010**, *23*, 675–682.
19. Nguyen, V.T.; Park, W.G. A free surface flow solver for complex three-dimensional water impact problems based on the VOF method International. *J. Numer. Methods Fluids* **2016**, *82*, 3–34. [CrossRef]
20. ANSYS Inc. *ANSYS Fluent User's Guide 2019R1*; ANSYS Inc.: Canonsburg, PA, USA, 2019.
21. Santner, T.J.; Williams, B.J.; Notz, W.I. *The Design and Analysis of Computer Experiments*; Springer: New York, NY, USA, 2003.

22. Stefano, M.; Bruno, S. UQLab: A Framework for Uncertainty Quantification in MATLAB. In Proceedings of the 2nd International Conference on Vulnerability and Risk Analysis and Management (ICVRAM 2014), Liverpool, UK, 13–16 July 2014. [CrossRef]
23. Chu, C.R.; Lin, Y.A.; Wu, T.R.; Wang, C.Y. Hydrodynamic force of a circular cylinder close to the water surface. *Comput. Fluids* **2018**, *171*, 154–165. [CrossRef]
24. Mckay, M.D.; Beckman, R.J.; Conover, W.J. A Comparison of Three Methods for Selecting Values of Input Variables in the Analysis of Output from a Computer Code. *Technometrics* **1979**, *21*, 239–245. [CrossRef]
25. Pebesma, E.J.; Heuvelink, G.B. Latin hypercube sampling of Gaussian random fields. *Technometrics* **1999**, *41*, 303–312. [CrossRef]
26. Iooss, B.; Boussouf, L.; Marrel, A.; Feuillard, V. Numerical study of algorithms for metamodel construction and validation. In *Safety, Reliability and Risk Analysis: Theory, Methods and Applications*; Martorell, S., Soares, C.G., Barnett, J., Eds.; Taylor & Francis Group: London, UK, 2009.

 *buildings*

Article

Behavior of Stiffened Rafts Resting on Expansive Soil and Subjected to Column Loads of Lightweight-Reinforced Concrete Structures

Mohamed H. Abu-Ali [1,*], Basuony El-Garhy [1], Ahmed Boraey [1], Wael S. Alrashed [1], Mostafa El-Shami [2], Hassan Abdel-Daiem [3] and Badrelden Alrefahi [1]

[1] Department of Civil Engineering, University of Tabuk, P.O. Box 741, Tabuk 71491, Saudi Arabia; belgarhy@ut.edu.sa (B.E.-G.); a.boraey@ut.edu.sa (A.B.); walrashed@ut.edu.sa (W.S.A.)
[2] Department of Civil Engineering, Faculty of Engineering, Menoufia University, Shibin El-Kom 32511, Egypt; mostafa.el-shami@protonmail.com
[3] Department of Electrical Engineering, University of Tabuk, P.O. Box 741, Tabuk 71491, Saudi Arabia; habdaldaiem@ut.edu.sa
* Correspondence: mabuali@ut.edu.sa

Abstract: An approach to estimate the behavior of stiffened rafts under column loads of a lightweight-reinforced concrete structure resting on expansive soils is presented in this paper. The analysis was conducted using the computer program SLAB97, which estimates the 3D distorted mound shape using the finite difference method by solving the transient suction diffusion equation in 3D and computing the corresponding soil movements. The interaction between the stiffened rafts and the 3D distorted mound shape is then analyzed using the finite element method. The SLAB97 program has been validated by comparing its results with the results of others that were shown to be valid. The goal of the study is to make the expansive soil structure interaction models that the previous researchers proposed more logical. Assuming the worst initial 3D distorted mound shapes of the two cases of edge heave and edge shrinkage, an upper-bound solution is obtained. Using the two scenarios of edge shrinkage and edge heave, the program was utilized in a parametric investigation to examine the impact of various parameters on the behavior of stiffened rafts on expansive soils. These parameters include the stiffening beam depth, the maximum differential movement of the distortion mound shape, and the raft dimensions. The behavior of the stiffened rafts subjected to concentrated column loads is concluded to be similar to that of the stiffened rafts subjected to uniform and perimeter line loads in both cases of distortion modes, with regard to the shape of raft deformation and distribution of the bending moments; however, the values of the design parameters such as maximum deflection, maximum differential deflection, and maximum moments are entirely different in these two situations.

Keywords: stiffened rafts; expansive soils; suction diffusion equation; lightweight-reinforced concrete structure; edge shrinkage; edge heave; distorted mound shape

Citation: Abu-Ali, M.H.; El-Garhy, B.; Boraey, A.; Alrashed, W.S.; El-Shami, M.; Abdel-Daiem, H.; Alrefahi, B. Behavior of Stiffened Rafts Resting on Expansive Soil and Subjected to Column Loads of Lightweight-Reinforced Concrete Structures. *Buildings* **2024**, *14*, 588. https://doi.org/10.3390/buildings14030588

Academic Editors: Zechuan Yu and Dongming Li

Received: 13 January 2024
Revised: 12 February 2024
Accepted: 18 February 2024
Published: 22 February 2024

1. Introduction

As is well known, the unsaturated expansive soils undergo movements (shrinkage or heave) in response to the changes in their moisture content (i.e., soil suction). The soil heaves and shrinks as its moisture content increases and decreases. The shrinkage or heave in expansive soils causes distortions in structures constructed over them, especially lightweight structures such as low to medium-rise buildings, commercial buildings, schools, hospitals, mosques, swimming pools, and rigid and flexible pavements. The changes in the soil suction may be caused by rainfall and evaporation, lawn irrigation, leaking from water or sewage pipelines, large trees, and raising or lowering in the groundwater table.

Lightweight structures constructed on expansive soils are frequently subjected to severe movements (shrink or heave) arising from non-uniform soil moisture changes, with consequent cracking and damage related to the distortion. The damage caused by expansive soils can range from minor cracking in walls and sidewalks to major cracking in the structural elements of the structures. Damage to lightweight structures constructed on expansive soils has been widely reported in many countries of the world [1,2]. In the United States alone, the annual loss due to structural damage caused by expansive soils exceeds that caused by earthquakes, hurricanes, and floods combined [3]. According to Krohn and Slosson [4], the annual cost of expansive soil damage in the United States exceeded $7.0 billion. In Australia, expansive soils cause structural cracks in nearly 50,000 houses each year [5]. According to Dafalla et al. [6], Saudi Arabia's expansive soils are believed to have caused economic losses in the hundreds of millions of US dollars.

Expansive soil is found in many parts of the world; however, the problems usually appear in the areas of semi-arid, arid, and severely arid climates such as Egypt and Saudi Arabia [7,8]. Only 41 years ago, Erol and Dhowian [9] conducted research on the swell and shrinkage behavior of active Madinash clays, which led to the discovery of expansive soil in Saudi Arabia. For the long-term monitoring of changes in soil suction and the consequent volume changes (shrink/heave) in expansive soil, Dhowian et al. [10] set up a primary field station in Al-Ghatt town in Saudi Arabia. Ruwaih [11] and Abduljauwad et al. [12] revealed the sites in Saudi Arabia where expansive soils have been discovered. Since there has been a growth in urban development in Saudi Arabia since 1998, it is expected that the expansive soils will be discovered in other areas of the country. For this reason, it is crucial for researchers and civil engineers to update the extensive soil presence places in Saudi Arabia.

Several scholars (e.g., [13–18]) have reported on the features of the expansive soils in various parts of KSA. Sabtan [19] and Dafalla and Al-Shamrani [18] reported the subsurface conditions and the geotechnical properties of the expansive soils in Tabuk City. Abduljauwad [16] reported the swelling properties of the expansive soils for different locations in the Eastern Province of Saudi Arabia. Abduljauwad and Ahmed [14] reported case histories in the Eastern Province of Saudi Arabia, including the residential district north of AI-Mubarraz city in the AI Hassa area and the A1-Qatif housing project.

The types of damage observed in structures constructed on expansive soils in Saudia Arabia include serious cracks, tilting, twisting, and sticking doors and windows [15,20]. The authors observed damage in a number of private and government lightweight structures in the Al Masif district of Tabuk City. Examples of the observed damage are shown in Figure 1. These cracks occurred shortly after the initial use of the structures as a result of moisture migration into the expansive soils.

Soil movements and improper foundation system selection and design are the causes of these cracks. Nelson et al. [2] state that grade beams on drilled piers, stiffened strip footing, and stiffened rafts are foundation systems that have been used successfully with expansive soils.

The performance and construction cost of lightweight structures built on expansive soil are influenced by the choice of foundation system, and the determination and categorization of the expansive soil, as well as the superstructure loads, are necessary factors for the proper foundation system selection [21]. Nelson et al. [2] reported the variables that affect how structures are designed on expansive soils, as well as the design options available to builders: (1) building the structure to withstand expansion-related movements and/or swell pressures; (2) treating the expansive soil or managing the supporting soil's environment by taking preventative measures to reduce the movements; (3) combining the two options.

Designing the structure to be stiff enough to accommodate soil movements is one way to reduce damage for structures built on expansive soil [2,22,23]. Stiffened rafts and strip footings and inverted T sections are the two foundation systems that are successfully used in various countries around the world (i.e., United States, Australia, South Africa, Saudia Arabia, and Egypt) to reduce or eliminate the damage of lightweight structures constructed on expansive soils [2,21,23–25]. Given Saudi Arabia's Vision-2030, urbanization is growing quickly, which makes it more likely that lightweight structures can be built on expansive soils in various parts of the country. This demonstrates the critical need for useful advice to assist civil engineers in minimizing or removing potential issues and damage to lightweight structures built on expansive soils, thereby lowering the loss of millions of dollars.

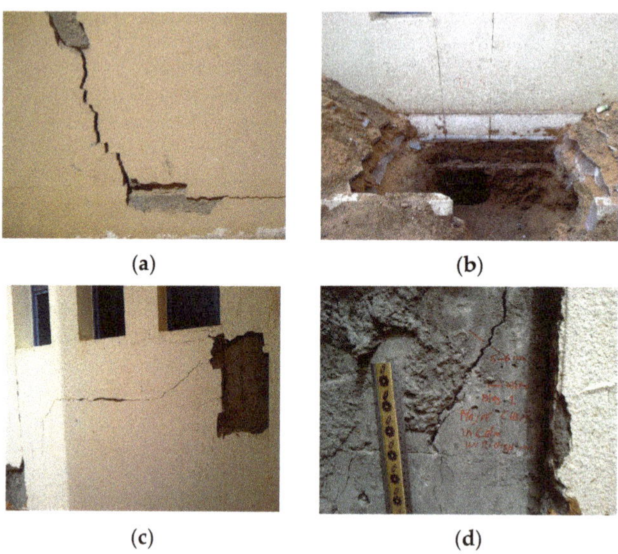

Figure 1. Examples of damage in lightweight structures constructed on expansive soils: (**a**) Cracks in a wall (Private building in Al Masif, Tabuk, Saudia Arabia), (**b**) Cracks in a ground beam (Government building in Al Masif, Tabuk, Saudia Arabia), (**c**) Cracks in a wall (Government building in Al Masif, Tabuk, Saudia Arabia), (**d**) Cracks in a column (Government building in Al Masif, Tabuk, Saudia Arabia).

Two primary items are presented in this study: Using the two scenarios of edge shrinkage, ES, and edge heave, EH, the program was utilized in a parametric investigation to examine the effect of various parameters on the behavior of stiffened rafts on expansive soils. Some of these parameters include the stiffening beam depth, the raft dimensions, and the mound shape's maximum differential movement. The three-story lightweight structure is made of reinforced concrete, which includes slabs, beams, columns, and a stiffened raft.

2. Analysis Method

2.1. Limitations in Existing Design Methods

Many design approaches have been documented and discussed in the literature for the design of a stiffened raft resting on expansive soils (e.g., [23,25–31]). Every one of the current design approaches has certain shortcomings that the new approach should address. The following limitations have been the subject of numerous studies (e.g., [23,32–34]).

The main flaw is that they require an estimate of the initial distorted mound shape and assume that the interaction is between the raft and an already-distorted mound shape. Mitchell [33] has noted that one of the key factors influencing the analysis of raft foundations resting on expansive soils in all of the design methods is the initial distorted mound shape. The mound exponent, m, the maximum differential movements, y_m, and the edge moisture variation distance, e_m, determine the initial distorted mound shape. One of the main issues is thought to be determining the e_m reliably. Additionally, estimating the initial distorted mound shape has the drawback of not accounting for the soil movements that result from various edge effects, such as ponded water, tree roots, and climate.

The current design methodologies made use of three different kinds of supporting soil models, including a rigid model (1), a Winkler and coupled spring model (2), and an elastic continuum model (3). Because of the soil's compressibility, the stiff model used by the BRAB method did not permit any decrease in the edge moisture variation distance, leading to extremely conservative structural design parameters. The soil was represented by independent springs in the Winkler model, each of which had a stiffness known as the modulus of subgrade reaction. According to Winkler's theory, the vertical displacement at a given point in the foundation is proportional to the contact pressure there. The primary limitation of the Winkler model is its inability to account for the soil's shearing resistance. The coupled spring model captures the soil's behavior in the absence of lateral displacement quite accurately. The model that best captures the behavior of the soil is thought to be the elastic continuum model [32].

The concentrated loads from columns are disregarded, and only two forms of structural loads (i.e., uniformly distributed loads and perimeter line loads) were taken into account by any of the approaches. Put another way, every design technique currently in use is appropriate for wall-bearing structures but inappropriate for use with reinforced concrete structures made up of slabs, beams, and columns.

A few of the current techniques were created for one-story residential buildings (e.g., [23]).

The majority of the design techniques currently in use ignored the third direction and instead focused on the interaction between the distorted mound shape and the stiffened raft in two dimensions. Put differently; they treat the raft similarly to a beam on an elastic foundation or a strip footing.

Pidgeon [35] advises using all of the current design methodologies with caution in areas other than their original development. The present study employs a design methodology that seeks to enhance the rationality of previous design approaches and to be suitable for the design of the stiffened raft under reinforced concrete, lightweight structures that are constructed on expansive soils in Saudi Arabia.

2.2. Computer Program SLAB97

A program named SLAB97 has been developed by the second author and his colleagues to analyze the stiffened raft–expansive soil interaction. The program is able to estimate the behavior of stiffened rafts subjected to concentrated column loads and resting on expansive soils. SLAB97 takes into account two problems that are solved separately: (1) analysis of the interactions between the stiffened raft and the 3D distorted mound shapes as a result of the soil movements and (2) 3D moisture movement (i.e., changes in soil suction) in the expansive soil underneath a flexible cover such as a raft and the associated soil movements. Since the vertical stress beneath the raft affects moisture movements in the supporting expansive soil, the analysis of these two problems actually had an impact on each other. The movements in the expansive soils beneath the stiffened raft or the outcome of the initial issue determine the structural analysis of the structure. To simplify computation, every issue is resolved independently. SLAB97 coupling between the finite element model was used to solve the second problem, and the 3D moisture diffusion and soil movements model was used to solve the first problem. A number of papers [24,36–38] have described and validated the two models. For background information, a brief description of the SLAB97 program is provided below.

The finite difference method, FDM, in the SLAB97 program, was used to solve the 3D suction diffusion equation, and the finite element method, FEM, which is based on the classical theory of thin plate on elastic half-space, was used to solve the stiffened raft–3D distorted mound shape interaction. Mitchell [39] developed the 3D suction diffusion equation, and Wray's model [40] served as the basis for the computation of soil movements, such as shrinkage or heave.

As illustrated in Figure 2, the raft and soil mass are discretized and represented as a grid of nodes. According to El-Garhy [41] and Abu Ali et al. [42], the size of the modeled soil mass beneath the stiffened raft is taken to be greater than the size of the raft by $B/4$ from each side. The advantage of symmetry can be considered in SLAB97 to spare time.

In the present study, for the solution of the 3D transient suction diffusion equation, the boundary conditions at the soil surface, outside the domain of the raft, are taken equal to $(\psi_e + \psi_0)$ and beneath the raft as well as at the end of the active zone is considered equal to ψ_e as shown in Figure 2. Whereas, at the corners of the soil mass and its sides (i.e., $x - z$ and $y - z$ planes), the boundary values of soil suction are obtained by solving the transient suction diffusion equation in 1D and 2D under the boundary conditions of $(\psi_e + \psi_0)$ at the top and ψ_e at the bottom, respectively.

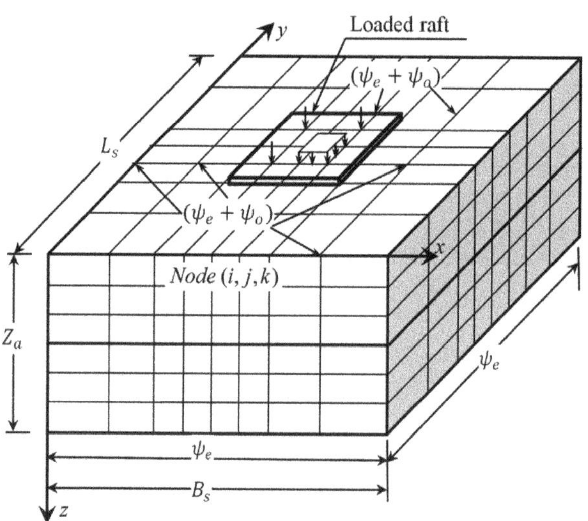

Figure 2. The discretization of the raft and the underneath soil mass for the SLAB97 program.

Some of the drawbacks of the current design methods are addressed by the program SLAB97. These include (1) predicting the response of the stiffened raft to changes in the climatic boundary conditions simply by stating the initial soil suction conditions in the soil mass; (2) taking into account the interaction between the stiffened raft and the distorted mound shape in three dimensions; (3) avoiding the need to prescribe the maximum differential movement and edge moisture variation distance as input values, as well as the locations of the points not in contact between the raft and the subgrade and the gap at these points. It does, however, necessitate estimating the worst-case scenario for the climatic boundary conditions, which could result in the worst possible distribution of soil suction throughout the expansive soil mass beneath the stiffened raft over the course of its lifetime.

SLAB97 is capable of estimating the 3D initial distorted mound shapes in addition to determining the maximum and minimum values of deflections, shears, and moments required for structural design, as well as the values induced in the stiffened raft. One of the benefits of the SLAB97 program's analysis of stiffened rafts is that it considers all of the variables that could influence the outcome, such as (1) load factors, (2) soil factors, (3) stiffened raft factors, and (4) climatic factors. The 3D initial distorted mound shape is produced by the interaction of the climatic and soil factors, and the deflections and internal forces induced in the stiffened raft are produced by the interaction of the stiffened raft and the initial distorted mound shape.

Validation of the Program SLAB97

El-Garhy et al. [36] reported the validation of the program SLAB97 against field measurements by comparing its results with the measured displacements of an actual raft resting on expansive soils at Sunshine, Melbourne, Australia. by Holland et al. [29]. In addition to previous validation, the problem of rafts resting on expansive soil has been solved by Shams et al. [43] resolved by SLAB97, and the results are compared for the purpose of more program validation. Shams et al. (2018) solved the raft problem by two different methods (i.e., ABAQUS program and AS2870-2011 [44]) for the two cases of ES and EH. The raft is of dimensions 20 m × 20 m and rests on a 4 m expansive soil layer. The thickness of the raft is 55 cm in the case of ES and 90 cm in the case of EH. The raft is subjected to a uniformly distributed load of 6.5 kPa plus its own weight and a perimeter line load of 10 kN/m. The elasticity modulus and Poisson's ratios of the raft and the expansive soil are 2.1×10^4 MPa, 0.16 and 80 MPa, 0.30, respectively.

The SLAB97 program's input parameters are chosen to generate a mound heave value that is similar to the mound heave employed by Shams et al. [43] in order to facilitate appropriate comparison. The values of the equilibrium soil suction, the amplitude of surface suction change, the diffusion coefficient of the soil, and the suction compression index of 2.0 p^F, 4.0 p^F 0.08 m²/day, and 0.0215, respectively, are used in the present analysis for the two cases of ES and EH. The SLAB97 program analyzes a 20 m × 20 m raft that is resting on an expansive soil mass with dimensions of 30 m × 30 m × 4 m. Only a quarter of the raft and the supporting soil mass are taken into account because of symmetry about the x and y axes.

Figures 3–6 show comparisons among the results of the SLAB97 program, ABAQUS program, and AS2870-2011 method. For the cases of ES, an excellent comparison is obtained between the distorted surface movements along the x axis of the SLAB97 program and the AS2870-2011 method, whereas the maximum differential movement of the SLAB97 program is greater than that of the ABAQUS program, as shown in Figure 3. The predicted deflection at the center of the raft by the SLAB97 program is slightly greater than the deflections predicted by the ABAQUS program and AS2870-2011 method, whereas, at the mid-edge point of the raft, the deflection of SLAB97 is equal to the deflection of the AS2870-2011 method and slightly smaller than the deflection of the ABAQUS program as shown in Figure 3. The predicted differential deflection along the x-axis by the SLAB97 program (i.e., 15.63 mm) is smaller than the differential deflections predicted by the ABAQUS program and AS2870-2011 method by about 18.4% and 53.6%, respectively. The maximum bending moment along the x-axis predicted by the SLAB97 program (i.e., 205.2 kN·m/m) is greater than that predicted by the ABAQUS program and AS2870-2011 method by about 13.3% and its location at 4 m from raft edge whereas, the location of the maximum moments of the other two methods at 5 m from the edge of the raft as shown in Figure 4.

For the case EH, the difference in the distorted surface mound shapes of the three methods is clear; however, the maximum differential movement is approximately equal, as shown in Figure 5. The predicted maximum deflection at the raft center by the SLAB97 program (i.e., 26.41 mm) is greater than the deflections predicted by the ABAQUS program and AS2870-2011 method by about 35.3% and 26.5%, respectively. However, the predicted differential deflection along the x-axis by the SLAB97 program (i.e., 16.79 mm) is slightly smaller than the differential deflection predicted by the ABAQUS program (i.e., 17.1 mm) and smaller than that predicted by the AS2870-2011 method by about 15.5% as shown in Figure 5. The maximum bending moment at the center of the raft predicted by the SLAB97 program (i.e., 578.6 kN·m/m) is slightly greater than that predicted by the ABAQUS program and AS2870-2011 method by about 11.5% and 8.96%, respectively, as shown in Figure 6.

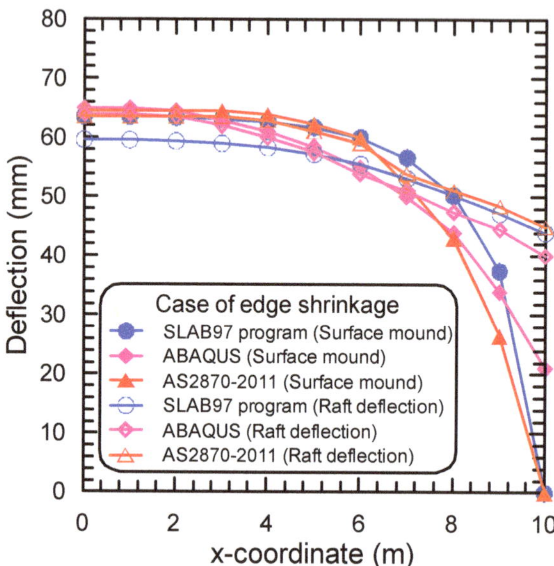

Figure 3. Comparisons between SLAB97 results and others for the surface mound shape and raft deflection (case of ES).

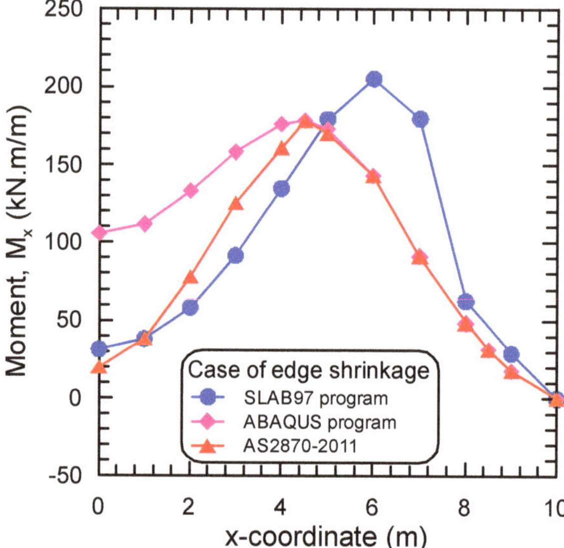

Figure 4. Comparisons between SLAB97 results and others for bending moments (case of ES).

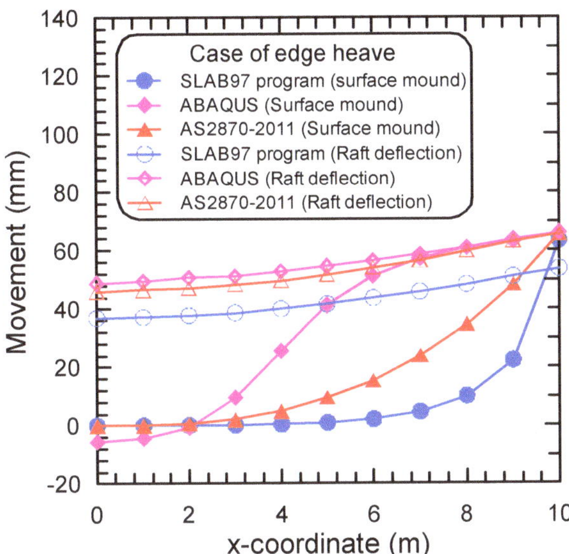

Figure 5. Comparisons between SLAB97 results and others for the surface mound shape and raft deflection (case of EH).

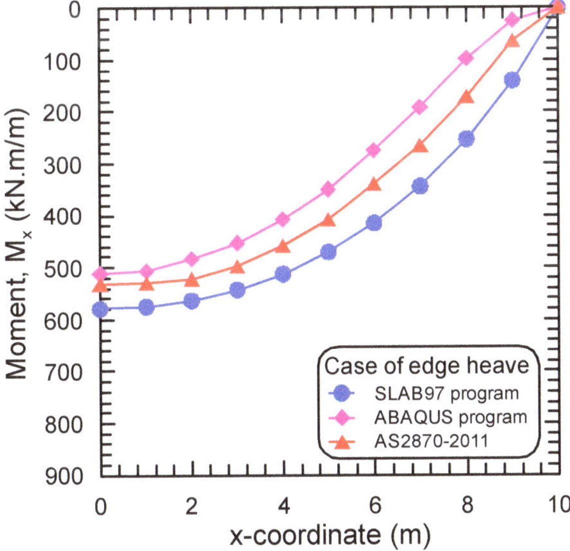

Figure 6. Comparisons between SLAB97 results and others for bending moments (case of EH).

3. Parametric Study

A parametric study is conducted using the SLAB97 program to examine the impact of different parameters on the stiffened rafts' behavior when they are built on expansive soils. These parameters are the raft's dimensions, L/B, the maximum differential movement of the mound shapes, y_m, and the depth of the stiffening beam d.

The following input parameters must be known at any site in order to use the SLAB97 program to calculate the structural design parameters (i.e., deflections, shears, and moments) of a stiffened raft resting on expansive soil: the equilibrium soil suction, ψ_e, the diffusion coefficient, α, the suction compression index, SCI, the amplitude of the surface suction change, ψ_0, and the depth of the active zone, Z_a, are the four initial parameters. Additionally, it is important to know the initial values of soil suctions in the expansive soil as well as the worst predicted weather scenarios that could result in the worst variations in soil suctions through the expansive soil beneath the raft over the course of its lifetime. Abu-Ali et al. [42] developed a rational procedure for estimating the climate-controlled soil parameters (i.e., α, ψ_0, Z_a, SCI, and ψ_0) from the results of routine geotechnical tests. They used their procedure and estimated the climate-controlled soil parameters for expansive soils at various locations in the KSA.

In this study, an equilibrium suction condition beneath the raft followed by a four-month dry spell (i.e., extended draught) to simulate the distortion mode of ES, and an equilibrium suction condition in the soil mass underneath the raft followed by a four-month wet spell (i.e., extended wet) to simulate the distortion mode of EH [3,38]. For the two instances of distortion modes with maximum differential movements, y_m, covering the soil classes taken into consideration in the Australian standard [44], it was found that a period of four months was sufficient to predict the distribution of wet and dry suction through the soil mass.

One of the climate-controlled parameters affecting the distorted mound shapes (i.e., the diffusion coefficient) is changed. In contrast, the other parameters are left unchanged to minimize the parameters affecting the analysis of the interaction between the stiffened rafts and the distorted mound shapes.

Because there are no measurements for the maximum differential movements, y_m, due to climatic changes at all locations of Saudia Arabia, and since the climatic conditions in the KSA are very close to the climatic conditions in large areas of Australia, the diffusion coefficient is changed to produce mound shapes of maximum differential movements, y_m, covering the soil classes (i.e., slightly reactive to extremely reactive) considered in the Australian standard [44]. The values of the climate-controlled parameters considered in the parametric study are selected to cover the range of those parameters in the different KSA locations, as shown in Table 1.

Table 1. Range of the parametric study's parameters.

Z_a (m)	ψ_e p^F (kPa)	ψ_0 p^F (kPa)	SCI	α (m^2/Day)	Raft Size $B \times L$ (m $\times$ m)	Aspect Ratio (L/B)	Stiffening Beam Depth (m)
5.0	4.0 (1000)	2.0 (10)	0.02	0.00144 0.00432 0.0104 0.0200 0.0360	8 × 8 12 × 16 16 × 28 20 × 40	1.00 1.33 1.75 2.00	0.30 0.60 0.90 1.20 1.50

The site classes recommended by the Australian standard [44] along with the values of the y_m considered in the parametric study for different rafts are shown in Table 2. As recommended by [44], the $y_m = 0.7\, y_s$ for footing design.

The modulus of elasticity of the soil, E_s, is taken at 60 MPa and 20 MPa for ES and EH, respectively, and its Poisson ratio is 0.4. The creep modulus of elasticity and Poisson's ratio of the raft are 12,000 MPa and 0.15, respectively. The raft thickness and stiffening beams width and spacing are kept constant and equal to 0.15 m, 0.30 m, and 4 m, respectively. The stiffening beam depth varies and includes the slab thickness.

The most common type of construction in the KSA is reinforced concrete structures, which consist of slabs, beams, columns, and footings. This study concerns the lightweight-reinforced concrete structures (e.g., Villas) consisting of three stories, which are usually susceptible to expansive soil problems. Unlike most of the previous studies, two types of loads are considered in the present study. These are column loads and a uniform distributed load representing the weight of the raft, the flooring, and the live loads.

Table 2. Site classes according to [44], along with the y_m values considered in the parametric study.

Site Classes According to AS2870-2011 [44]			Values of the y_m Considered in the Parametric Study (mm)			
			Raft size (m × m)			
Class	Site classification	y_s (mm)	8×8	12×16	16×28	20×40
S	Slightly reactive	$0 < y_s \leq 20$	20.14	20.52	20.52	20.52
M	Moderately reactive	$20 < y_s \leq 40$	31.37	32.88	32.88	32.88
H1	Highly reactive	$40 < y_s \leq 60$	45.60	48.52	48.52	48.51
H2	Highly reactive	$60 < y_s \leq 75$	58.54	63.68	63.76	63.75
E	Extremely reactive	$y_s > 75$	67.92	76.77	77.32	77.37

The columns' loads are calculated by the approximate tributary area method [45] based on an equivalent uniform load (i.e., dead load and live load) for each story = 11.63 kN/m^2. Figure 7 shows the dimensions of the studied stiffened rafts along with the columns' loads, columns dimensions, and spacing of stiffening beams.

The equivalent moment of inertia, I_{eq}, is used in the SLAB97 program to calculate the equivalent rigidity of the stiffened raft.

$$I_{eq} = \sqrt{I_x^2 + I_y^2} \tag{1}$$

where I_x and I_y are the moments of inertia per unit width of stiffened raft cross-section in x and y directions

To determine the equivalent thickness of the raft, the equivalent rigidity is equated to the rigidity of the constant raft thickness as follows.

$$h_{eq} = \sqrt[3]{12(1 - v_r^2)I_{eq}} \tag{2}$$

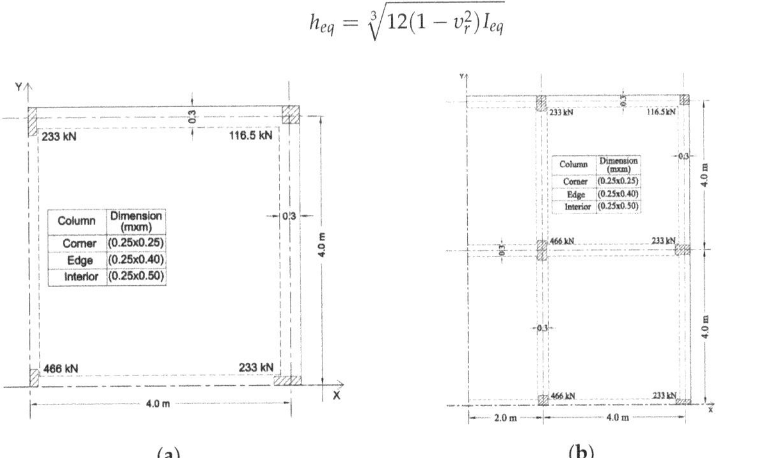

Figure 7. *Cont.*

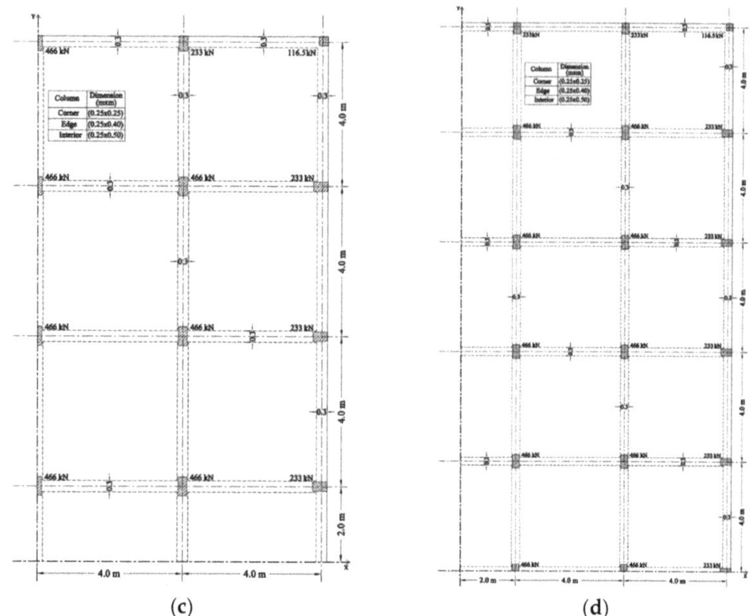

(c)   (d)

Figure 7. Dimensions and loads of the studied rafts: (**a**) raft size 8 m × 8 m, (**b**) raft size 12 m × 16 m, (**c**) raft size 16 m × 28 m, (**d**) raft size 20 m × 40 m.

4. Analysis of Results

The obtained results from the parametric study are used to investigate the effect of stiffening beam depth, the maximum differential movement of mound shape, and the raft dimensions represented by the aspect ratio, L/B, on the distribution of deflections and bending moments along the x and y axes. For space limitations, only the results of rafts of sizes 8 m × 8 m and 16 m × 28 m are presented here.

4.1. Effect of the Stiffening Beam Depth

The effect of the stiffening beam depth on the distribution of deflections and bending moments induced in the stiffened raft along the x and y axes for the rafts (i.e., 8 m × 8 m and 16 m × 28 m) at the maximum y_m shown in Table 2 (i.e., 67.92 mm and 77.32 mm) for two distortion modes of ES and EH are presented and discussed in this section.

For the case of ES, Figures 8–11 show the effect of beam depth on the deflections, bending moments, and the initial surface movements along the x and y axes for rafts of sizes 8 m × 8 m and 16 m × 28 m. As illustrated in these Figures, it is observed that: (1) The total and differential deflections (differential deflection is the difference between the deflections at the center of the raft and the mid-edge points on the x or y axes) decrease as the beam depth increases, (2) the length of the unsupported raft, L_{ur} (i.e., the distance inward from the edge of the raft to the point of separation between soil and raft) decreases as the beam depth increases in the x and y axes because of an increase in the raft deflections as a result of decreasing the beam depth (e.g., at the beam depths of 0.6, 0.9, and 1.2 m, the lengths of the unsupported raft are 0.80, 0.92, and 1.2 m for raft size 8 m × 8 m as shown in Figure 8 whereas, those lengths are 0.0, 0.88, and 0.96 m in the x direction and 0.84, 0.98, and 1.06 m in the y direction for raft size 16 m × 28 m as shown in Figure 10), (3) The distance from the edge of the raft to the point of separation between soil and raft, L_{ur}, is smaller than the distance from the edge of the raft to the point of the maximum moment in the two directions (e.g., at the beam depths of 0.6, 0.9, and 1.2 m, the lengths of L_{ur} are 0.84, 0.98, and 1.06 m for raft size 16 m × 28 m in the y direction whereas the distance to the points of maximum moments are 2.0, 2.3, and 5.0 m, respectively, as shown in Figure 11). A similar observation has been reported by Briaud et al. (2016) from the analysis of a stiffened raft on expansive soil subjected to uniform and perimeter line loads using ABAQUS software and they referred to the reason as that overhanging raft (i.e., raft on a dome shape mound) does not behave as a pure cantilever, (4) The distance from the raft edge to the points of maximum moments in the two directions increases as the beam depth increases (e.g., at beam depths of 0.6, 0.9, and 1.2 m, the distances to the points of maximum moments are 2.5, 3.0, and 3.5 m in x and y directions for raft size 8 m × 8 m as shown in Figure 9 and for raft size 16 m × 28 m the distances to the points of maximum moments are 1.0, 2.0, and 3.0 m in x direction and 3.0, 3.3, and 4.0 m in the y direction and as shown Figure 11), and (5) the bending moments induced in all stiffened rafts are negative moments because of the dome shape of the distorted mound and the bending moments along the x or y axes are not taken the same trends in some cases as shown in Figures 9 and 11 and this is may be due to the difference in the final contact area between the stiffened rafts and the distorted mound of expansive soils in the two directions and the columns concentrated loads.

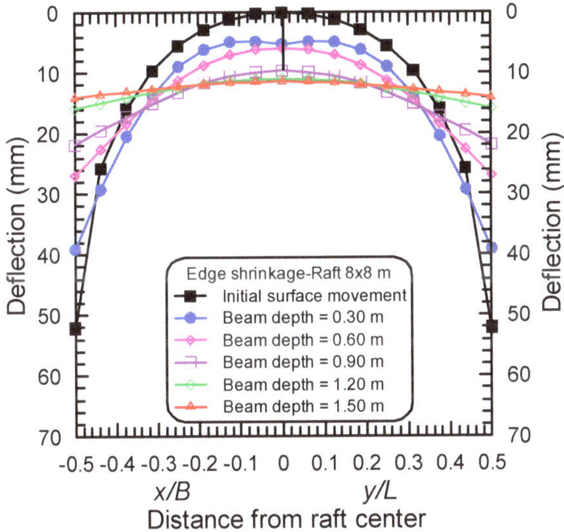

Figure 8. Effect of beam depth on the deflection along with the initial surface movement through the x and y axes for raft size 8 m × 8 m (case of ES).

For the case of EH, Figures 12–15 show the effect of beam depth on the distribution of deflections and bending moments along the x and y axes for rafts of sizes 8 m × 8 m and 16 m × 28 m. Referring to these Figures, it is shown that (1) generally, total and differential deflections decrease as the beam depth increases, and (2) two types of soil supports are observed; these are simply support conditions and multiple support conditions. For raft size 8 m × 8 m at beam depths of 0.6, 0.9, 1.2, and 1.5 m simply support occurs, and the raft behaves as a slab supported at its edges whereas, at beam depth of 0.3 m, multiple supports (i.e., support at the raft perimeter and support at the raft core area) are occurred as shown in Figure 12. For raft size 16 m × 28 m full contact occurs between the raft and soil in the x direction for all beam depths, whereas, in the y direction, full contact occurs except for cases of beam depths of 1.2 m and 1.5 m as shown in Figure 14. This explains the difference in the trends of the bending moments along the x or y axes, as shown in Figures 13 and 15. The two types of soil supports are observed by Shams et al. [23] from the analysis of stiffened rafts of single-story building subjected to a uniform load and a perimeter line load using ABAQUS software, (3) for raft size 8 m × 8 m because of the simple soil support condition, at beam depths of 0.6, 0.9, 1.2, and 1.5 m, the maximum moments occurred at the center of the raft and the moments taken the same trends except for the beam depth of 0.3 m the moments shows a different trend because of multiple soil support condition as shown in Figure 12 while, for raft size 16 m × 28 m, at all beam depths, the maximum moments occurred near the raft edges and shows similar trends in both x and y direction as shown in Figure 15 and this is attributable to the full contact between the rafts and soil, at all beam depths, because of the heavy interior columns loads, (4) for raft size 16 m × 28 m, the distance inward from raft edge to the point of maximum moment along the x or y axes increases as the beam depth increases as shown in Figure 15, (5) the moments induced in the rafts are positive moments due to the dish shape of the distorted mound.

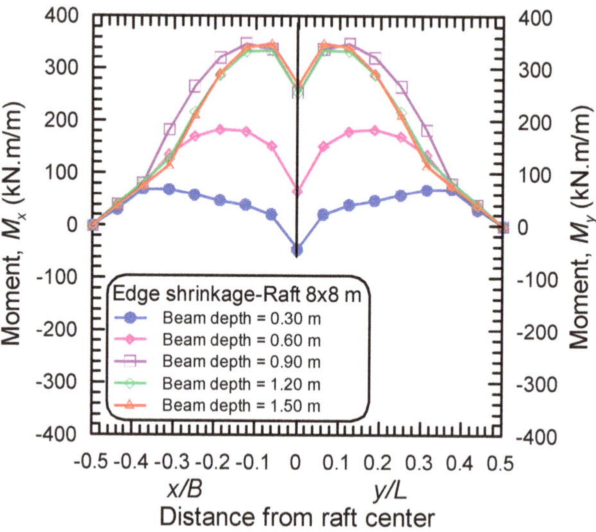

Figure 9. Effect of beam depth on the M_x and M_y through the x and y axes for raft size 8 m × 8 m (case of ES).

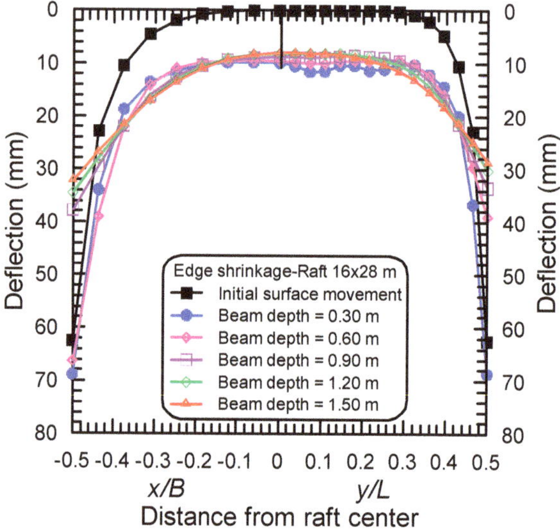

Figure 10. Effect of beam depth on the deflection along with the initial surface movement through the x and y axes for raft size 16 m × 28 m (case of ES).

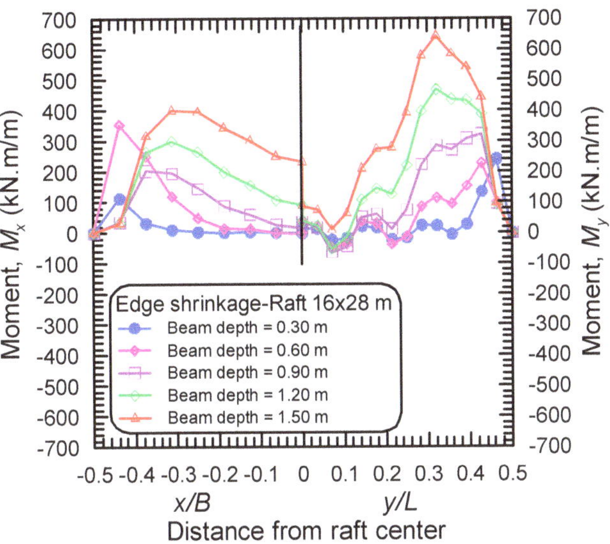

Figure 11. Effect of beam depth on the M_x and M_y through the x and y axes for raft size 16 m × 28 m (case of ES).

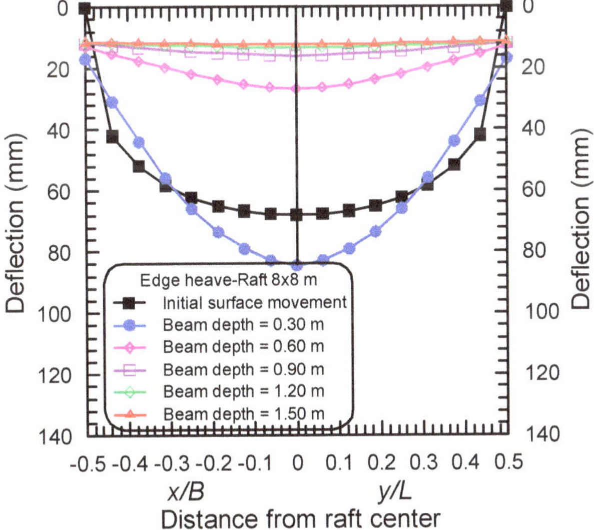

Figure 12. Effect of beam depth on the deflection along with the initial surface movement through the x and y axes for raft size 8 m × 8 m (case of EH).

The absolute values of maximum deflections and maximum bending moments along the x and y axes for the case of EH are greater than those in the case of ES for all the studied rafts except for raft size 8 m × 8 m, and this is attributable to two reasons: (1) the soil stiffness in case of EH is smaller than that in case of ES (i.e., $E_s = 20$ MPa in case of EH and equal to 60 MPa in case of ES) and (2) there is one interior column in case of raft size 8 m × 8 m and the rest of the columns on the raft perimeter that supported on the soil in the case of EH.

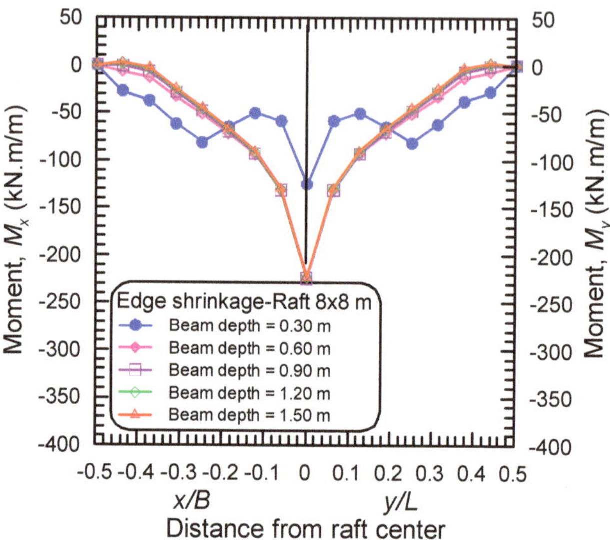

Figure 13. Effect of beam depth on the M_x and M_y through the x and y axes for raft size 8 m × 8 m (case of EH).

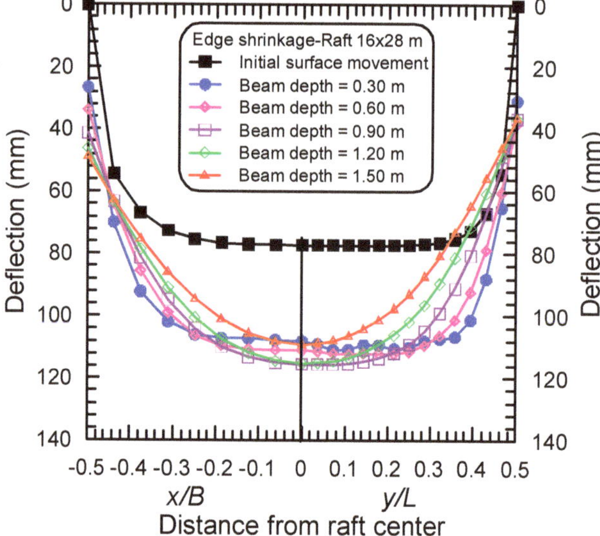

Figure 14. Effect of beam depth on the deflection along with the initial surface movement through the x and y axes for raft size 16 m × 28 m (case of EH).

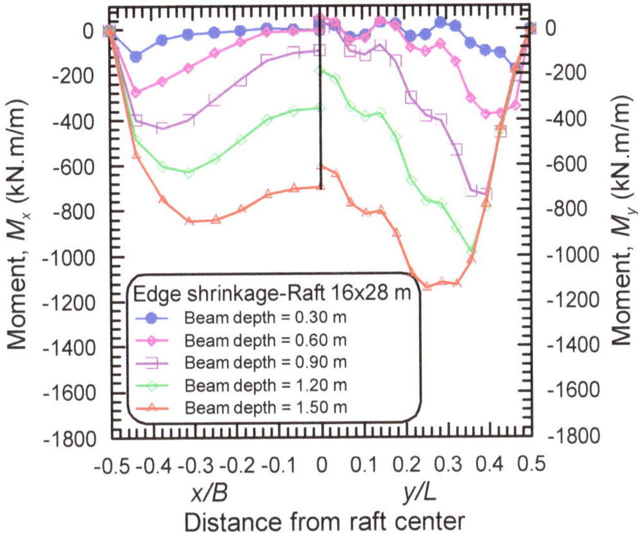

Figure 15. Effect of beam depth on the M_x and M_y through the x and y axes for raft size 16 m × 28 m (case of EH).

4.2. Effect of the Differential Movement of Mound Shape

The effect of the differential movements of mound shape, y_m, on the deflections and bending moments induced in the stiffened raft along the x and y axes, rafts of sizes 8 m × 8 m and 16 m × 28 m at beam depth of 1.2 m are presented and discussed.

For the case of ES, Figures 16–19 show the effect of the y_m on the deflections and bending moments along the x and y axes for rafts of sizes 8 m × 8 m and 16 m × 28 m. The deflections, differential deflections, and bending moments along the x and y axes increase as the y_m increases as shown in Figures 16–19, except for raft size 8 m × 8 m, the maximum differential deflections along the x and y axes at $y_m = 58.54$ mm is slightly greater than that at $y_m = 67.92$ mm by about 11.45%, and therefore, the maximum bending moments along the x and y axes at $y_m = 58.54$ mm is slightly greater than the maximum bending moments at $y_m = 67.92$ mm by about 7.7%, and this may be attributable to the maximum number of iterations (i.e., 10) considered by the program to achieve convergence, in other words, at $y_m = 67.92$ mm, the convergence condition may need iterations greater than 10, and this is considered a rare case. Also, from a loading point of view, a raft of size 8 m × 8 m is considered a special case because the columns' loads located on the perimeter of the raft represent 66.7% of building loads, and the load of the interior column represents 33.3% of the building loads unlike raft of size 16 m × 28 m the interior columns' loads represent 64.3% from the building loads and the columns' loads located on the perimeter of the raft represent a 35.7%. The distance inward from the raft edge to the point of maximum bending moment increases as the y_m increases as shown in Figures 17 and 19 (e.g., for raft size 8 m × 8 m, the distances from the raft edge to the points of maximum bending moments at the values of the $y_m = 37.37$ mm and 58.54 mm are 2.5 m and 3.0 m, respectively, and for raft size 16 m × 28 m, these distances along the y axis are 2.0 m and 5.0 m at the values of the $y_m = 32.88$ mm and 63.76 mm, respectively).

For the case of EH, Figures 20–23 show the effect of the y_m on the deflections and bending moments along the x and y axes for rafts of sizes 8 m × 8 m and 16 m × 28 m. For raft size 8 m × 8 m, the deflections, differential deflections, and bending moments along the x and y axes are the same at all values of the y_m as shown in Figures 20 and 21, and this is because the raft behaves as a simply supported plate at all values of the y_m because of three reasons: the high stiffness of the raft as the beam depth was 120 mm, the shortened length of the raft, and all the column loads located on the perimeter of the raft except one interior column load acting at the raft center. Whereas, for raft size 16 m × 28 m, the deflections, differential deflections, and bending moments along the x and y axes increase as the y_m increases, as shown in Figures 22 and 23. The distance inward from the raft edge to the point of maximum bending moment along the x and y axes are the same and equal to 2.0 m for all values of y_m except for the value of $y_m = 76.33$ mm. This distance along the y axis was increased to 2.5 m as shown in Figure 23.

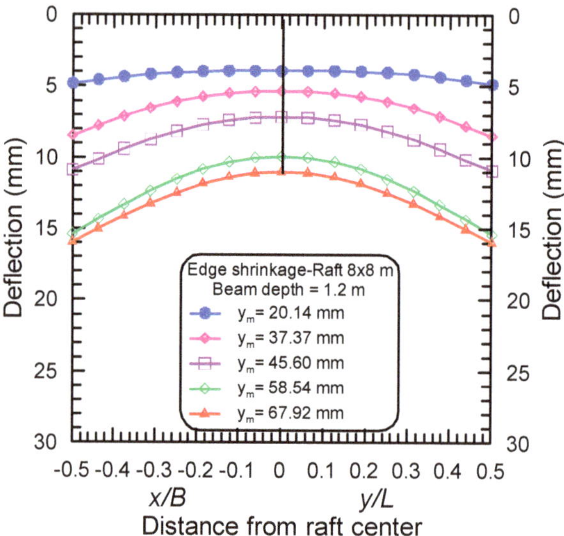

Figure 16. Effect of differential movement on the deflection through the x and y axes for raft size 8 m × 8 m (case of ES).

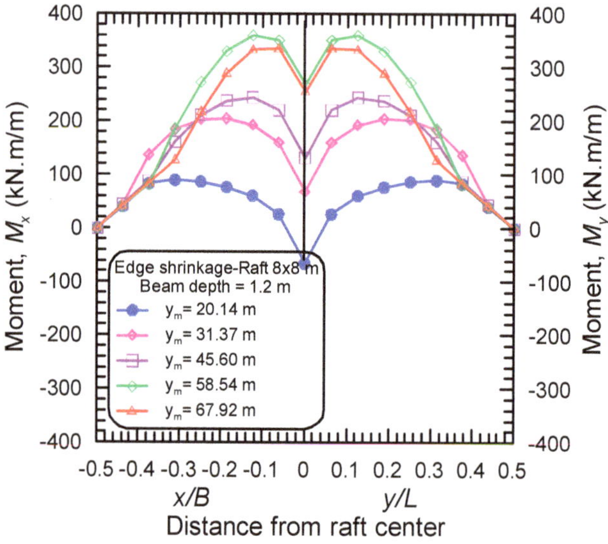

Figure 17. Effect of differential movement on the bending moments through the x and y axes for raft size 8 m × 8 m (case of ES).

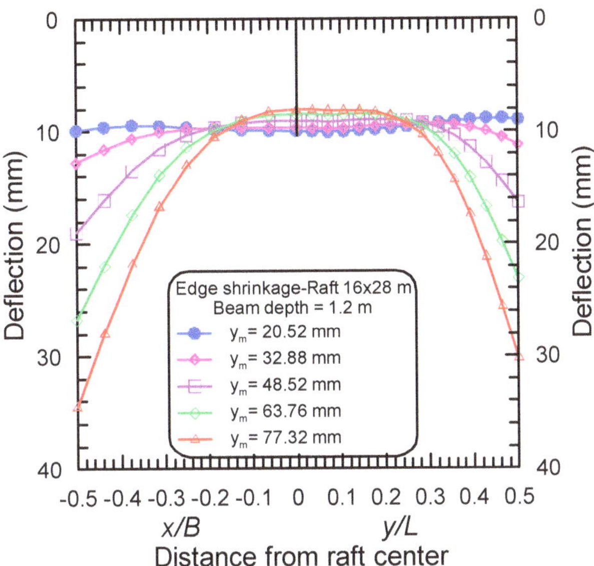

Figure 18. Effect of differential movement on the deflection through the x and y axes for raft size 16 m $\times$ 28 m (case of ES).

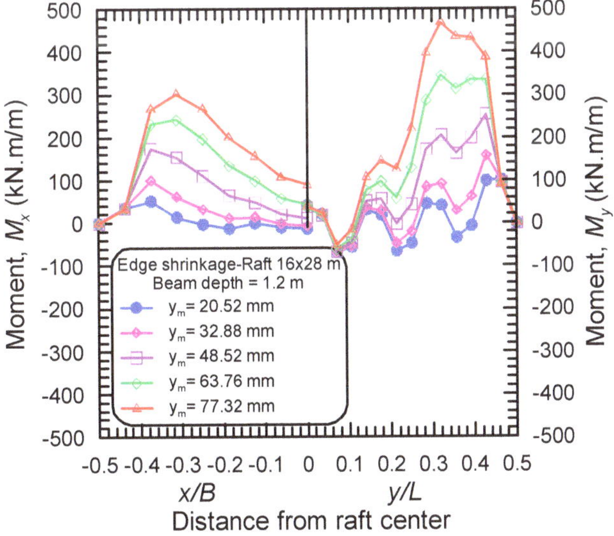

Figure 19. Effect of differential movement on the bending moments through the x and y axes for raft size 16 m $\times$ 28 m (case of ES).

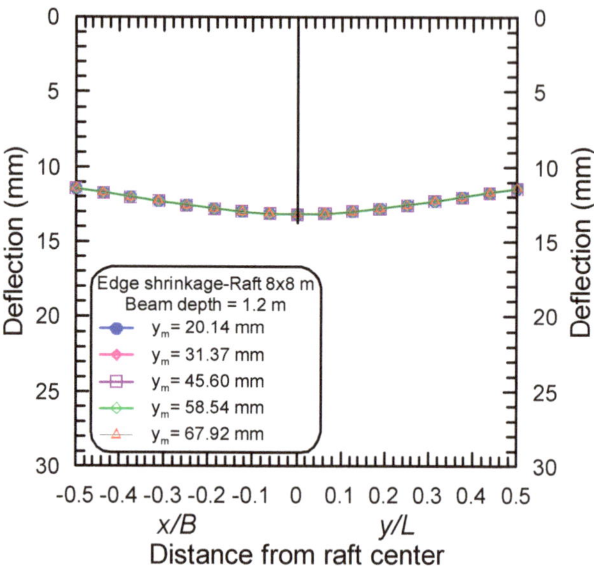

Figure 20. Effect of differential movement on the deflection through the x and y axes for raft size 8 m × 8 m (case of EH).

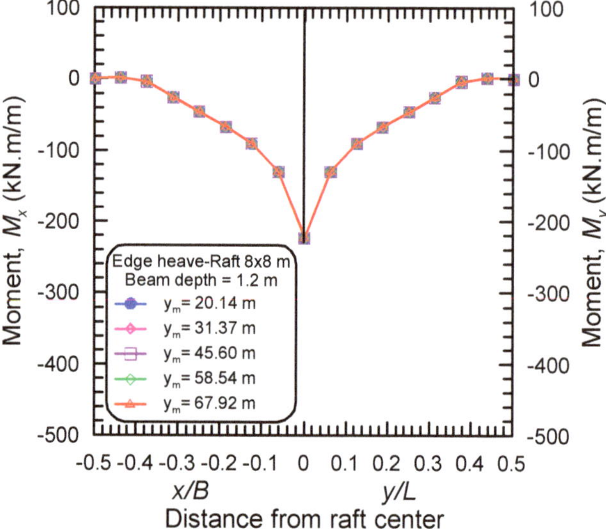

Figure 21. Effect of differential movement on the bending moments through the x and y axes for raft size 8 m × 8 m (case of EH).

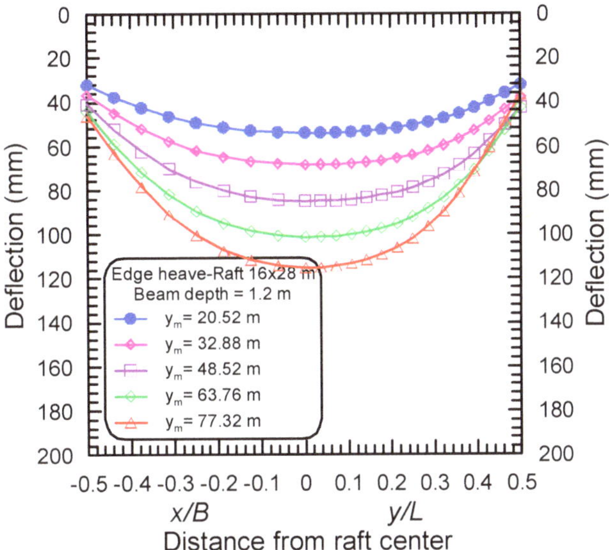

Figure 22. Effect of differential movement on the deflection through the x and y axes for raft size 16 m × 28 m (case of EH).

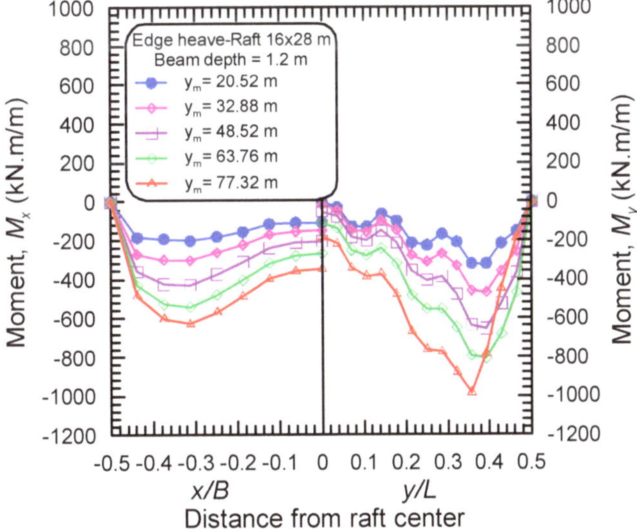

Figure 23. Effect of differential movement on the bending moments through the x and y axes for raft size 16 m × 28 m (case of EH).

4.3. Effect of Raft Dimensions

Figures 24–27 show the effect of raft dimension represented by the aspect ratio, L/B, on the deflections and bending moments induced in the raft along the x and y axes for all studied rafts at a diffusion coefficient of 0.036 m^2/day and stiffening beam depth of 1.2 m for the two cases of ES and EH. The deflections and maximum deflection along the x and y axes increase as the L/B ratio increases for the two cases of ES and EH, as shown in Figures 24 and 26. For the case of ES, the maximum bending moments along the x and y

axes increase as the L/B ratio increases, as shown in Figure 25. The distance from the raft edge to the point of maximum moments that occurred along the x and y axes decreases as the L/B ratio increases for the two cases of ES and EH, as shown in Figures 25 and 27.

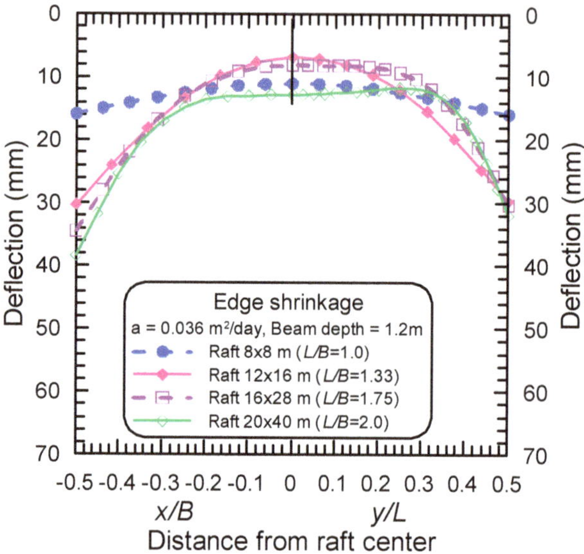

Figure 24. Effect of raft dimensions on the deflections through the x and y axes (case of ES).

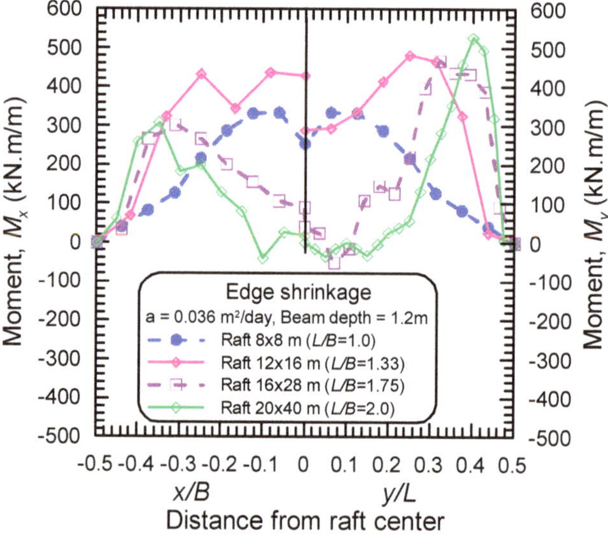

Figure 25. Effect of raft dimensions on the bending moments through the x and y axes (case of ES).

Not shown in this paper but discovered in the overall analysis is that for the studied rafts (i.e., 12 m × 16 m and 20 m × 40 m), the distribution of deflections and bending moments along the x and y axes are similar to that of raft size 16 m × 28 m with some small differences according to the condition of soil supports, the columns' loads, and the raft dimensions of each raft in both cases of ES and EH.

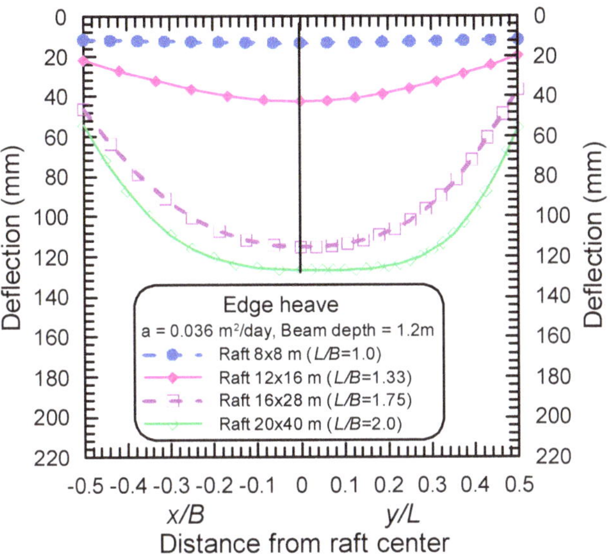

Figure 26. Effect of raft dimensions on the deflections through the x and y axes (case of EH).

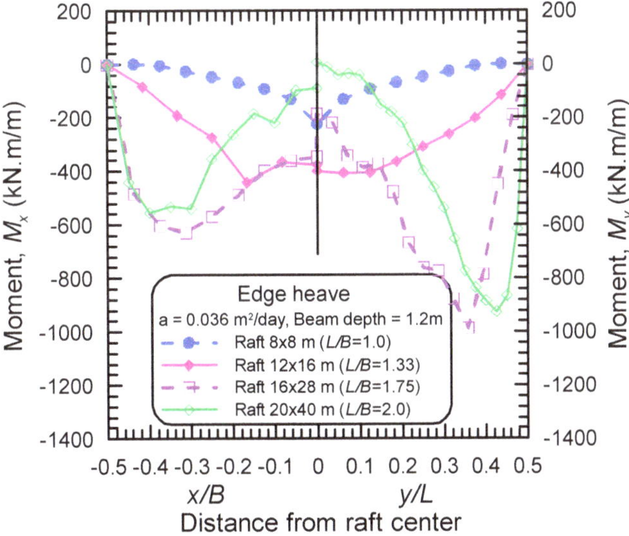

Figure 27. Effect of raft dimensions on the bending moments through the x and y axes (case of EH).

5. Conclusions

The behavior of stiffened rafts subjected to concentrated column loads and resting on expansive soils is calculated using a finite element program named SLAB97. The program can calculate the 3D distorted mound shape and examine how the stiffened raft interacts with the 3D distorted mound shape. The goal of the analysis technique presented in this work is to increase the earlier researchers' soil–raft interaction models' rationality. For both edge shrinkage and edge heave scenarios, the worst initial 3D distorted mound shapes are assumed to yield an upper-bound solution. The program has been validated by comparing its results with the results of others and shows good agreement. A parametric investigation is carried out by the program to study the impact of different parameters (i.e., the stiffening beam depth, the maximum differential movement of the mound shape, and the raft dimensions) on the behavior of stiffened rafts resting on expansive soils for the two cases of edge shrinkage and edge heave.

The following are some of the conclusions from this study:

1. The stiffened rafts subjected to concentrated columns' loads exhibit a similar shape of raft deformation and distribution of bending moments to those of the stiffened rafts subjected to uniform and perimeter line loads in both cases of distortion modes; however, the values of the design parameters (i.e., maximum deflection, maximum differential deflection, and maximum bending moments) are completely different.
2. For the case of EH distortion mode, two conditions of soil support were observed; these are simply-support condition and multiple-supports condition (i.e., support at the raft perimeter and support at the raft core area), depending on the stiffness of the raft.
3. The maximum bending moments in long and short directions, in both distortion modes, occur near the raft edge, and the distance from the raft edge to the locations of maximum moments depends on the stiffening beam depth, the maximum differential movement, and the aspect ratio of the raft.
4. The greatest amount of differential deflection was found to occur at the corner of the raft, and its value is dependent on the stiffening beam depth, the maximum differential movement, and the raft dimensions.
5. The stiffened raft components should be designed to withstand both negative moments and positive moments arising from both two distortion modes of ES and EH, respectively.

Author Contributions: Conceptualization, M.H.A.-A.; software, B.E.-G.; validation, B.E.-G., A.B. and W.S.A.; formal analysis, B.E.-G. and A.B.; investigation, B.E.-G. and A.B.; resources, B.E.-G. and W.S.A.; data curation, B.E.-G., A.B. and H.A.-D.; writing—original draft preparation, B.E.-G., W.S.A., A.B. and B.A.; writing—review and editing, B.E.-G., A.B. and W.S.A.; visualization, M.E.-S.; supervision, M.H.A.-A.; project administration, B.E.-G.; funding acquisition, M.H.A.-A. All authors have read and agreed to the published version of the manuscript.

Funding: This research was funded by the University of Tabuk (UT), Deanship of Scientific Research, under Grant No. S-1443-0090.

Data Availability Statement: The raw data supporting the conclusions of this article will be made available by the authors on request.

Acknowledgments: This research is supported by the University of Tabuk (UT), Deanship of Scientific Research, under Grant No. S-1443-0090. The author gratefully acknowledges this financial support.

Conflicts of Interest: The authors declare no conflicts of interest.

References

1. Li, J.; Cameron, D.A.; Ren, G. Case Study and Back Analysis of a Residential Building Damaged by Expansive Soils. *Comput. Geotech.* **2014**, *56*, 89–99. [CrossRef]
2. Nelson, J.D.; Chao, K.C.; Overton, D.D.; Nelson, E.J. *Foundation Engineering for Expansive Soils*; John Wiley & Sons, Inc.: Hoboken, NJ, USA, 2015.
3. Wray, W.K. *So Your Home Is Built on Expansive Soils: A Discussion of How Expansive Soils Affect Buildings*; American Society of Civil Engineers: Reston, VA, USA, 1995.
4. Krohn, J.P.; Slosson, J.E. Assessment of expansive soils in the United States. In Proceedings of the 4th International Conference on Expansive Soils, ASCE, Denver, CO, USA, 16–18 June 1980; pp. 596–608.
5. Robert, L.; Andrew, B.; Mary, L.C. *Soils Shrink, Trees Drink, and Houses Crack*; ECOS Magazine; The Science Communication Unit of CSIRO's Bureau of Scientific Services: Adelaide, Australia, 1980; pp. 13–15.
6. Dafalla, M.A.; Al-Shamrani, M.A.; Al-Mahbashi, A. Expansive soils foundation practice in a semiarid region. *J. Perform. Constr. Facil.* **2017**, *31*, 04017084. [CrossRef]
7. Fredlund, D.G.; Rahardjo, H. *Soil Mechanics for Unsaturated Soils*; Wiley: New York, NY, USA, 1993.

8. Fredlund, D.G. The Implementation of Unsaturated Soil Mechanics into Geotechnical Engineering. *Can. Geotech. J.* **2000**, *37*, 963–986. [CrossRef]
9. Erol, O.A.; Dhowian, A.W. Swell and shrinkage behavior of Madinish active clays. *J. Eng. Sci.* **1982**, 1–8.
10. Dhowian, A.W.; Ruwaih, I.A.; Erol, A.O. The distribution and evaluation of the expansive soils in Saudi Arabia. In Proceedings of the Second Saudi Engineers Conference, Dhahran, Saudi Arabia, 17–19 November 1985; Volume 1, pp. 308–326.
11. Ruwaih, I.A. Experiences with Expansive Soils in Saudi Arabia. In Proceedings of the 6th International Conference on Expansive Soils, New Delhi, India, 1–4 December 1987; pp. 317–322.
12. Abduljauwad, S.N.; Al-Sulaimani, G.J.; Al-Buraim, I.; Basunbul, I.A. Laboratory and field studies of response of structures to heave of expansive clay. *Géotechnique* **1998**, *48*, 103–121. [CrossRef]
13. Slater, D.E. Potential Expansive Soils in Arabian Peninsula. American Society for Civil Engineering. *Geotherm. Eng.* **1983**, *109*, 744–746.
14. Abduljauwad, S.N.; Ahmed, R. Expansive Soil in Al-Qatif Area. *Arab. J. Sci. Eng.* **1990**, *15*, 133–144.
15. Dhowian, A.W.; Erol, A.O.; Youssef, A. *Evaluation of Expansive Soils and Foundation Methodology in the Kingdom of Saudi Arabia*; General Directorate Research Grants Programs, King Abdulaziz City for Science and Technology: Riyadh, Saudi Arabia, 1990.
16. Abduljauwad, S.N. Swelling Behavior of Calcareous Clays from the Eastern Province of Saudi Arabia. *J. Eng. Geol.* **1994**, *27*, 333–351.
17. Al-Refeai, T.; Al-Ghamdy, D. Geological and geotechnical aspects of Saudi Arabia. *Geotech. Geol. Eng.* **1994**, *12*, 253–276. [CrossRef]
18. Dafalla, M.; Al-Shamrani, M. *Geocharacteristics of Tabuk Expansive Shale and Its Links to Structural Damage*; Geo-Congress 2014 Technical Papers; American Society of Civil Engineers: Reston, VA, USA, 2014; pp. 882–889.
19. Sabtan, A.A. Geotechnical Properties of Expansive Clay Shale in Tabuk, Saudi Arabia. *J. Asian Earth Sci.* **2005**, *25*, 747–757. [CrossRef]
20. Dafalla, M.A.; Al-Shamrani, M.A. "Performance-Based Solutions for Foundations on expansive Soils-Al Ghatt Region", Saudi Arabia. In Proceedings of the International Conference on Geotechnical Engineering, Chiangmai, Thailand, 10–12 December 2008.
21. Dafalla, M.A.; Al-Shamrani, M.A. Expansive Soil Properties in a Semi-Arid Region. *Res. J. Environ. Earth Sci.* **2012**, *4*, 930–938.
22. Dafalla, M.A.; Al-Shamrani, M.A.; Puppala, A.; Ali, H.E. *Use of Rigid Foundation System on Expansive Soils*; GSP (Geotechnical Special Publication); American Society for Civil Engineers (ASCE): Reston, VA, USA, 2010; p. 199.
23. Shams, M.A.; Shahin, M.A.; Ismail, M.A. Design of stiffened slab foundations on reactive soils using 3D numerical modeling. *Int. J. Geomech.* **2020**, *20*, 04020097. [CrossRef]
24. Wray, W.K.; El-Garhy, B.M.; Youssef, A.A. Three-Dimensional Model for Moisture and Volume Changes Prediction in Expansive Soils. American Society of Civil Engineering, ASCE. *J. Geotech. Geoenviron. Eng.* **2005**, *131*, 311–324. [CrossRef]
25. Briaud, J.L.; Abdelmalak, R.; Zhang, X.; Magbo, C. Stiffened slab-on-grade on shrink-swell soil: New design method. *J. Geotech. Geoenviron. Eng.* **2016**, *142*, 04016017. [CrossRef]
26. Building Research Advisory Board (BRAB). *Criteria for Selection and Design of Residential Slabs-on-Ground*; National Research Council, Academy of Sciences, Publication 1571; Building Research Advisory Board (BRAB): Washington, DC, USA, 1968.
27. Lytton, H. Observation Studies of Parent-Child Interaction, A Methodological Review. *Child Dev.* **1971**, *42*, 651–684. [CrossRef]
28. Walsh, D.S. *Papers in Pidgin and Creole Linguistics No. 1. (Pacific Linguistics A-54)*; Australian National University Press: Canberra, Australia, 1978; pp. 185–197.
29. Holland, J.E.; Pitt, W.G.; Lawrance, C.E.; Cimino, D.J. The Behavior and Design of Housing Slabs on Expansive Soils. In Proceedings of the Fourth International Conference on Expansive Soils, Denver, CO, USA, 16–18 June 1980; Volume 1, pp. 448–468.
30. Mitchell, P. A Simple method of design of shallow footings on expansive soil. In Proceedings of the 5th International Conference on Expansive Soils, Adelaide, SA, USA, 21–23 May 1984.
31. Post-Tensioning Institute (PTI). *Design and Construction of Post-Tensioned Slabs-on-Ground*, 2nd ed.; Post-Tensioning Institute: Phoenix, AZ, USA, 1996.
32. Wray, W.K. Development of a Design Procedure for Residential and Light Commercial Slabs-on-Ground Constructed over Expansive Soils. Ph.D. Dissertation, Texas A&M University, College Station, TX, USA, 1978.
33. Mitchell, P.W. The Design of Footings on Expansive Soils. Engineering Problems of Regional Soils. In Proceedings of the International Conference, Beijing, China, 8–12 August 1988; pp. 127–135.
34. Teodosio, B.; Baduge, K.S.K.; Pmendis, P. A review and comparison of design methods for raft substructures on expansive soils. *J. Build. Eng.* **2021**, *41*, 102737. [CrossRef]
35. Pidgeon, J.D. A Comparison of the suitability of two soils for direct drilling of spring barley. *Eur. J. Soil Sci.* **1980**, *31*, 581–594. [CrossRef]
36. El-Garhy, B.M.; Wray, W.K.; Youssef, A.A. Using Soil Diffusion to Design Raft Foundation on Expansive Soils. In Proceedings of the Sessions of Geo-Denver, Denver, CO, USA, 3–8 August 2000; Geotechnical Special Publication No. 99. ASCE: Reston, VA, USA, 2000; pp. 586–602.
37. El-Garhy, B.M.; Youssef, A.A.; Wray, W.K. Predicted and Measured Suction and Volume Changes in Expansive Soil. In Proceedings of the International Symposium on Suction, Swelling, Permeability and Structure of Clays, Shizuoka, Japan, 11–13 January 2001.
38. El-Garhy, B.M.; Wray, W.K. Method for Calculating the Edge Moisture Variation Distance. *J. Geotech. Geoenviron. Eng.* **2004**, *130*, 945–955. [CrossRef]

39. Mitchell, P.W. *The Structural Analysis of Footings on Expansive Soil*; Research Report No. I; Kenneth W. G. Smith and Associates: Adelaide, Australia, 1979.
40. Wray, W.K. *Using Soil Suction to Estimate Differential Soil Shrink or Heave*; Unsaturated Soil Engineering Practice, Geotechnical Special Publication No. 68; American Society of Civil Engineers: Reston, VA, USA, 1997; pp. 66–87.
41. E1-Garhy, B.M. Soil Suction and Analysis of Raft Foundation Resting on Expansive Soils. Ph.D. Dissertation, Civil Engineering Department, Menoufia University, Shebin EI-Koon, Egypt, 1999.
42. Abu-Ali, M.H.; El-Garhy, B.; Boraey, A.; Al-Rashed, W.S.; Abdel-Daiem, H. Estimating the Climate-Controlled Soil Parameters and the Distorted Mound Shape for Analysis of Stiffened Rafts on Expansive Soils. *Adv. Civ. Eng.* **2024**, *2024*, 5599356. [CrossRef]
43. Shams, M.A.; Shahin, M.A.; Ismail, M.A. Analysis and Modelling of Stiffened Slab Foundation on Expansive Soils. In *Numerical Analysis of Nonlinear Coupled Problems, GeoMEast 2017, Sustainable Civil Infrastructures*; Shehata, H., Rashed, Y., Eds.; Springer: Cham, Switzerland, 2018. [CrossRef]
44. *AS2870-2011*; Residential Slabs and Footings. Australian Standards: Sydney, NSW, Australia, 2011.
45. Hibbeler, R.C. *Structural Analysis*, 7th ed.; Prentice Hall: Saddle River, NJ, USA, 2011.

 *buildings*

Article

Fragility Analysis of Wind-Induced Collapse of a Transmission Tower Considering Corrosion

Chuncheng Liu and Zhao Yan *

School of Civil Engineering and Architecture, Northeast Electric Power University, Jilin 132012, China
* Correspondence: 2202000830@neepu.edu.cn

Abstract: To investigate the variation law of the wind-resistant performance of transmission towers during their operation, this paper proposes an evaluation method for the wind resistance of the transmission tower considering corrosion, and a 220-kV transmission tower is analyzed as an example. Considering the uncertainty of the material and geometric parameters, the wind-induced collapse of the transmission tower was analyzed, and the collapse wind speeds were obtained via pushover and incremental dynamic analyses. In addition, the sensitivity of the transmission tower to various parameters was analyzed. Based on the existing meteorological and corrosion data, corrosion prediction models were established using a back-propagation (BP) artificial neural network, and the mean relative error between the predicted and measured values of the test samples was 8.91%. On this basis, the corrosion depth of the tower members in the four regions was predicted, and the fragility of the transmission tower was analyzed considering the effects of corrosion and strong winds. The results show that the collapse wind speed of the transmission tower is most significantly affected by the thickness of the angle steel, followed by the elastic modulus and yield strength, and is less affected by the width of the angle steel. When the exposure time was 25 years, the wind-resistant performance of transmission towers in regions with severe acid rain and coastal industrial regions decreased by 10% to 20%. With an increase in exposure time, the failure mode of the transmission tower tended to be brittle failure.

Keywords: transmission tower; wind resistance; fragility analysis; BP artificial neural network; corrosion

Citation: Liu, C.; Yan, Z. Fragility Analysis of Wind-Induced Collapse of a Transmission Tower Considering Corrosion. *Buildings* **2022**, *12*, 1500. https://doi.org/10.3390/buildings12101500

Academic Editors: Dongming Li and Zechuan Yu

Received: 2 September 2022
Accepted: 18 September 2022
Published: 21 September 2022

Publisher's Note: MDPI stays neutral with regard to jurisdictional claims in published maps and institutional affiliations.

1. Introduction

A transmission tower is a tall and flexible structure, and it is significantly affected by wind load; therefore, wind load is very important in the structural design of transmission towers [1]. Wind tunnel tests and numerical simulation methods are often used in the research on the wind-resistant performance of transmission towers. Deng et al. [2] studied the dynamic characteristics and wind-induced vibration response of a tower-line system using wind-tunnel tests. Huang et al. [3] performed a numerical simulation and wind tunnel test of a transmission tower and compared the results of the test and simulation using the gust loading factors and gust response factors. The response and failure modes of transmission towers can be effectively predicted using nonlinear finite element analysis [4,5]. Zhang and Xie [6] used nonlinear buckling and dynamic analyses to evaluate the ultimate bearing capacity and vulnerable parts of a transmission tower. In the finite element analysis of transmission towers, more accurate results can be obtained by considering the coupling effects of the transmission tower and lines. Yasui et al. [7] simplified a transmission line as a truss element and studied the wind-induced vibration responses of different transmission towers. Battista et al. [8] calculated the response and stability of a transmission tower through time-domain and frequency-domain analyses.

Transmission towers are often affected by uncertain factors during their operation; therefore, it is more meaningful to evaluate the carrying capacity of transmission towers using probability analysis. The fragility analysis method is widely used to study the seismic performance of structures. Yazdani et al. [9,10] evaluated the seismic performance of plain concrete arch bridges under near-field and far-field earthquakes using an incremental dynamic analysis method. Chen et al. [11] revealed the potential failure modes of concrete gravity dams through incremental dynamic analysis. However, these studies only considered the randomness of seismic waves in their fragility analysis. So, Dolsek [12] conducted incremental dynamic analyses on four-story reinforced concrete frame models considering the uncertainty of the material properties. In recent studies, the probability analysis method has been applied to evaluate the wind resistance performance of transmission towers. Tian et al. [13] conducted fragility analysis of a transmission tower-line system considering the uncertainty of the wind load. Pan et al. [14] analyzed the sensitivity of transmission towers to earthquakes using the stripe analysis method. Fu et al. [15–18] conducted extensive research on the fragility analysis of transmission towers, proposed an uncertainty analysis method, and observed that deterministic analysis overestimated the wind-resistant performance of transmission towers; moreover, they confirmed that uncertainty analysis is effective in predicting the failure mode of the structure and performed fragility analysis of a transmission tower subjected to wind and rain loads. Based on the above research, we think it is necessary to consider the uncertainty of structure parameters and wind loads when conducting the fragility analysis of a transmission tower.

Research on the wind resistance of transmission towers has been conducted in-depth, but most analyses of transmission towers do not consider the effect of corrosion. As transmission towers are always in an outdoor atmospheric environment, they are vulnerable to atmospheric corrosion [19]. Corrosion causes mass loss and weakens the mechanical properties of steel [20,21], resulting in a decrease in the stability of the transmission tower. Therefore, the effect of corrosion should be considered when analyzing the bearing capacity of transmission towers. The degree of atmospheric corrosion of steel has a quantitative relationship with environmental factors and the chemical composition of steel [22–24]. Therefore, the corrosion rate of steel can be predicted according to the measured data. Zhi et al. [25] combined the nonlinear grey Bernoulli model with a genetic algorithm to predict the atmospheric corrosion rate of carbon steels; however, in the field of corrosion prediction, artificial neural network has broad application prospects. Song et al. [26] constructed four models to predict the corrosion rate of carbon steel in a dynamic atmospheric corrosion environment. Mohammed et al. [27] predicted the corrosion rate of medium carbon steel using an artificial neural network. In addition to the above research, the artificial intelligence algorithm can also be applied to the sensitivity analysis of steel corrosion. Li et al. [28] combined the mean impact value algorithm and back-propagation (BP) artificial neural network to evaluate the factors affecting the soil corrosion rate of Q235 steel. Cai et al. [29] conducted a sensitivity analysis of steel under atmospheric corrosion using an artificial neural network. Halama et al. [30] evaluated the effect of the SO_2 concentration on the atmospheric corrosion rate of carbon steel using an artificial neural network. Therefore, we think that the corrosion of steel can be accurately predicted if appropriate influencing factors can be selected.

Generally, strong winds are the main reason for the collapse of transmission towers, and steel corrosion is a hidden danger that affects their stability. However, the existing research did not give enough consideration to corrosion. Therefore, it is necessary to analyze the wind-resistant performance of transmission towers considering corrosion. In this paper, a method for evaluating the wind resistance of a transmission tower based on corrosion prediction and fragility analysis is proposed for the first time, and an uncertainty analysis of the collapse of a 220 kV transmission tower under the coupling effect of corrosion and strong wind is performed, which provides a valuable reference for the wind-resistant design of high-voltage transmission towers.

2. Probabilistic Evaluation Method for Wind Resistance of a Transmission Tower

The effect of corrosion on the wind resistance of transmission towers has not been considered in most studies. Therefore, in this paper, a method is proposed to evaluate the wind performance of a transmission tower by considering the effect of corrosion. As shown in Figure 1, the method is based on probability analysis and considers the uncertainty of the structural parameters of the transmission tower. The evaluation method includes three parts. First, after determining the probability distribution of parameters, a sensitivity analysis of the transmission tower is performed to obtain the collapse wind speed range of the structure and evaluate the impact of various parameters on the wind-resistant performance of the transmission tower. Subsequently, the Latin hypercube sampling method is used to sample each parameter, and the uncertainty models of the transmission tower are established. The fragility curves of the collapse wind speed and tower top displacement are obtained using pushover and incremental dynamic analyses, and the distribution of initial failure members in the transmission tower is obtained using nonlinear buckling analysis, which provides a comprehensive probabilistic assessment of the wind resistance performance of the transmission tower in the initial state.

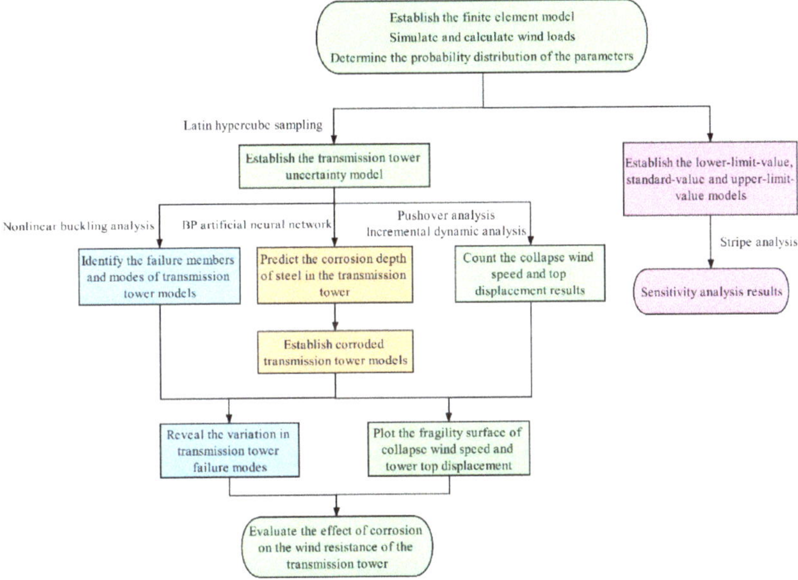

Figure 1. Proposed procedure of the wind resistance evaluation method for the transmission tower.

Based on the existing measured steel corrosion data, the corrosion depth prediction models of steel are obtained using a BP artificial neural network, and corroded transmission tower models are established. Combined with the above analysis methods, the corresponding fragility surfaces are obtained, and the variation rules for the failure modes and members of the transmission tower are identified. Based on the analysis results, the impact of corrosion on the wind-resistant performance of the transmission tower is evaluated.

3. Sensitivity Analysis of the Tower-Line System under Wind Loads

3.1. Finite Element Model

A latticed 220 kV transmission tower was investigated in this study. The nominal height was 30 m. The parameters of the conductor and the ground wire are listed in Table 1. The horizontal span was 410 m. The tower members were composed of Q235 and Q345 angle steels. The segmentation of the tower and parameters of the main leg members are shown in Figure 2.

Table 1. Material parameters of the conductor and ground wire.

Parameters	2 × LGJ-400/35	JLB20A-150
Diameter (mm)	26.82	15.75
Elastic modulus (GPa)	65	147.2
Cross-sectional area (mm^2)	425.24	148.07
Weight (per unit length) (kg/km)	1349	989.4
Tensile breaking force (N)	103900	178570

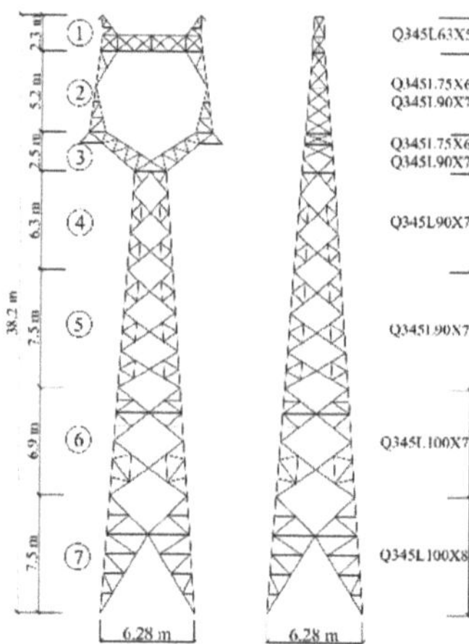

Figure 2. Segmentation of the tower and parameters of main leg members.

The finite element models of the transmission tower and tower line system were established using the Abaqus 2020 software. The B31 element was used to simulate the tower members, and the T3D2 element was used to simulate both the transmission line and the insulator. Fixed constraints were applied to the bottom of the transmission tower. Hinge restraints were used at the ends of the transmission line and at the connection of the insulator to the transmission tower and conductor. A bilinear isotropic hardening plasticity model was used to simulate the constitutive model of the steel material, as shown in Figure 3. The finite element model of the structure is shown in Figure 4. The modal analysis results of the transmission tower were as follows: the first-order natural frequencies of lateral bending (the direction perpendicular to the transmission line), longitudinal bending (the direction parallel to the transmission line), and torsion were 2.095, 2.107, and 4.856 Hz, respectively.

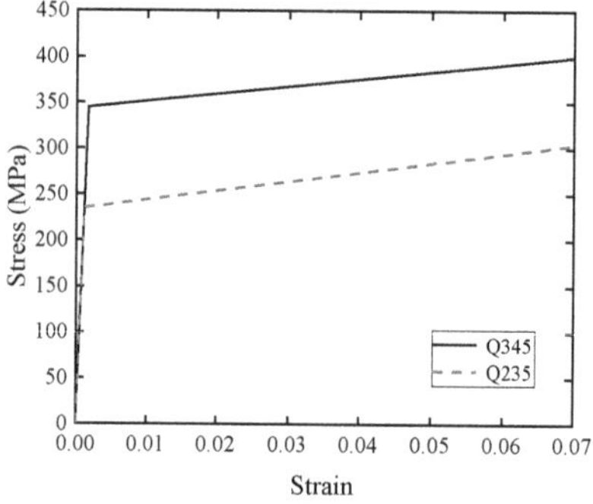

Figure 3. The stress–strain relationship for angles in the FEM models.

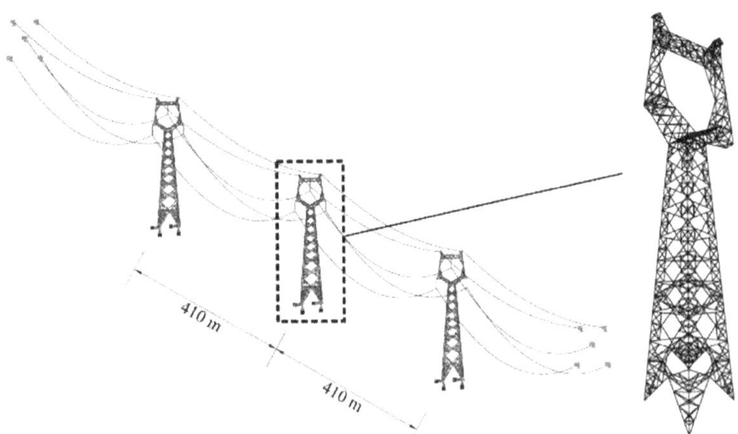

Figure 4. Finite element model of the structure.

3.2. Uncertainty of Material and Geometric Parameters

Transmission tower members are inevitably affected by external factors during their production and transportation, resulting in random variations in their parameters, which also affects the wind-resistant performance of transmission towers. Therefore, in this study, the uncertainty of the six parameters in the structure was considered based on existing research [15]. The material property parameters included the yield strength of Q235 steel, yield strength of Q345 steel, elastic modulus, and Poisson's ratio. The geometric parameters included the width and thickness of the angle steel. The probability distributions of different parameters are listed in Table 2. According to the unified standard for the reliability design of building structures [31], the standard values of the elastic modulus and Poisson's ratio took the 0.5 quantile value of the probability distribution, and the standard values of material strength took the 0.05 quantile value of the probability distribution. Therefore, the mean value of each material-property parameter was obtained. The mean values of the geometric parameters were obtained from the statistical results of relevant research [32]. In addition, the mean value of the geometric parameters in Table 2 was equal to that of the measured results divided by the standard value.

Table 2. Probability distributions of material and geometric parameters.

Parameter	Variable	Mean Value	Coefficient of Variation	Distribution Type
Elastic modulus (GPa)	E_s	206	0.03	Lognormal
Poisson ratio	ν	0.3	0.03	Lognormal
Yield strength of Q235 steel (MPa)	f_{y_Q235}	263.7	0.07	Lognormal
Yield strength of Q345 steel (MPa)	f_{y_Q345}	387.1	0.07	Lognormal
Width of the angle steel	b	1.001	0.008	Normal
Thickness of the angle steel	t	0.985	0.032	Normal

Considering that the calculation cost of the fragility analysis for multiple groups of transmission tower models with different corrosion degrees was relatively large, the sample size in this paper was determined to be 20 based on relevant research [13], and the accuracy of analysis can be evaluated as follows [33]:

$$N > \frac{-\ln(1-K)}{P_f} \tag{1}$$

where K is the confidence level, P_f is the failure probability, and N is the sample size. By calculation, the accuracy of probability analysis in this paper is close to 90%.

The above parameters were sampled using the Latin hypercube sampling (LHS) method, which has an advantage over Monte Carlo sampling in that the sampling effect is good even when the sample size is low. LHS is a stratified sampling method, that is, the research object is divided into multiple parts with equal probability, and then the sample proportion is determined according to the sample size, and each part is sampled according to this proportion [34]. The sampling results are shown in Figure 5. Uncertainty models of the transmission tower were established based on the sampling results.

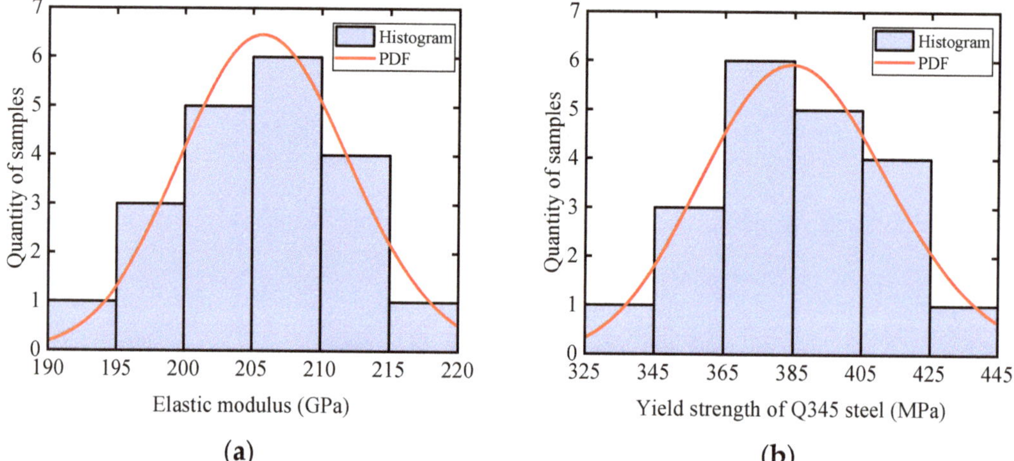

(a) **(b)**

Figure 5. Sample distributions of uncertainty parameters. (**a**) Sampling result of elastic modulus. (**b**) Sampling result of yield strength of Q345 steel.

3.3. Simulation of Wind Load

The simulation of the wind load on a structure is the basis of nonlinear dynamic analysis. The transmission tower was divided into seven panels vertically, and the transmission line was divided into ten parts longitudinally. The simplified wind load points of the tower-line system are shown in Figure 6.

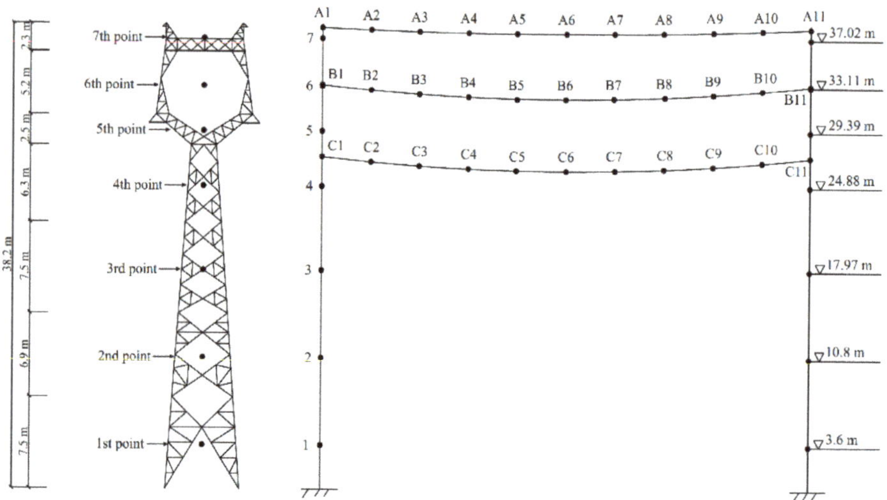

Figure 6. Schematic diagram of simplified wind load points.

Atmospheric boundary-layer wind consists of static and fluctuating winds. The static wind speed was calculated according to the exponential law, and the fluctuating wind speed with the Davenport spectrum was simulated using the harmonic superposition method. The harmonic synthesis method uses spectral decomposition and trigonometric series superposition to realize the numerical simulation of random process samples [17]. The total time of the wind speed time series was 300 s. The time interval was 0.125 s, and the cutoff frequency was 4 Hz. Taking a basic wind speed of 25 m/s as an example, the wind speed at the top of the tower is shown in Figure 7a. In addition, Figure 7b shows that the simulation spectrum was consistent with the target spectrum, indicating that the simulated wind speed can be used for nonlinear dynamic analysis.

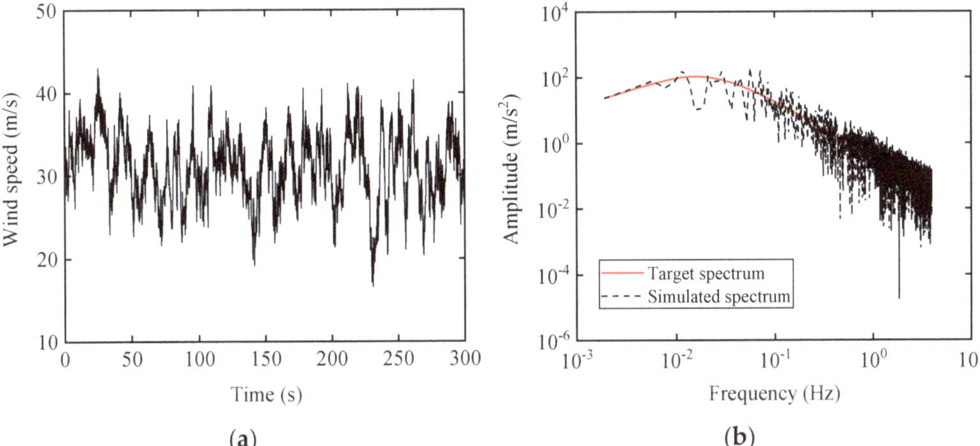

Figure 7. Simulated wind speed results. (**a**) Wind speed at the top of the tower. (**b**) Comparison of the simulated and target spectrum.

3.4. Sensitivity Analysis

Different material and geometric parameters have different degrees of impact on transmission towers. Using the basic collapse wind speed of the transmission tower as the criterion, the lower-limit-value, standard-value, and upper-limit-value models of the tower-line system were established. The strip analysis method is widely used in the sensitivity analysis of structures under earthquake [14], so in this paper, we studied the sensitivity of the transmission tower under wind load using this method.

The lower and upper limit values of each parameter were taken as the 0.05 and 0.95 quantiles of its probability distribution, respectively. To better reflect the impact of each parameter on the transmission tower, according to the control variable method [34], we changed only one parameter of the structure, and the other parameters still had their standard values when the model was established. The most unfavorable wind incidence angle for this tower was 90° (the direction perpendicular to the transmission line); therefore, the response of the transmission tower at this wind incidence was selected for research.

Multiple basic collapse wind speeds can be obtained by an extensive dynamic analysis of the models, and the log mean and log-standard deviation of the collapsed wind speeds for each group of models can be calculated using the maximum likelihood estimation method:

$$\begin{cases} \hat{\mu} = \frac{1}{n} \sum_{i=1}^{n} \ln X_i \\ \hat{\sigma}^2 = \frac{1}{n} \sum_{i=1}^{n} \left(\ln X_i - \frac{1}{n} \sum_{i=1}^{n} \ln X_i \right)^2 \end{cases} \tag{2}$$

where $\hat{\mu}$ and $\hat{\sigma}$ are the maximum likelihood estimators of the parameters in the lognormal distribution. The results of stripe analysis for the standard-value model are shown in

Figure 8. The log mean and log-standard deviation of the basic collapse wind speeds for the standard-value model were 3.412 and 0.0397, respectively.

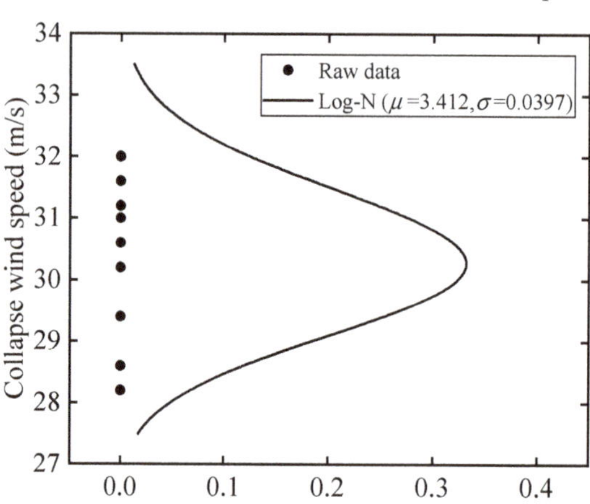

Figure 8. Stripe analysis results.

The log mean interval of the basic collapse wind speed corresponding to each parameter is shown in Figure 9a. The vertical dotted line represents the log-mean value of the collapsed wind speed of the standard value model, the end of each bar represents the analysis result of the lower- or upper-limit-value model, and the length of the bar reflects the impact level of its corresponding parameter on the basic collapse wind speed of the transmission tower. The bar length of the angle steel thickness was the largest, as shown in Figure 9a, indicating that the angle steel thickness had the greatest impact on the collapsed wind speed when the tower was in operation. The yield strength of Q345 steel and elastic modulus also had a significant effect on the collapse of the transmission tower. The width of the angle steel had a slight effect on the wind-resistant performance of the tower because the measured values of the angle steel width were close to the standard value according to the statistical results of the relevant research [32]. The bar length of the yield strength of Q235 steel was the minimum, which was due to the stress and deformation of the diagonal members of the transmission tower in this paper were relatively small. However, the results will be different for transmission towers with large stress on diagonal members.

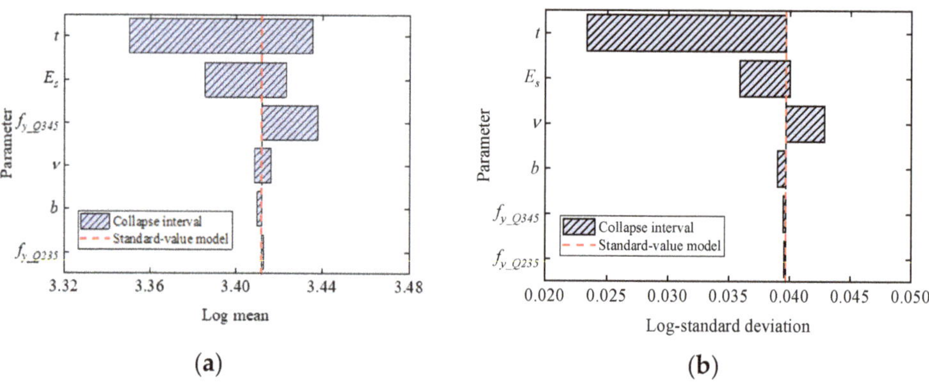

Figure 9. Tornado chart of collapse wind speeds. (**a**) Log mean of basic collapse wind speeds. (**b**) Log-standard deviation of basic collapse wind speeds.

The log-standard deviation reflects the dispersion degree of the collapsed wind speed results. Figure 9b shows that the change in geometric parameters reduced the log-standard deviation of the collapsed basic wind speeds, which meant that the collapsed basic wind speed of the transmission tower became more concentrated. The variations in the elastic modulus and Poisson's ratio made the basic collapse wind speed results more dispersed; however, the strength of the steel did not significantly affect the dispersion of the basic collapse wind speed results.

4. Fragility Analysis of Transmission Tower Considering Structural Uncertainty

Based on the established uncertainty models, pushover analysis for the transmission tower and incremental dynamic analysis for the tower-line system were performed. The calculation results of the two analysis methods were compared to study the collapse resistance of the transmission tower, and the calculation accuracy of the pushover analysis was evaluated using probability analysis.

4.1. Pushover Analysis

The simplified wind load points of the transmission tower are the same as those in the previous section, and the wind incidence angle was still determined to be $90°$ when calculating the equivalent static loads. The equivalent static wind loads of the structure were calculated according to the load code for the design of overhead transmission lines [35]. The wind load on the transmission tower panels can be calculated using the following equation:

$$P_T = W_0 \cdot \mu_Z \cdot \mu_S \cdot \beta_Z \cdot A_S \tag{3}$$

where W_0 is the standard value of the reference wind pressure (kN/m^2); μ_Z is the coefficient of wind pressure variation with height; μ_S is the shape coefficient of the tower; A_S is the calculated value of the projected area; and β_Z is the wind vibration coefficient, calculated as follows [35]:

$$\beta_{Zi} = 1 + 2g \cdot \varepsilon_t \cdot I_{10} \cdot B_{Zi} \cdot \sqrt{1 + R^2} \tag{4}$$

$$B_{Zi} = \frac{m_i \phi_{1i}}{\mu_{Si}\mu_{Zi}A_i} \cdot \frac{\sqrt{\sum_{j=1}^{n} \sum_{j'=1}^{n} (\mu_{Sj}\mu_{Zj}\phi_{1j}\bar{I}_{Zj}A_j)(\mu_{Sj'}\mu_{Zj'}\phi_{1j'}\bar{I}_{Zj'}A_{j'})coh_Z(z_j, z_{j'})}}{\sum_{j=1}^{n} m_j \phi_{1j}^2} \tag{5}$$

$$R^2 = \frac{\pi}{6\zeta_1} \frac{x_1^2}{\left(1 + x_1^2\right)^{4/3}} \tag{6}$$

$$x_1 = \frac{30 f_1}{\sqrt{k_w W_0}} \tag{7}$$

where g is the peak factor equal to 2.5; ε_t is the reduction coefficient of the wind load fluctuation; I_{10} is the nominal turbulence intensity at a 10 m height, which is 0.14 for class B ground roughness; B_{Zi} is the background factor; R^2 is the resonance factor; ϕ_1 is the first-order mode coefficient of the structure; $coh_Z(z_j, z_{j'})$ is the vertical coherence function; ζ_1 is the first-order damping ratio of the structure; and f_1 is the first-order natural frequency of the structure.

The wind load on the transmission line can be calculated using the following equation:

$$P_D = \beta_C \cdot \alpha_L \cdot W_0 \cdot \mu_Z \cdot \mu_{SC} \cdot d \cdot L_P \tag{8}$$

where β_C is the gust coefficient of the conductor and ground wire; α_L is the span reduction factor; μ_{SC} is the shape coefficient of the transmission line, which is 1.0 when the diameter is greater than 17 mm and otherwise equal to 1.1; d is the diameter of the transmission line; and L_P is the horizontal span.

The wind load of the insulator string can be calculated using the following equation:

$$P_J = n \cdot \lambda_1 \cdot W_0 \cdot \mu_Z \cdot \mu_{S1} \cdot A_1 \tag{9}$$

where n is the number of insulator strings perpendicular to the wind direction; λ_1 is the shielding reduction factor of the wind load on the insulator string; μ_{S1} is the shape coefficient of the insulator, equal to 1.0; and A_1 is the calculated value of the wind-pressure bearing area of the insulator string.

The calculated results for the wind load on the structure are shown in Tables 3 and 4.

Table 3. Wind load of each panel (basic wind speed at 25 m/s, wind incidence angle at 90°).

Panel Number	Height above Ground (m)	Height of Panel (m)	Wind Pressure Height Variation Coefficient μ_z	Wind Vibration Coefficient β_z	Shape Coefficient μ_S	Projected Area A_S (m^2)	Standard Value of Reference Wind Pressure W_0 (kN/m^2)	Standard Value of Wind Load (kN)	Design Value of Wind Load (kN)
1	37.0	2.3	1.481	5.860	2.217	0.528	0.391	3.432	4.805
2	33.1	5.9	1.430	1.537	2.217	2.154	0.391	4.100	5.740
3	29.4	1.8	1.380	2.642	2.217	1.596	0.391	3.441	4.817
4	24.9	6.3	1.308	1.316	2.356	3.635	0.391	5.758	8.061
5	18.0	7.5	1.190	1.116	2.356	6.018	0.391	7.352	10.293
6	10.8	6.9	1.021	1.059	2.356	7.120	0.391	7.083	9.156
7	3.6	7.5	1.000	1.012	2.356	9.467	0.391	8.818	12.345

Table 4. Wind load of conductor, ground line, and insulator.

Wind Speed (m/s)	Wind Load Design Value of Middle Conductor P_{D1} (N)	Wind Load Design Value of Side Conductor P_{D2}, P_{D3} (N)	Wind Load Design Value of Ground Wire P_B (N)	Wind Load Design Value of Middle Insulator P_{JD1} (N)	Wind Load Design Value of Side Insulator P_{JD2}, P_{JD3} (N)
15	4567	4344	1497	314	297
20	8118	7720	2662	558	528
25	12,679	12,056	4158	871	826
26	13,712	13,039	4497	942	893
27	14,786	14,059	4850	1016	963
28	15,900	15,118	5216	1093	1036
30	18,249	17,351	5987	1255	1189

The calculated wind loads were applied to the corresponding nodes on the transmission tower in the form of concentrated forces. The weights of the conductor, ground wire, and insulator were also simplified as a concentrated force and applied at the corresponding position of the transmission tower. A pushover analysis was performed on the uncertainty models. By increasing the basic wind speed continuously to cause the collapse of the transmission tower models, the pushover curves of the uncertainty models were obtained (Figure 10).

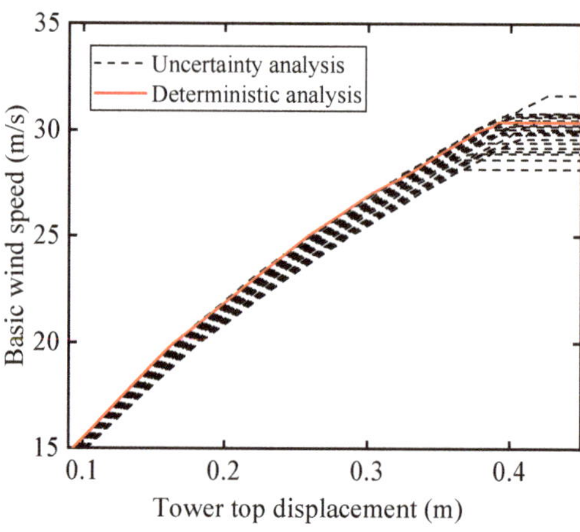

Figure 10. Pushover curves for transmission tower models.

We observed that the displacement response of the deterministic model lay between those of the uncertainty models. The basic collapse wind speed of the deterministic model was 30.4 m/s, and the collapsed-tower top displacement was 0.391 m. According to the analysis results of the uncertainty models, multiple collapse wind speeds and tower-top displacements were obtained. The fragility curve of the uncertainty analysis for the transmission tower was obtained by fitting the collapse data to the cumulative function of the lognormal distribution.

The fragility curves of the basic collapse wind speed and tower top displacement are shown in Figure 11. The value corresponding to a 10% probability in the fragility curve is frequently used as the critical collapse datum; thus, the collapse wind speed calculated by the uncertainty analysis was 28.68 m/s and the ultimate displacement at the top of the tower was 0.368 m. Compared with the results of the deterministic analysis, the collapse wind speed and ultimate tower top displacement of the uncertainty analysis were 5.66% and 5.88% lower, respectively, which indicated that the results of the deterministic analysis overestimated the bearing capacity of the structure.

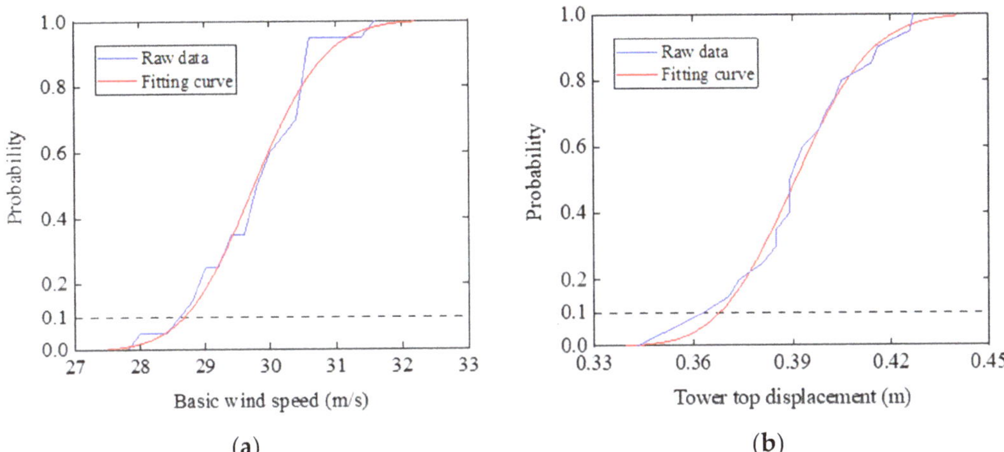

Figure 11. Fragility curve for transmission towers. (**a**) Fragility curve of the basic collapse wind speed. (**b**) Fragility curve of collapsed-tower top displacement.

4.2. Incremental Dynamic Analysis

The collapsed tower top displacement of each sample was determined using the nonlinear static analysis results. These displacements were taken as the critical values of the transmission tower models for incremental dynamic analysis.

When calculating the response of the tower-line system under different wind speeds, multiple wind speed time histories were generated by the harmonic superposition method to consider the uncertainties in the wind load. In total, 20 groups of wind speeds were gradually increased from 20 m/s, and the increment in the basic wind speed was 0.2 m/s, when calculating the response of the tower-line system under different wind speeds. By capturing the maximum top displacement of the transmission tower at each basic wind speed, the variation curve could be obtained, as shown in Figure 12. The red line in Figure 12a is the collapsed tower top displacement of the transmission tower model. Only part of the data are shown in Figure 12b to avoid overlapping curves, and each curve represents the result of the incremental dynamic analysis.

In the incremental dynamic analysis of the tower-line system, we considered that the basic wind speed corresponding to the tower top displacement that exceeded the critical displacement value for the first time was the basic collapse wind speed. The collapse wind speed for each sample was obtained and compared with that calculated using pushover analysis. As shown in Figure 13a, the collapse wind speed values obtained using the two methods were similar, with a maximum relative error of 5.7% and a mean relative error of 3.18%, indicating that the mass and coherence functions were considered when calculating the equivalent wind load according to the load code for the design of the overhead transmission line [35]; therefore, accurate response results can be obtained using pushover analysis.

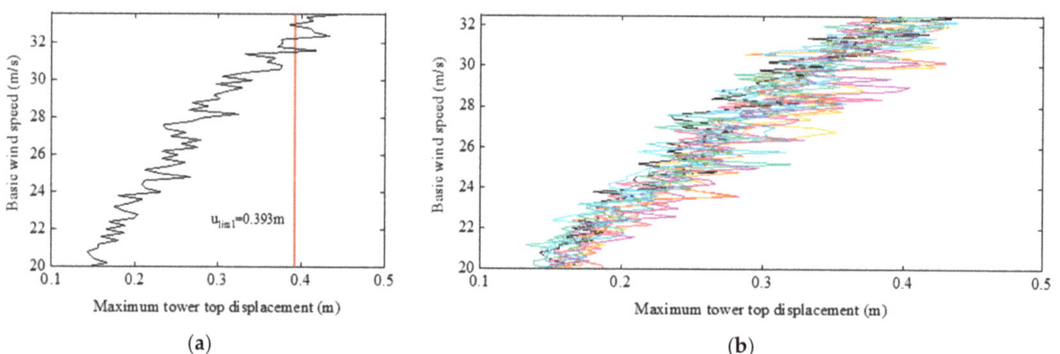

(a)

(b)

Figure 12. Variation curve of the maximum tower top displacement. (**a**) One of the incremental dynamic analysis results. (**b**) Multiple incremental dynamic analysis results.

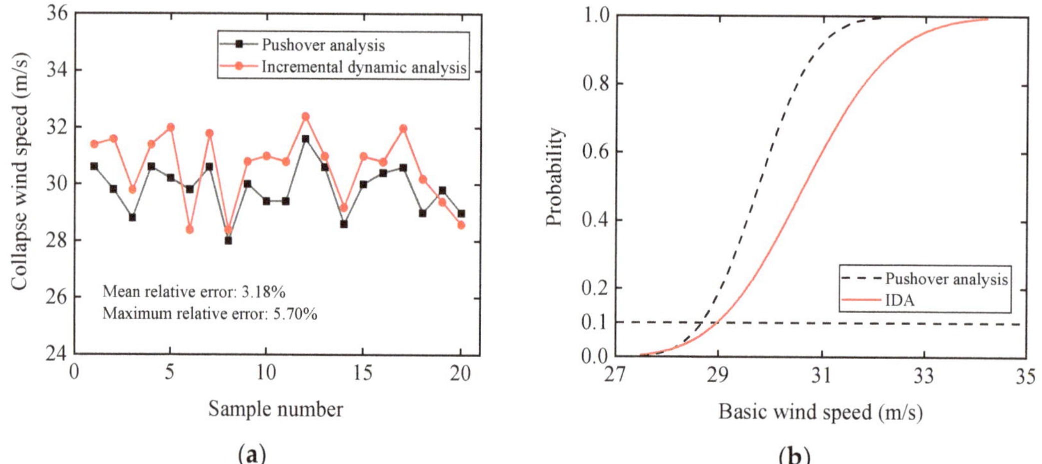

(a)

(b)

Figure 13. Comparison of pushover analysis and IDA results. (**a**) Comparison results of the basic collapse wind speeds. (**b**) Comparison results of the basic collapse wind speed fragility curves.

Based on the collapse wind speed of each sample, the fragility curve corresponding to the incremental dynamic analysis was fitted. The comparison results of the collapse wind speed fragility curves obtained using the two methods are shown in Figure 13b. We observed that the end position of the fragility curve corresponding to the incremental dynamic analysis was farther from the starting position, which was due to the uncertainty of the wind load being considered in the incremental dynamic analysis. The curve of the pushover analysis was on the left side of the curve of the incremental dynamic analysis, and the starting positions of the two curves were close. In the fragility curve corresponding to the incremental dynamic analysis, the basic collapse wind speed corresponding to the 10% probability was 28.96 m/s, which is only 0.97% different from the result of the pushover analysis, which further indicated that the collapse wind speed obtained using the pushover analysis was accurate.

5. Prediction of Corrosion Depth Based on the BP Artificial Neural Network

The BP artificial neural network is a widely used network model. Its construction concept is as follows: first, a part of the measured data is input into the neural network as the training set, the weights in the network are adjusted through the back-propagation algorithm until the error meets the requirement, and then the test set is calculated using the generated model to evaluate the prediction accuracy of the artificial neural network.

5.1. Generation of the Artificial Neural Network Model

In this study, a three-layer neural network was constructed based on the corrosion data from the National Materials Corrosion and Protection Data Center [36]. As shown in Figure 14, the connections between adjacent layers of the neural network were fully connected. The data selection for the neural network prediction model is shown in Table 5. There were 15 steel materials involved in the model construction, and the steel types were carbon steel and low-alloy steel. Meteorological data and atmospheric corrosion data during model training were obtained from the test stations in six regions. The meteorological data are presented in Table 6. The exposure times of the specimens were 1, 2, 4, 8, and 16 years. The input factors of the neural network model included material parameters, meteorological factors, and exposure time. The number of nodes in the input layer was 15, and the parameter in the output layer was the corrosion rate of steel.

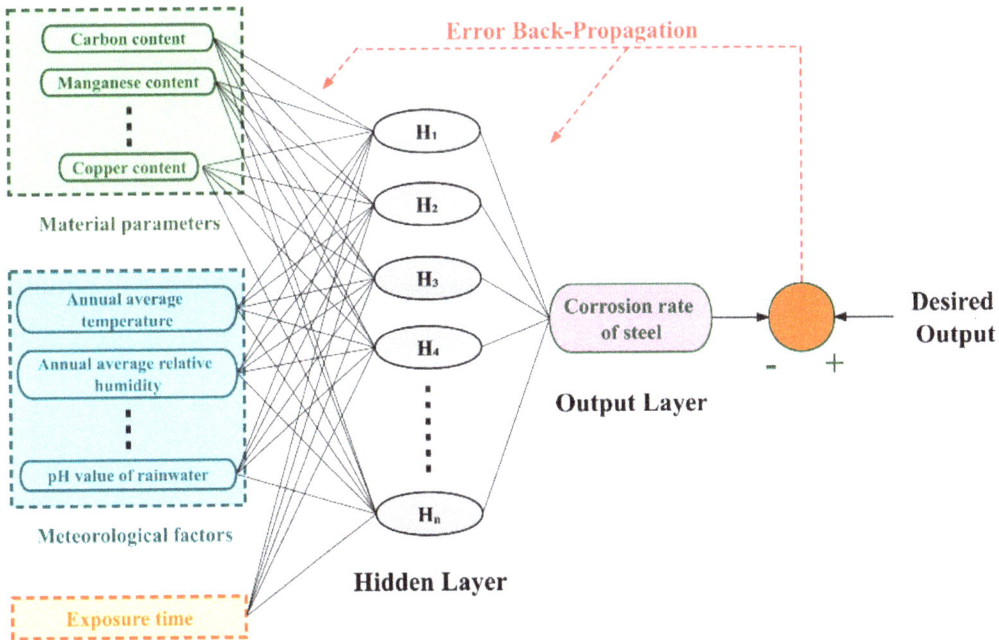

Figure 14. Schematic diagram of the BP artificial neural network.

The transfer function of the network was a sigmoid function with a learning rate of 0.05. The interval of the optimal number of hidden layer nodes was determined using existing research [37]. After testing, the number of hidden layer nodes was nine, and a neural network prediction model was obtained through training.

Table 5. Data selection for the artificial neural network model.

Network Parameter	Data Selection
Steel material	Carbon steel: 3C, 20, 15MnMoVN, 14MnMoNbB, 09MnNb(S), 08Al, 12CrMnCu, Q235; Low-alloy steel: 16MnQ, 10CrMoAl, 10CrCuSiV, 09CuPTiRE, 09CuPCrNi, 09CuPCrNiA, Q345
Region	Beijing, Qingdao, Jiangjin, Guangzhou, Wuhan, Qionghai
Input factor	Material parameters: Content of carbon, manganese, sulfur, phosphorus, silicon, and copper Meteorological factors: annual average temperature, annual average relative humidity, annual sunshine hours, annual precipitation, SO_2 concentration, Cl^- concentration, NO_2 concentration, pH value of rainwaterExposure time

Table 6. Meteorological data of each region.

Region	Annual Average Temperature (°C)	Annual Average Relative Humidity (%)	Annual Sunshine Hours (h)	Annual Precipitation (mm)	SO_2 Concentration (mg/cm³)	Cl^- Concentration (mg/cm³)	NO_2 Concentration (mg/100 cm²/d)	pH Value of Rainwater
Beijing	12.8	55	2368.6	578.7	0.06	0.85	0.11	6.52
Qingdao	12.8	70	2199.9	582.6	0.05	0.11	0.08	5.42
Jiangjin	19.8	77	1369.3	998	0.22	0.00	0.08	5.44
Guangzhou	21.5	81	1582.9	2095.4	0.06	0.03	0.08	6.68
Wuhan	17.1	77	2092.5	1434.2	0.08	0.02	0.14	6.81
Qionghai	24.6	82	1743.1	2506.1	0.02	0.05	0.01	6.38

5.2. Corrosion Depth Prediction Results

Test samples were calculated using the generated model to evaluate the prediction accuracy of the artificial neural network. As shown in Figure 15a, the predicted results of the test samples were close to the measured values. The error analysis results for the test set are shown in Figure 15b to reflect the prediction accuracy of the model more intuitively. We observed that the relative error of almost all predicted values was less than 20%, the mean relative error between the predicted and measured values was 8.91%, and the correlation coefficient was 0.9849, which indicated that the prediction accuracy of the corrosion prediction model established by the BP artificial neural network was high.

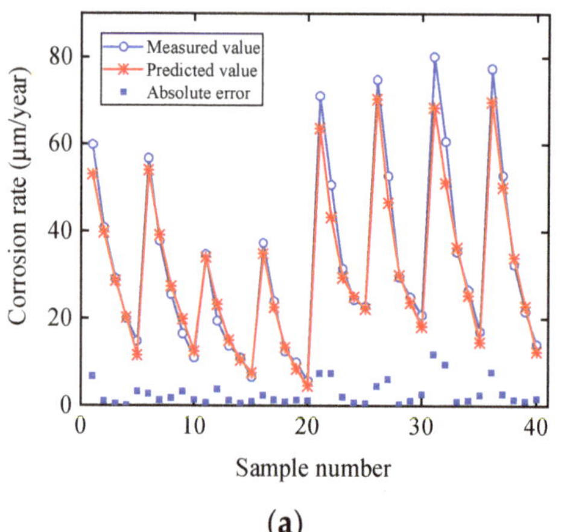

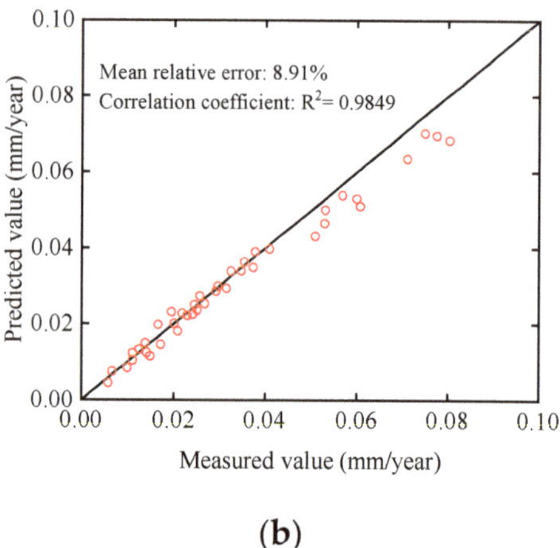

Figure 15. Evaluation results of the test set. (**a**) Comparison between measured values and predicted values. (**b**) Error analysis results of the test set.

Owing to the randomness of the prediction model generated by the BP neural network, 20 prediction models were generated in this study, and the average of the 20 prediction results was considered the final result. The transmission tower contained Q235 and Q345 steels; therefore, the corrosion rates of Q235 and Q345 steels in Beijing, Qingdao, Jiangjin, and Guangzhou were predicted in this study.

Through multiple attempts, we observed that the change trends of the prediction curves within 25 years were the same, and the results were relatively dense. However, when the exposure time exceeded 25 years, the dispersion of the predicted results was larger; therefore, the maximum exposure time was determined to be 25 years. Figures 16 and 17 show the corrosion rate prediction results for Q235 and Q345 steels, respectively, in the four regions.

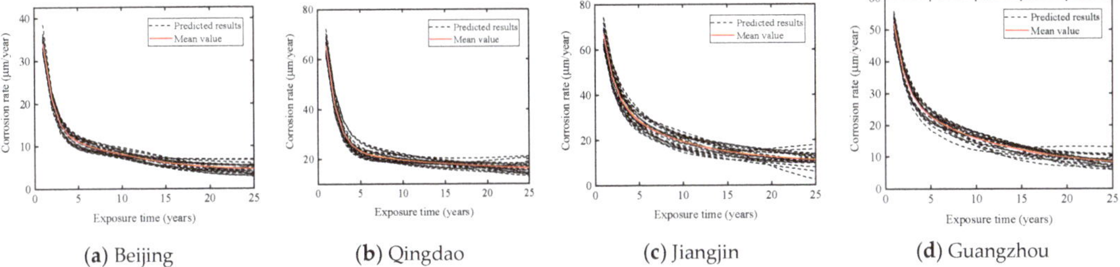

(**a**) Beijing   (**b**) Qingdao   (**c**) Jiangjin   (**d**) Guangzhou

Figure 16. Predicted results of the Q235 steel corrosion rate.

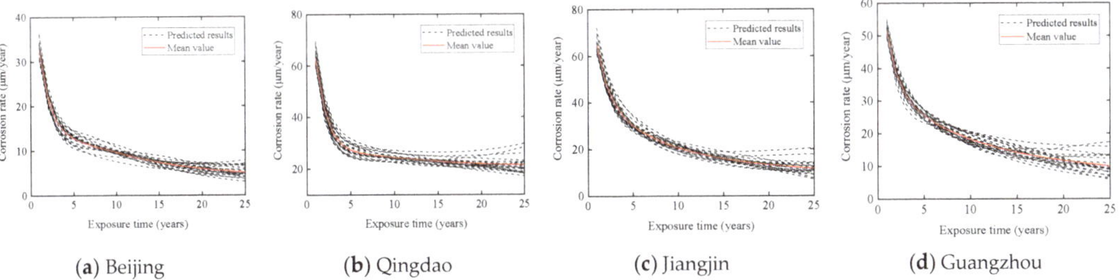

(**a**) Beijing   (**b**) Qingdao   (**c**) Jiangjin   (**d**) Guangzhou

Figure 17. Predicted results of the Q345 steel corrosion rate.

Based on the predicted corrosion rate, the corresponding variation law of the corrosion depth was obtained using integration, that is, the corrosion depth in the t year is the sum of the corrosion rates in the previous t years. The fitting function of the mean curve of the corrosion depth can be expressed as follows:

$$D = At^n \tag{10}$$

where D is the corrosion depth in t years, A is the corrosion depth in the first year, and n reflects the changing trend of the curve. The fitting results for the corrosion depth are listed in Table 7. The R-squared value represents the correlation between the power function and the predicted value. The R-squared value of each fitting result was close to 1, indicating that the effect of fitting the corrosion depth by the power function was good. The fitting curves for the corrosion depths are shown in Figure 18. We observed that the corrosion depth in Beijing was smaller, and the corrosion depths in Qingdao and Jiangjin were larger. By comparing the meteorological factors in different regions, we can consider that the relative humidity and SO_2 concentration have a significant impact on the corrosion rate of steel. In addition, the corrosion depth of Q235 steel is greater than that of Q345 steel in a short exposure time, but with an increase in exposure time, the corrosion depth of Q345 steel exceeds that of Q235 steel. However, Q345 angle steel is used as the main leg member in the transmission tower; therefore, the section dimensions of Q345 angle steel are relatively large. Therefore, although the corrosion depth of Q345 steel is larger, its mass loss ratio is generally smaller than that of Q235 steel, indicating that the mass loss ratios of the diagonal members are greater during the operation of the transmission tower.

Table 7. Fitting results of corrosion depth prediction values.

Steel Material	Region	*A*	*n*	*R*²
Q235	Beijing	34.03	0.5951	0.9953
Q235	Qingdao	65.37	0.6581	0.9988
Q235	Jiangjin	66.78	0.6592	0.9887
Q235	Guangzhou	51.89	0.6675	0.9886
Q345	Beijing	32.94	0.6444	0.9944
Q345	Qingdao	63.42	0.7219	0.9985
Q345	Jiangjin	65.20	0.6817	0.9877
Q345	Guangzhou	51.81	0.7044	0.9894

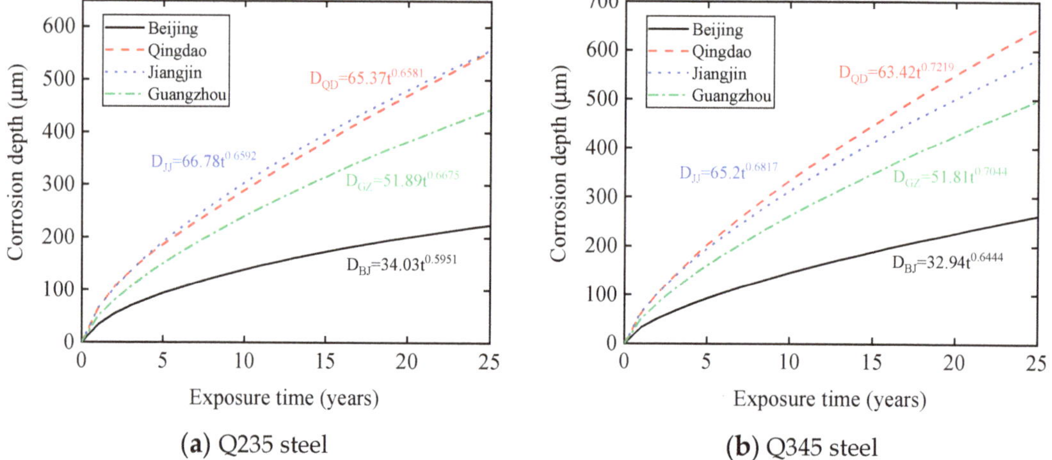

(**a**) Q235 steel (**b**) Q345 steel

Figure 18. Fitting curves of corrosion depth.

6. Uncertainty Analysis of Transmission Tower Considering Corrosion and Strong Wind Effects

6.1. Results of Wind Resistance Degradation

According to statistics [38], the steel protective layer of the transmission tower in regions with severe acid rain and coastal regions will become invalid within a few years, and the transmission tower will be completely corroded. The material properties and geometric parameters of the steel also decrease noticeably because of corrosion. Therefore, in this paper, we converted the corrosion depth into the mass loss ratio of steel and analyzed the wind-resistant performance of corroded transmission towers. The mechanical properties of Q235 and Q345 steels before and after corrosion were compared in relevant research [20,21], and the variation law of the mechanical properties of Q235 and Q345 steels with the mass loss ratio was revealed based on the statistical results. The formula for the mechanical property degradation of the Q235 and Q345 steel is as follows:

$$\frac{p'}{p}=1-c\eta \tag{11}$$

where p and $p\prime$ are the mechanical property of the steel before and after corrosion. η is the mass loss ratio. c is the value of the reduction coefficient of mechanical properties as shown in Table 8.

Table 8. Value of the reduction coefficient of mechanical properties.

Steel Type	Yield Strength	Ultimate Strength	Elastic Modulus
Q235	0.875	0.894	0.88
Q345	0.96	0.99	0.98

The uncertainty of geometric parameters will affect the calculation results of the mass loss ratio, and further, it will affect the decline of material parameters. Therefore, it is necessary to consider the uncertainty of its structural parameters when analyzing the wind-resistant performance of corroded transmission towers. According to the obtained corrosion depth prediction results, the mass loss ratios of angle steel with different section dimensions in each uncertainty model were calculated, and the material properties of each angle steel after corrosion were calculated using the degradation formula. Considering the degradation of the mechanical properties and geometric parameters, an uncertainty analysis of the wind-induced collapse of the transmission tower was performed.

Figures 19 and 20 show the relationship between the collapse probability of the transmission tower and the basic wind speed and the relationship between the collapse probability and tower top displacement in Beijing, Qingdao, Jiangjin, and Guangzhou at different exposure times, respectively. Tables 9 and 10 show the basic wind speeds and tower top displacements corresponding to a 10% probability of the collapse fragility surface for different exposure times. The collapse wind speed and tower top displacement decreased significantly with the intensification of corrosion. Beijing and Jiangjin are both inland regions, but the collapse wind speed of the transmission tower in Jiangjin decreased faster than that in Beijing, which was due to severe acid rain caused by the high SO_2 concentration in the atmosphere of Jiangjin. The decline in wind speeds in Qingdao and Guangzhou, which are coastal regions, was also greater than that in ordinary inland regions. In addition, the decline in tower top displacements in regions with severe acid rain and coastal regions was also greater than that in ordinary inland industrial regions.

To reflect the declining trend in the collapse wind speed and tower top displacement of the transmission tower with exposure time more intuitively, Figure 21 shows the decay curves of the wind-resistant performance of the transmission tower. The decrease ratio in Figure 21 was calculated as the reduction value of the collapse wind speed or collapsed tower top displacement divided by its initial value. We observed that the decay curves of the collapse wind speed and tower top displacement in the same region were almost coincident, indicating that the decrease ratios of the collapse wind speed and tower top displacement were synchronized. When the exposure time was less than 5 years, the wind resistance performance of the transmission tower decreased rapidly. With increasing exposure time, the decay rate gradually slows. When the exposure time was 25 years, the wind resistance performance of the transmission tower in the Beijing region could still be maintained at more than 90%. However, the wind resistance performance of transmission towers in regions with severe acid rain and coastal industrial regions decreased by 10% to 20%. In addition, the decrease ratios of the collapsed wind speeds of the transmission towers in Qingdao, Jiangjin, and Guangzhou were all greater than 10% when the exposure time was 10–15 years.

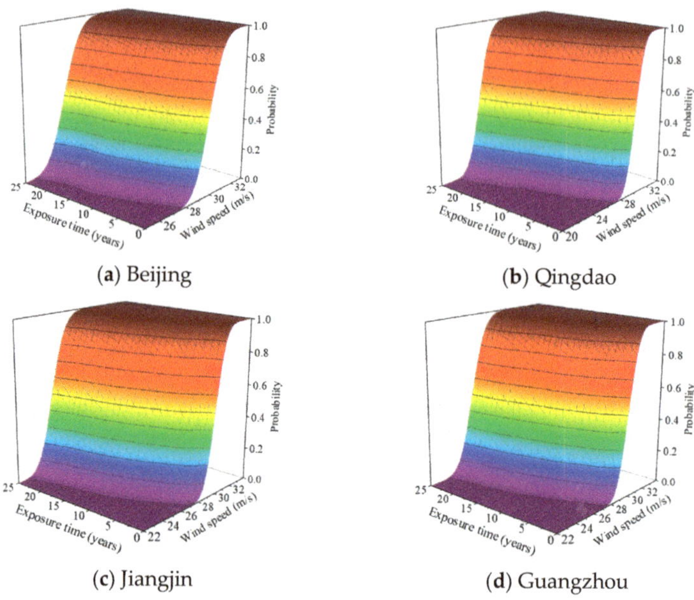

(**a**) Beijing

(**b**) Qingdao

(**c**) Jiangjin

(**d**) Guangzhou

Figure 19. Fragility surface of the collapsed basic wind speed in each region.

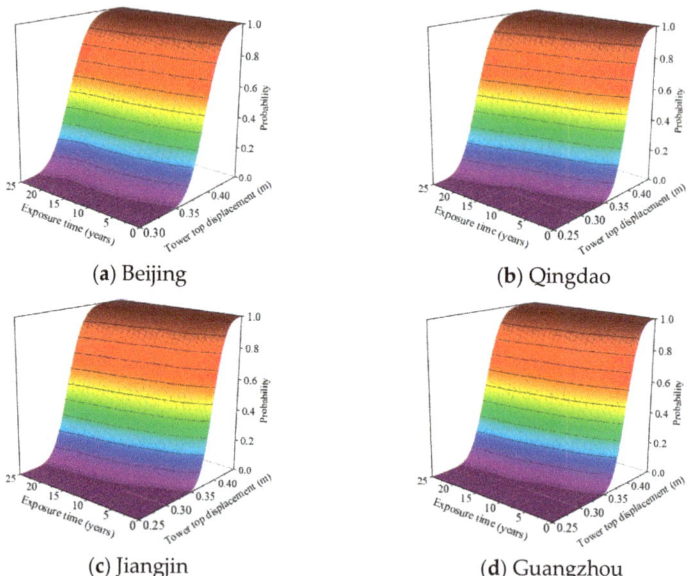

(**a**) Beijing

(**b**) Qingdao

(**c**) Jiangjin

(**d**) Guangzhou

Figure 20. Fragility surface of the collapsed-tower top displacement in each region.

Table 9. Collapsed basic wind speed statistics at different exposure times in each region.

Exposure Time (Years)	V_{BJ} (m/s)	V_{QD} (m/s)	V_{JJ} (m/s)	V_{GZ} (m/s)
0	28.68	28.68	28.68	28.68
5	27.96	27.00	26.79	27.31
10	27.47	26.09	25.82	26.45
15	27.12	25.23	25.16	25.78
20	27.06	24.22	24.75	25.35
25	26.67	23.45	24.19	24.96

V_{BJ}, V_{QD}, V_{JJ}, and V_{GZ} are the critical basic collapse wind speeds of the transmission towers in Beijing, Qingdao, Jiangjin, and Guangzhou, respectively.

Table 10. Collapsed-tower top displacement statistics at different exposure times in each region.

Exposure Time (Years)	U_{BJ} (m)	U_{QD} (m)	U_{JJ} (m)	U_{GZ} (m)
0	0.368	0.368	0.368	0.368
5	0.359	0.346	0.346	0.349
10	0.353	0.336	0.332	0.340
15	0.350	0.326	0.323	0.330
20	0.350	0.310	0.318	0.326
25	0.342	0.301	0.313	0.321

U_{BJ}, U_{QD}, U_{JJ}, and U_{GZ} are the critical collapse tower top displacements of the transmission towers in Beijing, Qingdao, Jiangjin, and Guangzhou, respectively.

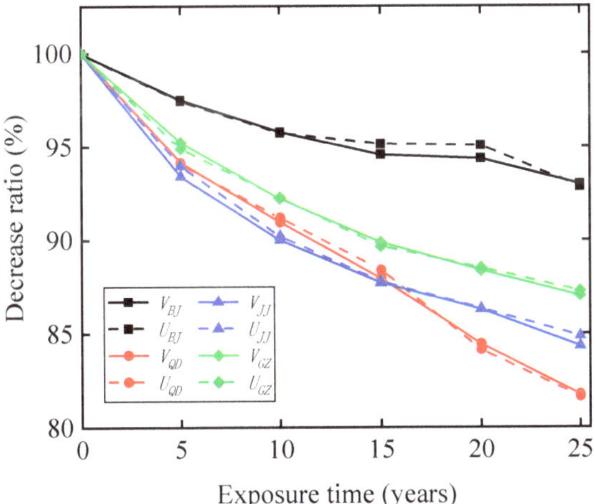

Figure 21. Decay curves of the wind-resistant performance of the transmission tower.

6.2. Variation in Transmission Tower Failure Modes

When tower members are corroded, in addition to the smaller collapse wind speed and tower top displacement, the failure mode of the tower also changes. Therefore, in this study, the failure members and modes of the transmission tower models with different exposure times were analyzed to study the impact of corrosion on the transmission tower more comprehensively.

The failure mode of the tower was observed through the nonlinear buckling analysis of the finite element model of the transmission tower. Firstly, the equivalent loads of the transmission tower and lines were applied to the corresponding nodes, and the eigenvalue buckling analysis was performed on the transmission tower model. Secondly, the vibration mode result of buckling analysis was applied to the tower as an initial defect. Finally, the nonlinear buckling analysis was carried out on the tower after updating the model. The results showed that vulnerable members in the studied tower complied with buckling failure. Additionally, the possible failure members of the transmission tower can be obtained through uncertainty analysis. As shown in Figure 22a, the possible failure members of the tower are the main leg members in the 5th and 6th panels and the failure mode of tower is shown in Figure 22b.

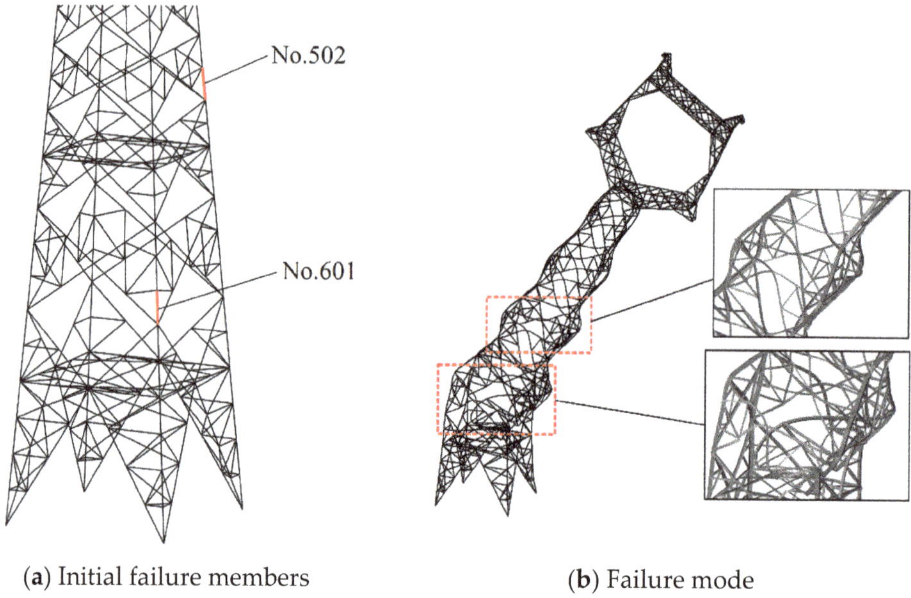

(**a**) Initial failure members   (**b**) Failure mode

Figure 22. Initial failure members and failure mode of tower.

The changes in the failure members and modes of transmission towers in different regions with the exposure time are listed in Table 11. Without considering the effect of corrosion, when the transmission tower collapses, there was a 70% probability that the failure member was No. 502, and a 30% probability that the failure member was No. 601. The failure mode of the transmission tower is plastic instability, that is, the transmission tower undergoes plastic deformation before collapsing. However, with an increase in the exposure time, the failure mode of the structure changes, and the transmission tower is prone to elastic instability, which indicates that the failure mode of the transmission tower tends to be brittle failure. In addition, the failure probability of each member in the transmission tower also changes with increasing corrosion of the tower members. The failure probability of member No. 502 decreased, and the failure probability of member No. 601 increased. From the failure modes of transmission towers in different regions, we observed that in the Beijing region, elastic instability would not occur in the transmission tower until the exposure time is 25 years; however, the transmission towers in the Qingdao, Jiangjin, and Guangzhou regions may experience elastic instability when the exposure time is 10 years, which indicates that in ordinary inland industrial regions, the failure modes of transmission towers are less affected by corrosion. However, in regions with severe acid rain and coastal industrial regions, the collapse failure modes of transmission towers will change significantly with the increase in the exposure time.

Table 11. Collapse failure probability of the transmission tower in each region.

Exposure Time	Initial Failure Member	Failure Probability (%)	Probability of Plastic Instability (%)	Probability of Elastic Instability (%)
Initial state	No. 502	70	70	0
	No. 601	30	30	0
5 to 20 years in Beijing 5 years in Qingdao 5 years in Jiangjin 5 years in Guangzhou	No. 502	70	70	0
	No. 601	30	30	0
25 years in Beijing 10 to 15 years in Qingdao 10 to 15 years in Jiangjin 10 to 20 years in Guangzhou	No. 502	70	65	5
	No. 601	30	30	0
20 years in Jiangjin 25 years in Guangzhou	No. 502	70	55	15
	No. 601	30	25	5
20 years in Qingdao 25 years in Jiangjin	No. 502	65	45	20
	No. 601	35	30	5
25 years in Qingdao	No. 502	65	45	20
	No. 601	35	20	15

7. Conclusions

In this paper, a wind resistance evaluation method for a transmission tower is proposed and taking a 220 kV transmission tower as an example, the sensitivity of the transmission tower to various uncertain parameters was studied. The fragility curves for the transmission tower were obtained by pushover and incremental dynamic analyses, respectively. The variation curves of the steel corrosion depth in the transmission tower with exposure time were obtained based on a BP artificial neural network, and by taking the collapse wind speed and tower top displacement as the evaluation indicators, the decline trend of the wind-resistant performance of the transmission tower with the increase in corrosion was evaluated. The conclusions drawn from this study are summarized as follows:

(1) The sensitivity analysis of the transmission tower shows that the angle steel thickness has the greatest impact on the wind-resistant performance of the transmission tower when the tower is in operation, and the yield strength of Q345 steel and the elastic modulus also have a significant impact on the collapse wind speed. The change in geometric parameters reduces the log-standard deviation of the basic collapse wind speeds, and the variations in the elastic modulus and Poisson ratio can make the basic collapse wind speed results more dispersed. Therefore, more attention should be paid to the thickness of steel when designing and manufacturing transmission towers.

(2) The collapse wind speed results obtained using the pushover analysis based on the load code for the design of the overhead transmission line [35] were close to those obtained by the incremental dynamic analysis. The maximum relative error was 5.7% and the mean relative error was 3.18%. The starting positions of the fragility curves obtained by the two methods were almost coincident, and the basic collapse wind speed results corresponding to a 10% probability differed by only 0.97%. Therefore, probability analysis method can improve the accuracy of the results.

(3) The accuracy of predicting the steel corrosion rate using a BP artificial neural network was high. The mean relative error between the predicted and measured values was 8.91% and the correlation coefficient was 0.9849. The mass loss ratios of the diagonal members were greater than those of the main leg members during the operation of the transmission tower, so in engineering design, it is feasible to use the artificial neural network method for corrosion prediction.

(4) The corrosion of tower members will reduce the basic collapse wind speed of the tower and collapsed-tower top displacement, particularly in regions with severe acid rain and coastal industrial regions, and will result in variations in the failure mode and members of the transmission tower. With the increase in the exposure time, the possibility of brittle failure of the transmission tower increases; therefore, the

transmission tower should be maintained in time according to the corrosion degree of the tower members in different regions.

This study evaluated, for the first time, the wind-induced collapse of a transmission tower with different corrosion degrees using the fragility analysis method, but we did not consider the texture characteristics of the steel surface after corrosion. In future research, more measured data and artificial intelligence methods should be combined to study the corrosion of transmission towers.

Author Contributions: C.L.: conceptualization, project administration, supervision, writing—review and editing, and funding acquisition. Z.Y.: methodology, investigation, software, data curation, and writing original draft. All authors have read and agreed to the published version of the manuscript.

Funding: This research was funded by the National Natural Science Foundation of China, grant number 51278091.

Institutional Review Board Statement: Not applicable.

Informed Consent Statement: Not applicable.

Data Availability Statement: The authors confirm that the data supporting the findings of this study are available within the article.

Acknowledgments: The authors sincerely appreciate the funding provided by the National Natural Science Foundation of China (Grant No. 51278091).

Conflicts of Interest: The authors declare that they have no known competing financial interests or personal relationships that could have influenced the work reported in this study.

References

1. Roy, S.; Kundu, C.K. State of the art review of wind induced vibration and its control on transmission towers. *Structures* **2021**, *29*, 254–264. [CrossRef]
2. Deng, H.Z.; Si, R.J.; Hu, X.Y. Wind Tunnel Study on Wind-Induced Vibration Responses of a UHV Transmission Tower-Line System. *Adv. Struct. Eng.* **2013**, *16*, 1175–1185. [CrossRef]
3. Huang, M.F.; Lou, W.J.; Yang, L. Experimental and computational simulation for wind effects on the Zhoushan transmission towers. *Struct. Infrastruct. E.* **2012**, *8*, 781–799. [CrossRef]
4. Prasad Rao, N.; Kalyanaraman, V. Non-linear behaviour of lattice panel of angle towers. *J. Constr. Steel Res.* **2001**, *57*, 1337–1357. [CrossRef]
5. Zheng, H.D.; Fan, J. Progressive collapse analysis of a truss transmission tower-line system subjected to downburst loading. *J. Constr. Steel Res.* **2022**, *188*, 107044. [CrossRef]
6. Zhang, J.; Xie, Q. Failure analysis of transmission tower subjected to strong wind load. *J. Constr. Steel Res.* **2019**, *160*, 271–279. [CrossRef]
7. Yasui, H.; Marukawa, H.; Momomura, Y. Analytical study on wind-induced vibration of power transmission towers. *J. Wind Eng. Ind. Aerod.* **1999**, *83*, 431–441. [CrossRef]
8. Battista, R.C.; Rodrigues, R.S.; Pfeil, M.S. Dynamic behavior and stability of transmission line towers under wind forces. *J. Wind Eng. Ind. Aerod.* **2003**, *91*, 1051–1067. [CrossRef]
9. Yazdani, M.; Jahangiri, V.; Marefat, M.S. Seismic performance assessment of plain concrete arch bridges under near-field earthquakes using incremental dynamic analysis. *Eng. Fail. Anal.* **2019**, *106*, 104170. [CrossRef]
10. Moazam, A.M.; Hasani, N.; Yazdani, M. Incremental dynamic analysis of small to medium spans plain concrete arch bridges. *Eng. Fail. Anal.* **2018**, *91*, 12–27. [CrossRef]
11. Chen, D.H.; Yang, Z.H.; Wang, M. Seismic performance and failure modes of the Jin'anqiao concrete gravity dam based on incremental dynamic analysis. *Eng. Fail. Anal.* **2019**, *100*, 227–244. [CrossRef]
12. Dolsek, M. Incremental dynamic analysis with consideration of modeling uncertainties. *Earthq. Eng. Struct. D* **2009**, *38*, 805–825. [CrossRef]
13. Tian, L.; Zhang, X.; Fu, X. Fragility analysis of a long-span transmission tower-line system under wind loads. *Adv. Struct. Eng.* **2020**, *23*, 2110–2120. [CrossRef]
14. Pan, H.Y.; Tian, L.; Fu, X. Sensitivities of the seismic response and fragility estimate of a transmission tower to structural and ground motion uncertainties. *J. Constr. Steel Res.* **2020**, *167*, 105941. [CrossRef]
15. Lei, X.; Fu, X.; Xiao, K. Failure Analysis of a Transmission Tower Subjected to Wind Load Using Uncertainty Method. *Proc. CSEE* **2018**, *38*, 266–274. [CrossRef]
16. Fu, X.; Wang, J.; Li, H.N. Full-scale test and its numerical simulation of a transmission tower under extreme wind loads. *J. Wind Eng. Ind. Aerod.* **2019**, *190*, 119–133. [CrossRef]

17. Fu, X.; Li, H.N.; Li, G. Fragility analysis of a transmission tower under combined wind and rain loads. *J. Wind Eng. Ind. Aerod.* **2020**, *199*, 104098. [CrossRef]
18. Fu, X.; Li, H.N.; Li, G. Fragility analysis and estimation of collapse status for transmission tower subjected to wind and rain loads. *Struct. Saf.* **2016**, *58*, 1–10. [CrossRef]
19. Kenny, E.D.; Paredes, R.; Lacerda, L. Artificial neural network corrosion modeling for metals in an equatorial climate. *Corros. Sci.* **2009**, *51*, 2266–2278. [CrossRef]
20. Kong, Z.Y.; Jin, Y.; Sabbir, H. Experimental and theoretical study on mechanical properties of mild steel after corrosion. *Ocean Eng.* **2022**, *246*, 110652. [CrossRef]
21. Jia, C.; Shao, Y.S.; Guo, L.H. Incipient corrosion behavior and mechanical properties of low-alloy steel in simulated industrial atmosphere. *Constr. Build. Mater.* **2018**, *187*, 1242–1252. [CrossRef]
22. Mccuen, R.H.; Albrecht, P. Effect of Alloy Composition on Atmospheric Corrosion of Weathering Steel. *J. Mater. Civil Eng.* **2005**, *17*, 117–125. [CrossRef]
23. Wang, Z.F.; Liu, J.R.; Wu, L.X. Study of the corrosion behavior of weathering steels in atmospheric environments. *Corros. Sci.* **2013**, *67*, 1–10. [CrossRef]
24. Feliu, S.; Morcillo, M. The prediction of atmospheric corrosion from meteorological and pollution parameters —I. Annual corrosion. *Corros. Sci.* **1993**, *34*, 415–422. [CrossRef]
25. Zhi, Y.J.; Fu, D.M.; Yang, T. Long-term prediction on atmospheric corrosion data series of carbon steel in China based on NGBM (1,1) model and genetic algorithm. *Anti. Corros. Method. Mater.* **2019**, *66*, 403–411. [CrossRef]
26. Song, X.X.; Wang, K.Y.; Zhou, L. Multi-factor mining and corrosion rate prediction model construction of carbon steel under dynamic atmospheric corrosion environment. *Eng. Fail. Anal.* **2022**, *134*, 105987. [CrossRef]
27. Mohammed, A.; Amer, M.; Ahmad, A. Predicting the Corrosion Rate of Medium Carbon Steel Using Artificial Neural Networks. *Prot. Met. Phys. Chem.* **2022**, *58*, 414–421. [CrossRef]
28. Li, J.Y.; Men, C.; Qi, J.F. Impact factor analysis, prediction, and mapping of soil corrosion of carbon steel across China based on MIV-BP artificial neural network and GIS. *J. Soil. Sediment.* **2020**, *20*, 3204–3216. [CrossRef]
29. Cai, J.P.; Cottis, R.A.; Lyon, S.B. Phenomenological modelling of atmospheric corrosion using an artificial neural network. *Corros. Sci.* **1999**, *41*, 2001–2030. [CrossRef]
30. Halama, M.; Kreislova, K.; Van Lysebettens, J. Prediction of Atmospheric Corrosion of Carbon Steel Using Artificial Neural Network Model in Local Geographical Regions. *Corrosion-US* **2011**, *67*, 065004. [CrossRef]
31. National Standards of the People's Republic of China. *Unified Standard for Reliability Design of Building Structures GB50068-2018*; China Architecture & Building Press: Beijing, China, 2018.
32. Chen, G.X.; Li, J.H. Statistical Parameters of material strength and geometric properties of shapes for steel members. *J. Chongqing Jianzhu Univ.* **1985**, *1*, 1–23.
33. Alembagheri, M.; Seyedkazemi, M. Seismic performance sensitivity and uncertainty analysis of gravity dams. *Earthq. Eng. Struct. Dyn.* **2015**, *44*, 41–58. [CrossRef]
34. Xu, Z.; Zhang, T.; Ge, X.D. Uncertainty analysis of transmission tower-line system under wind load. *J. Shandong Univ.* **2021**, *51*, 99–105.
35. Power Industry Standards of the People's Republic of China. *Load Code for the Design of Overhead Transmission Line DL/T 5551-2018*; China Planning Press: Beijing, China, 2018.
36. National Materials Corrosion and Protection Data Center. Available online: https://www.corrdata.org.cn/index.php (accessed on 2 March 2022).
37. Wang, R.B.; Xu, H.Y.; Li, B. Research on Method of Determining Hidden Layer Nodes in BP Neural Network. *Comput. Technol. Dev.* **2018**, *28*, 31–35.
38. Chen, Y.; Qiang, C.M.; Wang, G.G. Corrosion and Protection of Transmission Towers. *Elec. Power Constr.* **2010**, *31*, 55–58.

 *buildings*

Article

Numerical Strategy for Column Strengthened with FRCM/SRG System

Salvatore Verre

Department of Civil Engineering, University of Calabria, Via P. Bucci Cubo 39B, Arcavacata di Rende, 87036 Cosenza, Italy; salvatore.verre@unical.it; Tel.: +39-0984494055

Abstract: The use of fabric-reinforced cementitious mortar (FRCM) or steel-reinforced grout (SRG) is now recognized to be effective in enhancing the axial capacity of masonry columns when confinement is achieved. Numerous experimental tests demonstrated the symbiotic role of the fabric and the inorganic matrix. An open issue is still related to the numerical simulation. In fact, if the compressive behavior by the numerical simulation of the unreinforced and reinforced masonry columns confined by a FRCM/SRG jacket may follow different approaches. The inorganic matrix transfers the stresses from the substrate to the fabric differently, depending on the presence or absence of cracks. The fabric consists of an open grid whose yard could be differently stressed after the matrix damage because of the occurrence of a possible slippage at the fabric–matrix interface. Definitely, these aspects are difficult to numerically predict. The paper herein is devoted to the assessment of different numerical approaches for the FRCM/SRG confinement of masonry columns by considering data from the literature and varying the parameters related to the matrix, the fabric, and the masonry itself. The goal is to best fit the experimental outcomes (from different available sources) with different strategies based on a finite element (*FE*) modeling. The results show good matching between the experimental and theoretical curves for the different FRCM/SRG systems. The results evidenced that the accuracy of the experimental versus the numerical curves match is met for the different FRCM/SRG systems.

Keywords: FRCM systems; SRG systems; masonry columns; numerical modeling

Citation: Verre, S. Numerical Strategy for Column Strengthened with FRCM/SRG System. *Buildings* **2022**, *12*, 2187. https://doi.org/10.3390/buildings12122187

Academic Editors: Dongming Li and Zechuan Yu

Received: 29 October 2022
Accepted: 25 November 2022
Published: 9 December 2022

Publisher's Note: MDPI stays neutral with regard to jurisdictional claims in published maps and institutional affiliations.

1. Introduction

An important part of existing structures is mainly made out of masonry. Indeed, most historical buildings consist of monumental masonry structures (churches, temples, towers, etc.), as well as most ordinary buildings. There are substantial differences that exist between monumental and ordinary buildings in terms of geometry and structural details. Various strategies for the prediction and the assessment of the structural behavior of masonry buildings by a numerical model have been developed in recent decades. Numerical models have been favorably developed and preferred over analytical approaches, given the complex mechanical response of masonry and the irregular geometries of historic masonry buildings. The numerical strategies are subdivided into four classes: block-based models (BBM), continuum models (CM), macro-element models (MM), and geometry-based models (GBM). The strategy of *BBM* models is based on masonry heterogeneity by the assembly of the blocks with mortar joints. Through this strategy, the real texture of the masonry structure can be described, while the individual mechanical properties can be evaluated through experimental tests on small-scale samples. In addition, by means of models, it is possible to simultaneously describe both the out-of-plane and in-plane behavior of masonry walls [1,2]. Interaction is the fundamental part of the *BBM* strategy; it depends on the type of interaction. In the technical literature, there are five categories: (i) interface element-based approaches [1–4], (ii) contact-based approaches [5,6], (iii) textured continuum-based approaches [7–9], (iv) block-based limit analysis approaches [10–13], and (v) extended finite element (XFEM) approaches [14,15]. The drawbacks of this strategy are the massive computational burden of solving the numerical model, and the time dedicated to modeling the block and assembling it [16–18]. The strategy of the continuum model (CM) is based on the use of a deformable continuous body. In particular, the mesh does not always have to respect the real texture but can be much bigger. Through this characteristic, the computational burden is less than that of *BBM* models. However, the assumption of an appropriate constitutive law that is due to the properties of the masonry itself can be calibrated through two strategies: direct approach and homogenization procedures and multiscale approaches. The strategies calibrate the constitutive laws directly on the results of experimental tests. There are two types of direct approaches in the literature. With the introduction of FE and the use of the complementary energy theorem, it was possible to obtain a solution via of minimization of a quadratic function with equality and inequality constraints [19]. Other continuum-directed approaches base their nonlinear constitutive laws on theories of fracture or damage mechanics and/or plasticity. The use of these models has been shown to be favorable for assessing the structural performance of historic monumental masonry buildings, particularly because of the limited computational demands of these models and their ease in representing complex geometries [20–22], churches and temples [23–25], palaces [26–28], and bridges [29,30]. The second strategy introduces a homogenization strategy that connects the structural scale model to a scale model of the material and its heterogeneities. Homogenization procedures are generally based on accurate modeling strategies of an *RVE* (representative volume element). In particular, the *RVE* must statically represent the heterogeneity of the masonry under study. Three models based on this strategy can be identified in turn in the technical literature: (i) a priori homogenization approaches, (ii) step-by-step multiscale approaches, and (iii) adaptive multiscale approaches. The first model uses the RVE technique to initially define the homogenized material properties and then use them in the structural-scale model, while using homogenized properties [31–35]. In the second model, the structural response is evaluated for each point in the structural model of a boundary value problem on the *RVE* [36–39]. Finally, in the adaptive multiscale approaches one uses, through adaptability, the material-scale model in the structural-scale one [40–42]. The MM strategy generalizes the structure with panel-scale behavior—in other words, as a macro-element [43]. The two main elements are pier (vertical elements) and spandrel, which are horizontal portions of the structure between two openings aligned along the height. The global behavior of the structure under seismic action depends on the panel response and, consequently, on the load redistribution given by the diaphragms [44]. In fact, the MM strategy does not predict out-of-plane ruptures and would lead to an overestimation of the capacity of the structure [2]. Moreover, in severely irregular structures, subdivision appears complicated and, in some cases, impossible. The advantages are easy discretization of the model and definition of mechanical properties. This strategy is employed through two approaches: equivalent beam and spring based. The first approach is based on modeling the masonry structure by schematization in equivalent frame models. A first model, called the POR (pushover response) method, is based on the simplified elasto-plastic relationships to describe the beam nonlinearity connected by rigid links [45–47]. Recently, an advanced equivalent beam-based macro-element was developed in [48] for the nonlinear static and dynamic simulation of masonry structures. The beam mechanical description conceived axial, bending, and shear deformation within the Timoshenko beam theory. The second approach is based on nonlinear springs within an equivalent frame to simulate the in-plane nonlinear behavior of masonry walls [49]. In [50], the authors developed a spring-based approach where piers and spandrels are conceived as equivalent arrangements of nonlinear springs, while in [51], each masonry facade was conceived as an integral unit, instead of one of piers and spandrels. As a result, masonry behavior is described by two vertical springs and one horizontal spring for the shear behavior of the wall. Finally, the last strategy (GBM) conceives the structure as a rigid body, the only input being the geometry of the structure and the loading conditions. Through this approach, it is possible to analyze structural equilibrium or evaluation of a possible collapse through static or film theorems, both based on limit analysis. There are several approaches based on the static theorem in the technical literature. For example, thrust-network analysis (TNA), reported in [52,53], is based on the duality between geometry and in-plane forces in networks, and plausible funicular solutions under gravitational loading within a defined envelope are studied. A further approach to thrust networks was proposed in [54], where the equilibrium of masonry vaults was analyzed using polyhedral stress functions, while the discrete singular stress network is calculated based on Airy's stress formulation [55]. Solutions based on kinematic approaches are based on discretization into rigid blocks according to the collapse mechanisms observed in earthquake-affected structures. Various strategies have been proposed, including [56] a discontinuous upper boundary analysis tool with sequential linear programming and mesh fitting to investigate the actual collapse mechanisms of double-curved masonry structures and a recently developed tool based on genetic algorithms for upper boundary analysis of masonry vaults [57]. Application of the modeling approaches described above normally each extend to an experimental case. Often in a strengthening or rehabilitation action for masonry buildings, it is necessary to improve the capacity in plane (under compression load). Next-generation systems used for reinforcing masonry structures are composed of an inorganic matrix combined with an ultra-strong fiber fabric. In this so-composed system, the fibers have the task to bear the tensile load while the matrix has the task of protecting and transferring the compression load between the masonry substrate and the fiber. The types of inorganic matrices could be cement-based or lime-based. In the technical literature, there are many acronyms attributable to the new generation of reinforced systems based on the type of fiber used in the reinforcement system: basalt B-FRCM [58–62], poliparafenilenbenzobisoxazole [63,64] PBO-FRCM, glass G-FRCM [61,65,66], steel SRG [59,60,62,66–68], and carbon C-FRCM [69–73]. It was also observed that, depending on the reinforcement system adopted or better depending on the fiber used, different modes of failure were observed. The most recurrent ones are the opening of the reinforcement jacket near the overlap zone or the rupture of the fabric, always in the overlap zone. The purpose of this paper is to show several different types of modeling on unreinforced masonry (URM) under compressive loads while providing a general and different approach of modeling reinforced masonry with the FRCM or SRG system that combines versality and moderate computational cost.

2. Numerical Model

2.1. Modeling of Unreinforced Masonry (URM)

The performance of the numerical model depends on the experimental description of the heterogenous behavior of the masonry structure through the adequate constitutive laws. The type of experimental method used for evaluating the material properties and the boundary conditions are of critical importance for the numerical model and results. However, particular attention should also be given to the geometrical aspects of the masonry structure: unit dimensions, type and quality of mortar joint, and unit surface conditions. The variety of clay brick units and mortar types and typology methods of construction do not allow development of unified constitutive laws. In fact, several codes and standards prioritize experimental characterizations of material properties for design and numerical simulations [74,75]. The set of mechanical properties used for the masonry numerical model depends both on the accurate description (elastic and inelastic) of material and on the adopted modeling approach. Generally, the elastic range can be defined through the modulus of elasticity and the compressive strength. When a description of cracking is necessary, the nonlinearity effects are necessary through additional mechanical properties such as shear strength, tensile strength, and fracture energies. The extent of knowledge required on material properties depends on the modeling strategy, where in technical literature [76] there are three different approaches aimed at the modeling of a masonry column: (i) macro-modeling, (ii) micro-modeling, and (iii) simplified micro-modeling. In Figure 1, the schematic models used for a small-scale column are reported.

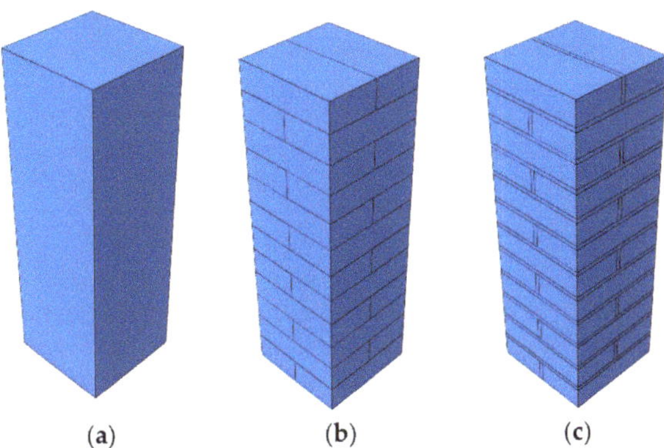

Figure 1. Numerical approaches: (**a**) macro; (**b**) simplified micro; (**c**) micro.

2.1.1. Macro-Modeling (MA-Approach)

This is the simplest of the three approaches, where the single brick and the mortar joint merge into a unique element. This approach does not take into account the interaction between the two constituents of the column. In the literature, there exists two different strategies where the internal structure of masonry cannot be described explicitly and the damage within the masonry is evaluated through a continuous medium. The first strategy most widely used is considered as a homogenous and anisotropic material using plasticity or another macro-scale constitutive relationship. The main advantage consists of the reduced computational burden (time-consuming). Moreover, this method is the most used by researchers; in fact, in the literature there are different studies that take into account different strategies based on this technique. In [49,77–81], the research developed a shear wall or masonry column such as a continuous element where the mortar unit of clay brick was described through an average function of the mechanical properties of the single parts. The second strategy is based on the reduction in the elasticity modulus by increasing the load in the model, which represents the propagation of cracks in the elements. This strategy is generally used in reinforced concrete.

2.1.2. Micro-Modeling (MI-Approach)

This approach is more realistic, and it takes into account the behavior of the single constituents of the column such as continuous and discontinuous elements. In fact, it is also known as the heterogeneous approach, while in terms of time it is uneconomic and inefficient because of the many parameters and various interactions between the individual parts. It introduces two important steps to create a numerical model, which are the local and global steps of the unreinforced masonry. The local view refers to the single details in terms of geometrical and mechanical characteristics, respectively, while the global view refers to the definition of contact surface (plane) and interaction (Mohr–Coulomb [82,83] law) among all the parts. Therefore, this approach is not applicable if applied on a realistic scale of masonry structures. Moreover, to avoid this drawback, in the literature there are two strategies. Both strategies focus their attention on how to represent and how to model the mortar joint. The first strategy considers the masonry made only of bricks while modeling the vertical and horizontal mortar joints as an interaction surface [84,85] with a zero thickness. The merging of the clay brick and the mortar joint (as interaction surface) is called the unit to represent the continuum elements. Important research conducted in [84] followed this approach, where the mortar joints (vertical and horizontal) are modeled through the elastic–plastic behavior of the interface. The obtained numerical curves fit the experimental results satisfactorily. The second approach consists of representing the mortar joint with its real geometric thickness [85], thus modeling the entire masonry 3D structure, including both the brick and the mortar joints. In particular, the modeling of mortar joints requires particular attention, such as isolating a part of the structure for particular boundary conditions [85] or inhibiting any interaction with the external reinforcement [33]. Furthermore, in terms of mortar strength, there is a reduction and the introduction of penalties at the interaction surface in terms of tangential behavior. In addition, this approach with respect to the first one allows the occurrence of cracks in the mortar joints and in the brick unit.

2.1.3. Simplified Micro-Modeling (SIMI-Approach)

This approach is an intermediate one with respect to macro- and micro-modeling, which is generally used in real-scale structures. In the literature, there are two strategies for using this modeling. The first strategy consists of those mortar joints that are clamped into the unit/mortar interface as a discontinuous element. Expanded units, up to half of the mortar thickness in vertical and horizontal directions, were simulated by continuum elements as reported in [13,86–88]. In this approach, the mortar joints are considered as the weakest elements and modeled by an elastic–plastic interface behavior. The obtained results showed that the strategy was able to reproduce the experimental response and evaluate the cracks inside the expand unit. The second strategy, called the homogenization approach in the literature, is reported in [56,89]. It is made up of periodic units. Through the periodic units, it is possible to model heterogeneous masonry structures, reducing the number of material parameters and by avoiding independent modeling of all mortar joints. The use of these strategies permits the modeling of masonry structures, reducing the computational burden and, at the same time, the material parameters in input. The obtained results showed that the strategy was able to reproduce the experimental response and evaluate the cracks.

2.2. Constitutive Laws of: Unreinforced Masonry (URM), Clay Brick and Mortar Joints

A largely adopted constitutive law is based on Feenstra [90]. It considers a three-branch behavior in compression. For it, the macro and simplified micro approaches were used. The geometrical and the mechanical parameters that govern the model are compressive strength (f_{cm}), elastic modulus (E_{cm}), fracture energy (G_{cm}), and mesh size (h).

$$
f_c = \begin{cases}
-f_{cm}\dfrac{1}{3}\dfrac{\varepsilon_j}{\varepsilon_{\frac{c}{3}}} & \varepsilon_{\frac{c}{3}} < \varepsilon_j \leq 0 \\[2ex]
-f_{cm}\dfrac{1}{3}\left(1 + 4\left(\dfrac{\varepsilon_j - \varepsilon_{\frac{c}{3}}}{\varepsilon_c - \varepsilon_{\frac{c}{3}}}\right) - 2\left(\dfrac{\varepsilon_j - \varepsilon_{\frac{c}{3}}}{\varepsilon_c - \varepsilon_{\frac{c}{3}}}\right)^2\right) & \varepsilon_c \leq \varepsilon_j \leq \varepsilon_{\frac{c}{3}} \\[2ex]
-f_{cm}\left(1 - \left(\dfrac{\varepsilon_j - \varepsilon_c}{\varepsilon_u - \varepsilon_c}\right)^2\right) & \varepsilon_{cu} < \varepsilon_j \leq \varepsilon_c \\[2ex]
0 & \varepsilon_j < 0
\end{cases}
\tag{1}
$$

where, particularly, the first branch is assumed linear, while the other two, after the elastic range and the peak stress, are nonlinear. The behavior is defined through three characteristic strain values. The first strain $\varepsilon_{c/3}$ where the linear branch ending is expressed:

$$
\varepsilon_{\frac{c}{3}} = -\frac{1}{3}\frac{f_{cm}}{E_{cm}}
\tag{2}
$$

while the relative strain at the peak stress is expressed as:

$$
\varepsilon_c = -\frac{5}{3}\frac{f_{cm}}{E_{cm}}
\tag{3}
$$

and the ultimate strain where the *URM* has terminated the softening compression is expressed as:

$$
\varepsilon_{cu} = \varepsilon_c - \frac{3}{2}\frac{G_{cm}}{h\,f_{cm}}
\tag{4}
$$

The ultimate strain depends on two parameters; fracture energy (G_{cm}) and the characteristic element length (h). Fracture energy (G_{cm}) was computed as being equal to the area under the softening third branch (Equation (1)), while the h is evaluated as the cubic root of the masonry column volume.

$$G_{cm} = h \int_{\varepsilon_c}^{\varepsilon_{cu}} \sigma(\varepsilon_c) d\varepsilon_c \tag{5}$$

$$h = \sqrt[3]{Volume\ of\ the\ masonty\ column} \tag{6}$$

Other researchers used a modified constitutive law based on that of reinforced concrete [91], described by nonlinear equations. The simplicity of this last approach is due to the parameters involved, i.e., the stress (f_{cm}) and the relative strain (ε_c) at peak. The constitutive tensile law was modeled according to the following branches. The first branch is linear elastic until the peak tensile stress (f_{ct}), while the second branch is expressed by:

$$f_t = f_{ct}\, e^{-\frac{\varepsilon_t}{\varepsilon_{tu}}} \tag{7}$$

where ε_t is the crack strain and ε_{tu} is the ultimate crack strain. The softening branch also depends on the fracture energy (G_{fm}) and the characteristic element length. The G_{fm} is evaluated according to:

$$G_{fm} = h \int_{\varepsilon_t=0}^{\varepsilon_{tu}=\infty} \sigma(\varepsilon_t) d\varepsilon_t \tag{8}$$

while h is evaluated according to Equation (6). Therefore, the ultimate crack strain is:

$$\varepsilon_{tu} = \frac{G_{fm}}{f_{ct}\, h} \tag{9}$$

However, the post-peak branch is exponential and, to avoid snap-back phenomena, the parameter h is evaluated according to the expression:

$$h \leq \frac{G_{fm}\, E_{tm}}{f_{ct}^2} \tag{10}$$

where E_{tm} is the initial tangent Young's modulus. The constitutive law in tension and in compression were reported in Figure 2, which is possible to use by the internal functions of the commercial software adopted [92], called Elastic (E) and Concrete Damage Plasticity (CDP).

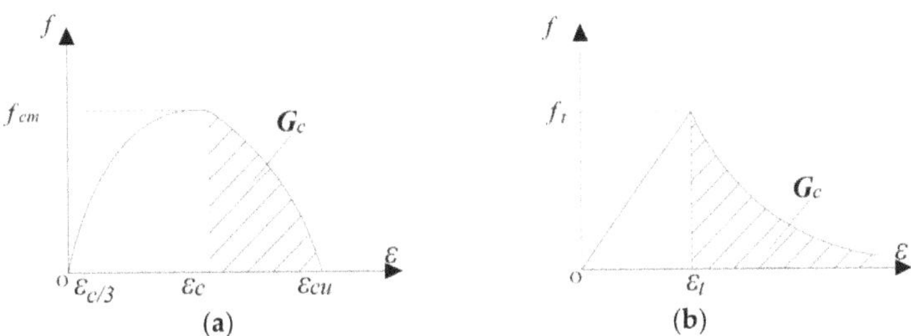

Figure 2. Material constitutive law: of masonry in (**a**) compression and (**b**) tension.

The elastic branch of the constitutive law (compression and tension) was modeled in E by parameters such as density (ρ), both elastic modulus in compression and in tension (E_{cm} and E_{ct}), the Poisson ratio (v), and the maximum stress (f), which were kept constant during the analysis. The nonlinear branches, in terms of two main failure mechanisms, which are compressive crushing and tensile cracking, were modeled through the CDP. The evolution of the yield surface is controlled by two hardening variables called equivalent plastic strains in compression and in tension. Consequently, in the CDP model, one must necessarily describe the behavior in uniaxial terms outside the elastic branch. Moreover, the CDP entails the nonlinear branch through the following parameters, namely, the yield stress damage parameter and the relative strain inelastic and cracking for the masonry compression and tension damage, respectively. The CDP function, as understood by its name, is a concrete material function, but by means of the following parameters it was possible to adopt this function for the masonry column as reported in [77,93–96] and for other quasi-brittle materials:

- Dilation angle (DA): 30° angle measured in the meridional plane between the failure surface and the hydrostatic axis;
- Plastic potential eccentricity (PPE): 0.1 due to a non-associated potential plastic flow and it is a length's segment between the vertex of the hyperbola and the asymptotes with respect to the center of the hyperbola;
- Ratio between the initial biaxial and yield compressive stress: 1.16;
- Viscosity parameter (VP): 0.0 visco-plastic regularization.

In the SIMI approach, the periodic unit was modeled in compression using Equations (1)–(6). Through the micro approach, it was possible to independently describe both the single clay brick and the mortar joints. In particular, the clay brick was modeled by linear elasticity until failure by function E through the mechanical parameters (f_{brick} and E_{brick}), generally evaluated by [95,96], while the mortar joints were modeled in both compressive and tensile behavior, through function CDP. The compressive constitutive law was modeled with a nonlinear model suggested in [77,92,93,97–99], while the equations are reported by the following:

$$f_c = f_{cmat} \left[2\left(\frac{\varepsilon_j}{\varepsilon_{cmat}} \right) - \left(\frac{\varepsilon_j}{\varepsilon_{cmat}} \right)^2 \right] \tag{11}$$

where f_{cmat} is the peak compression stress and the relative strain. The strain is evaluated by Equation (12) and the Young's modulus (E_{cmat}):

$$\varepsilon_{cmat} = 2 \, \frac{0.85 \, f_{cmat}}{E_{cmat}} \tag{12}$$

In addition, the tensile constitutive law was modeled with a bilinear model [77,92,93,97–99] with a μ factor equal to 25.

$$f_t = \begin{cases} E_{tmat} \, \varepsilon_{tmat} & \varepsilon_j \leq \varepsilon_{tmat} \\ f_{tmat} - \frac{f_{tmat}}{\mu \, \varepsilon_{tmat}} \left(\varepsilon_j - \varepsilon_{tmat} \right) & \varepsilon_{tmat}(1 - \mu) \geq \varepsilon_j \geq \varepsilon_{tmat} \end{cases} \tag{13}$$

In both micro- and simplified micro-modeling strategies, an important step is the definition of contact surfaces and the type of interaction. The contact surfaces between expansion cells by a standard contact including surface-to-surface and self-contact was used to avoid penetration among them. The surfaces were assumed to be zero thickness; therefore, hard contact for normal behavior of contact was assigned. Hard contact refers to an interaction without any softening to avoid no penetration of the surfaces, which can occur in the model. Another mechanical characteristic was assigned in terms of tangential behavior, called the friction coefficient. Generally, the most common friction coefficient of concrete/masonry, which is set equal to 0.67, is reported in [100]. Meanwhile, in the micro-modeling, all nodes are degree-of-freedom and the bond slip is not considered between brick and mortar. In other words, the perfect bond between the vertical and horizontal mortar joints and between single clay bricks and the mortar joints were considered. This strategy, because of the not easy estimation of bond parameters and the related slip law between the single clay brick and the mortar joints [83], is hard to apply. In all of the three approaches, it is possible to introduce the cracking phenomenon to simulate the experimental behavior of the masonry or concrete structures being tested. This phenomenon is introduced through the nonlinear behavior of the material in compression and in tension, as reported in these studies [77,78,80,94]. The results furnished are in good agreement in the comparison between the numerical model and the experimental one. Moreover, the authors emphasize that the efficiency is related to the accurate definition of the equations that describe the behavior of the unreinforced column and the nonlinear solution strategy.

2.3. FRCM–SRG (Macro and Micro-Approach) and Interface Modeling

The numerical simulation of the *FRCM* or *SRG* system for the confinement of masonry columns is poorly investigated in the literature. All the available studies [58–73,77] carried out by the authors are devoted, as their main goal, to the investigation of the structural or the debonding problem and to model the strengthened systems on the basis of the experimental results that refer to the specific type of FRCM or SRG systems. A clear distinction within the FRCM systems is crucial for proper selection of the numerical strategies. The acronym FRCM includes all of the several types of fibers, namely metallic and nonmetallic ones, their different behavior should be underlined. All nonmetallic reinforcements are characterized by a negligible stiffness, except under tensile stresses in the fiber's direction; contrarily, the metallic reinforcement, known also as SRG when combined with the mortar matrix, presents a significant stiffness under a different stress state. In the case of confinement, this specific feature often involves a failure by the opening of the jacket after the matrix damage, instead of fibers breaking [68,77]. For the above reason, the two kinds of composites (metallic and nonmetallic) are herein modeled with different strategies. The presence of the metallic reinforcements provides evidence of only one failure type that was possible to observe in the confinement action, independent of the matrix (cement or lime-based) used [67,68]. Consequently, the matrix was not physically modeled, but the low influences were considered in the mechanical values of system characterization by tensile tests on *SRG* specimens made up of fiber and matrix [68]. The above-stated and the effect of the matrix in the FRCM system, based on the experimental observation, has a crucial role for the initial tensile stiffness of the composite, while the fabric mainly affects the strength and the post-cracking stiffness. In the relatively few works present in the technical literature [77,93,101,102] the matrix was excluded from the numerical model. The performance and effects of the matrix were considered in this model by assessing the mortar cracking. Generally, the ductility is related to the damage evolution depending on both the matrix and the fabric and on their interactions. In fact, the nonlinear models for the matrix and the fabric and matrix modeled separately were introduced. The behavior of external reinforcement was described by two approaches: macro approach (MA) and micro approach (MI). The first approach (MA) was used for the SRG system, where the strengthened system is modeled without distinguishing the matrix (lime- or cement-based) and the fabric by an element shell. This approach is made possible by similar values between the mechanical value of the dry fibers such as steel fibers (with different steel density) and the mechanical values obtained by tensile tests on the *SRG* specimen (steel fibers and matrix), while the second approach (MI) was used for the *FRCM* system, where the external reinforced (matrix) was modeled with real thickness (t_{fm}) and the fabric was modeled with the equivalent thickness (t_f) for both reinforcement systems. The compressive constitutive law of the matrix was modeled with a nonlinear model (see Figure 3), while the tensile constitutive law was modeled with a bilinear model. The equations used are (11)–(13). The behavior in compression and in tensile was described by the *CDP* function. A similar modeling technique was used in [75–91], where the external *FRCM* reinforcement was used to strengthen the unreinforced masonry columns. In both approaches for the two external reinforcement systems, the fabric was linear-elastic-until-failure modeled (see Figure 3) and was described by the internal function E.

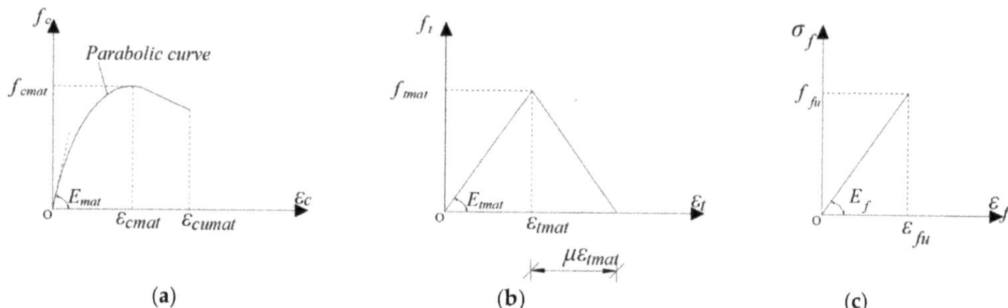

(a)   (b)   (c)

Figure 3. Stress–strain relationship: (**a**) compression, (**b**) tension for the mortar, and (**c**) linear elastic for fabric.

Moreover, even if the adhesion of the composite with the masonry substrate is generally not relevant in the case of confinement, as demonstrated for the case of FRP [103], a perfect bond was considered in the proposed numerical simulations to make the computation easier and more robust. The same assumption was also imposed for the bond between the matrix and the fabric interaction for the FRCM system. On the basis of the failure mode [58–73], in particular in the overlap zone, the presence of a greater quantity of fibers was considered through the equivalent thickness parameter (t_f and t_{mat}), modifying it appropriately.

The interaction between the steel fabric and the overlap layer was used as an interface cohesive surface and a different interaction between the masonry substrate and the external reinforcement was considered (see Figure 4). The bond slip law adopted is reported in [104]; in particular, it is a bilinear model. The bond slip law was evaluated by statistical studies and a meso-scale finite element model on a large database on the single-lap direct shear test. The latter one was used to evaluate the initial stiffness of the bond slip curve. This bilinear model is illustrated in Figure 5.

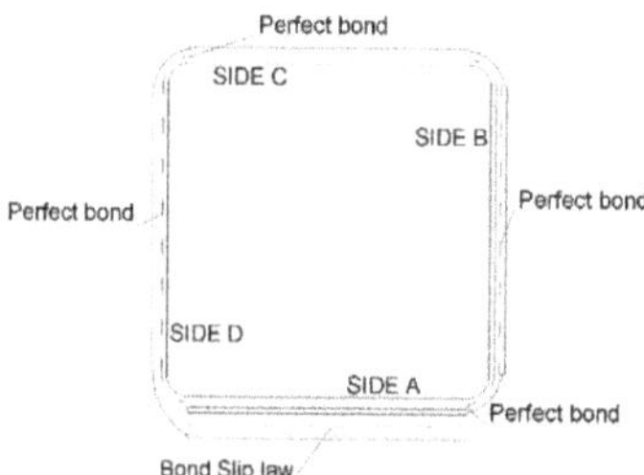

Figure 4. Scheme of interaction.

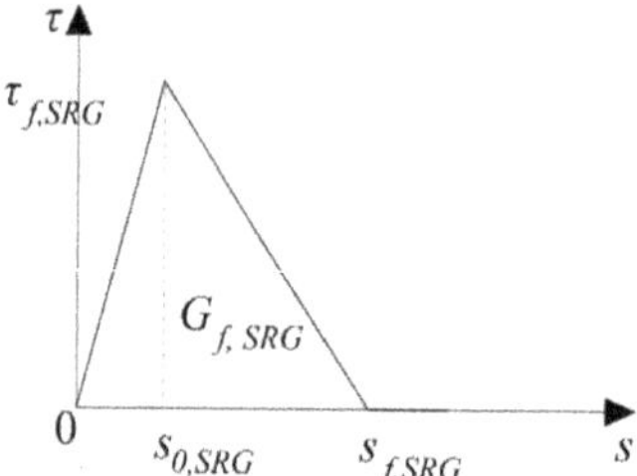

Figure 5. Interface modeling.

To evaluate parameter k_0, many experimental results are needed. This is due to the limited available test results relative to the considered confining systems. Consequently, the value of $k_{0,SRG}$ was assumed unchanging with respect to that reported in [104]. The parameter $G_{f,SRG}$ was calculated as:

$$G_{f,SRG} = \int_0^{s_{f,SRG}} \tau_f\, ds \tag{14}$$

Finally, the parameters $s_{0,SRG}$ and sf, SRG are evaluated by the following equations:

$$s_{0,SRG} = 0.0195\, \beta_w\, f_{tmat} \tag{15}$$

$$s_{f,SRG} = \frac{2\, G_{f,SRG}}{\tau_{f,SRG}} \tag{16}$$

$$\tau_{f,SRG} = 1.5\, \beta_w\, f_{tmat} \tag{17}$$

$$G_{f,SRG} = 0.308\, \beta_w{}^2\, \sqrt{f_{tmat}} \tag{18}$$

and they depend on the geometric parameter β_w, the fractures energy ($G_{f,SRG}$), and tensile strength of the mortar (f_{tmat}). The bond slip law adopted depends on the three failure modes that are due to the opening and sliding associated with the normal and shear stress, respectively. The initialization and evaluation of the damage were evaluated by the quadratic function reported in a previous numerical work on the columns strengthened with an *SRG* system [77,92].

2.4. Geometry, Boundary Conditions, and Solution Technique

The all-masonry columns were modeled in three dimensions (3D) through the three types of modeling strategies described in Section 2. The element used to model the masonry column, single clay brick, mortar joints, and external matrix of FRCM system is the linear tetrahedral four node C3D4 element with constant stress. The FE element used in the MA for the SRG system and fibers of the *FRCM* system is a two-dimensional shell element called S4R. This element is used to model mono-dimensional structures with small thickness. The equivalent thickness of the fibers (t_f) adopted for the *S4R* element is equal to the values of the fabric mesh considered, while the matrix thickness (t_{mat}) is equal to the matrix layer adopted in the tests. Moreover, the masonry column equipped with externally reinforced corners was rounded to avoid stress concentration. All numerical tests in displacement control were conducted through the enforcement of a displacement $-\lambda u$ along the y-axis. The displacement on the entire surface at the top column was applied, while all nodes on the surface of the column bottom (translations and rotations) were blocked. To solve the nonlinear equations associated to the numerical problem focused on in this work, a dynamic approach was used. Generally, this approach is not used to solve a quasi-static problem because of the parameters involved. The first users of this technique were Chen et al. [105]; they suggested paying particular attention to two parameters, providing their values to obtain only the static solution. The first parameter is the variable mass scaling and the value used is equal to 0.00005. The role of this parameter is to scale the mass of all (or single macro-element) the elements at the beginning of a step and periodically during the displacement phase. The second parameter is the ratio between the kinetic and total energy of the model. The value of this ratio is less than 5% during the entire analysis, except at the first displacement increment.

3. Inventory of Experimental Data and Results

3.1. Unreinforced Masonry (URM)

The unreinforced masonry considered in this numerical work was reported in [65,68]; in particular, both experimental campaigns reported that all masonry columns had a square-type cross-section of 250×250 mm with a horizontal and vertical thickness of mortar joints of 10 mm. The current columns have different heights equal to 500 and 720 mm, respectively. Finally, two and three unreinforced columns in [65,68] were used as reference columns, respectively.

Table 1 and Figure 6 show all mechanical and geometrical parameters regarding the unreinforced masonry columns investigated in [65,68].

Table 1. Statistical values of the masonry's constituents.

ID		Brick			Mortar	
		Compressive Strength f_{brick} (MPa)	Flexural Strength - (MPa)	Elastic Modulus E_{brick} (MPa)	Flexural Strength f_{tmat} (MPa)	Compressive Strength f_{cmat} (MPa)
[65]	Average (CoV %)	12.43 (8%)	1.91 (17%)	1625 (3%)	0.83 (1%)	1.89 (12%)
[68]		20.8 (18.4%)	-	-	0.55 (13.4%)	4.3 (7.6%)

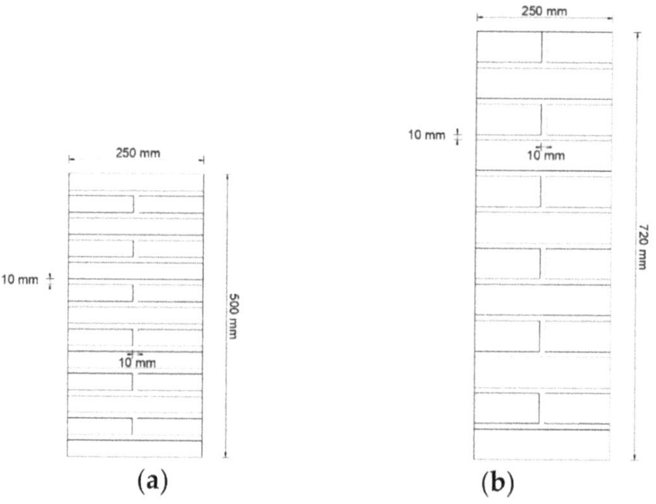

Figure 6. Geometrical details of stress–strain relationship: (**a**) Cascardi et al. [65] and Sneed at al. [68].

The columns were tested under axial compression, and the typical failure observed was masonry crashing (brittle failure) through a vertical crack in the mortar joints that was then propagated at the single clay brick units. In Table 2, a compressive strength f_{cm} (peak stress) and the elastic modulus in compression E_{cm} were reported. The value of E_{ct} (elastic modulus in tension) was set equal to compression.

Table 2. Test result of unreinforced masonry.

ID		*Fcm* (MPa)		*Ecm* (MPa)	
[65]	U1	8.08	7.61	-	250.33
	U2	7.15			
[68]	UC-1	8.7	7.36	-	2953.21
	UC-2	6.6			
	UC-3	6.8			

To perform the analyses of the available case studies, it was necessary to assume values for the missing parameters. The available literature overviewed in the present work along with the guidelines [106] offers an adequate amount of information upon which to base these assumptions. The average values of Poisson's ratio v was equal to 0.15, 0.20, and 0.3 for the masonry, mortar joint, and clay brick, respectively; similar values were adopted in [78,81,85,94,107–109]. The values used for the density ρ was suggested by the standard [106] and it was equal to 18 (kN/m³). The elastic modulus for mortar joints was calculated according to Eurocode 2 [110]. Finally, parameter f_{ct} was set equal to $1/3 f_{cm}$.

The parameter is that the softening branches of the constitutive laws of materials depend on fracture energies (G_{cm} and G_{fm}) and these were calculated by the equations reported in [100]. The values adopted are 0.784 and 0.25, respectively, for the experimental campaign reported in [65]. While for the [68], they were equal to 0.760 and 0.025, respectively. Finally, for both [65,68], the parameter h was set equal to 10, while in Figure 7 the finite-element resolution is shown.

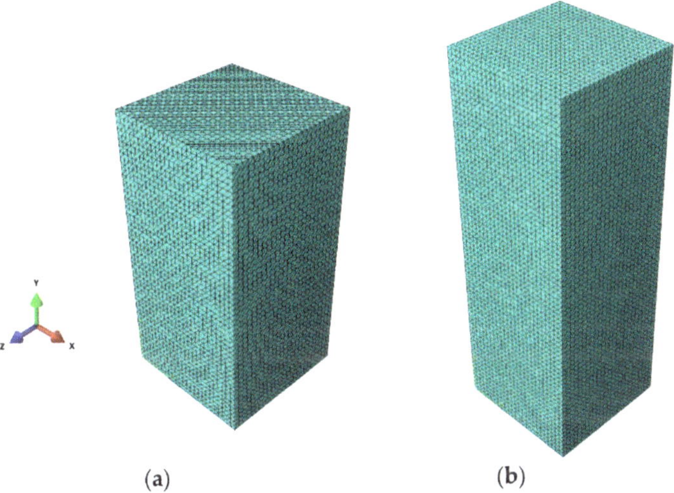

(a)   (b)

Figure 7. Geometrical modeling and finite-element resolution for URM: (**a**) Cascardi et al. [65] and (**b**) Sneed et al. [68].

3.2. FRCM and SRG System

In [65], nine columns were divided into three groups (three columns for each group); they were reinforced with the G-FRCM system using glass fiber combined with a different strengthening mortar. As described in the section, the FRCM external reinforcement was modeled through the MI approach. The matrix was modeled in 3D with its real thickness, while the fiber was modeled in 2D through its equivalent thickness. Specifically, the mortars presented a difference in terms of compression strength (f_{cmat}), and in Table 3 the flexural, compressive strength and the elastic modulus are reported (E_{mat}). The thickness (t_{mat}) adopted per all types of mortar is equal to 5 mm, while the density (ρ) and Poisson's ratio v were assumed to be 20 (kN/m³) and 0.15, respectively. In Figure 8, the finite-element resolution for external reinforcement is shown.

Table 3. Statistical values of the mortar and the results of reinforced columns reported in [65].

Mortar				Reinforced Columns	
	Flexural Strength f_{tmat} (MPa)	Compressive Strength f_{cmat} (MPa)	Elastic Modulus E_{mat} (MPa)	Peak Axial Strength f_{cmax} (MPa)	Average f_{cmax} (MPa)
M4	0.83	4.15	16,898 *	7.54 8.28 8.34	8.06
M7	1.46	7.26	19,984 *	10.31 8.34 11.37	10.01
M23	4.61	22.93	28,219 *	15.65 12.21 14.73	14.20

* Note: evaluated according to [110].

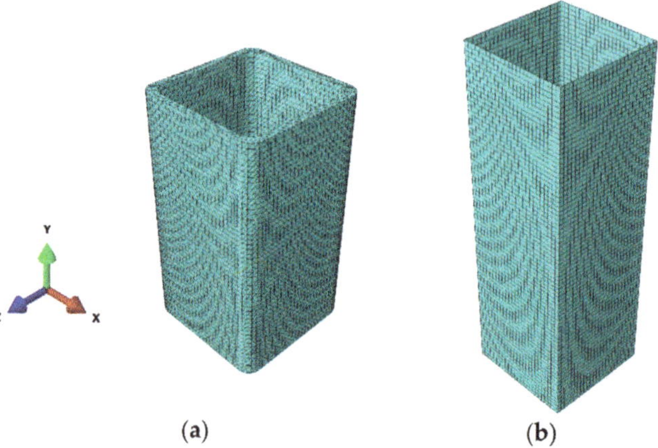

Figure 8. Finite-element resolution for (**a**) FRCM and (**a**) SRG.

All columns were reinforced with a single layer of continuous reinforcement over the entire height of the column, while the length of the overlap was equal to the width of the column. The mechanical properties of glass fibers in terms of tensile strength (f_{fu}) and elastic modulus (E_f) were equal to 742.40 (CoV 9%) and 37,120 (11%) MPa, respectively. Meanwhile, the equivalent thickness (t_f) was equal to 0.046 mm. In addition, round corners equal to 30 mm were in the four column corners to avoid stress concentration.

The typical failure observed was the brittle failure type, while a knife effect was observed for the reinforced ones, i.e., opening of a large vertical crack in the corner zone (overlap zone). In [68], the masonry columns were reinforced by an *SRG* system; in particular, three groups of four reinforced columns were considered. The reinforced system consisted of two types of inorganic matrices (hydraulic lime mortar and cementitious mortar matrix) with compressive strengths and different steel density.

The first approach used with the SRG system is *MA* without distinguishing the mortar and the steel fabric, and it was modeled in 2D. The equivalent thickness value used depends on the steel fibers' density (Table 4) and the constitutive law was linear elastic until failure. The mechanical parameters adopted in terms of tensile strength (f_{SRG}), ultimate strain (ε_{SRG}), and cracked modulus (E_{SRG}) were suggested by the manufacturer [111], and they were obtained by tensile tests on the SRG specimen. In addition, the density used is equal to 7.8 g/cm^3. The mechanical parameters of the bilinear model are summarized in Table 5.

Table 4. Statistical values of the external reinforcement and its constituents for columns reported in [68].

Group	Steel Fabric	Mortar					SRG Specimen		
	Round Corner r (mm)	Steel Density p (g/m^2)	Equivalent Thickness t_f (mm)	Flexural Strength f_{tmat} (MPa)	Compressive Strength f_{cmat} (MPa)	Tensile Strength f_{SRG} (MPa)	Elastic Modulus E_{mat} (MPa)	Ultimate Strain ε_{SRG} (-)	Cracked Modulus E_{SRG} (GPa)
1	0	670	0.084	1.5	13.0	3060	23801	0.010	156.0
2	0	670	0.084	4.4	47.1	2900	35021	0.018	160.0
3	9.5	1200	0.169	4.4	47.1	3060	35021	0.021	170.0

Table 5. Lu's parameter values.

	Interface Modeling	
	Group 1/3	Group 2
$k_{0,SRG}$ [N/mm^2]	76.92	76.92
$\tau_{f,SRG}$ [N/mm^2]	1.66	4.88
$G_{f,SRG}$ [N/mm]	0.21	0.37

The typical failure observed, independent of the types of steel density and inorganic matrix used, was located at the overlap zone with the opening of the reinforcement jacket. It should be noted that by increasing the steel density it was possible to observe a decrease in terms of peak axial strength. Table 6 shows the peak axial strength values for each type of column analyzed in [68].

Table 6. Results of reinforced columns reported in [68].

Group	Reinforced Columns	
	Peak Axial Strength f_{cmax} (MPa)	Average f_{cmax} (MPa)
1	10.3 9.5 9.1 8.5	9.3
2	9.1 10.1 9.4 8.7	9.3
3	10.7 11.1 10.1 10.1	10.5

3.3. Experimental versus Numerical Results

3.3.1. Unreinforced Columns

Figure 9 shows the comparison between the stress–axial-strain curves obtained from experimental tests and the numerical results obtained by the *MA* for the experimental campaign reported in [65,68]. The numerical branch before the peak stress is in good agreement with the experimental curves, while the softening branch presents higher scatter. In terms of peak axial stress, the numerical curves present an error between 2 and 3%. Furthermore, in terms of peak axial stress, the error of the curves shown in Figure 9 was evaluated by means of Equation (19) and summarized in Table 7.

$$\Delta_{err}[\%] = \frac{f_{cm\,(Num)} - f_{cm\,(Exp)}}{f_{cm\,(Exp)}} \, 100 \tag{19}$$

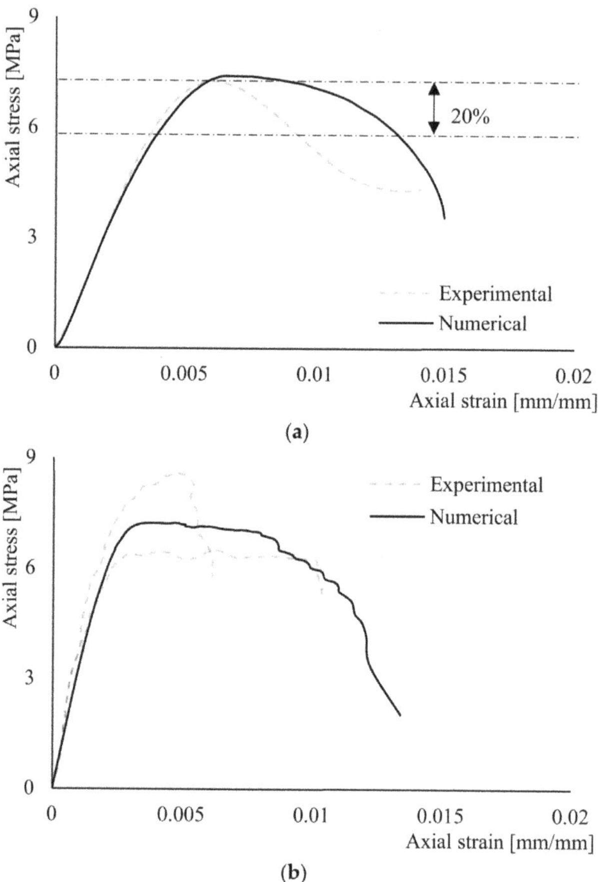

Figure 9. Comparison results for unreinforced column between the numerical and experimental curves reported in (**a**) Cascardi et al. [65] and (**b**) Sneed et al. [68].

Table 7. Results of comparison.

	Cascardi et al. [65]	Sneed et al. [68]
$\Delta_{err}[\%]$	2.1	2.8

The numerical model curve is affected by the type of approach used; in particular, the experimental curves have a more rapid descending post-peak branch until final collapse.

Figure 10 shows the comparisons between the experimental curve reported in [65] and the curves obtained from the numerical models by using the MI and SIMI approaches. In terms of peak stress, lower values are reached compared with the experimental and with higher values of axial deformation in correspondence to the peak stress. However, in the first branch, the numerical curve differs from the experimental one and exhibits nonlinear behavior before reaching the peak stress. In addition, it is possible to observe drops near the peak stress because of the type of interaction adopted. The differences in drops are more pronounced in the SI approach. Finally, for the assumptions present in the interaction between bricks and mortar joints and between the mortar joints themselves, the compressive and tensile strength of the vertical mortar joints was reduced by 50% as suggested in [81] to not harden the numerical model and cause failure in the vertical mortar joints. The softening branch is in good accordance until the reduction of 20% in the peak stress. In terms of peak axial stress, the numerical curves present an error of less than 1% for both the approaches.

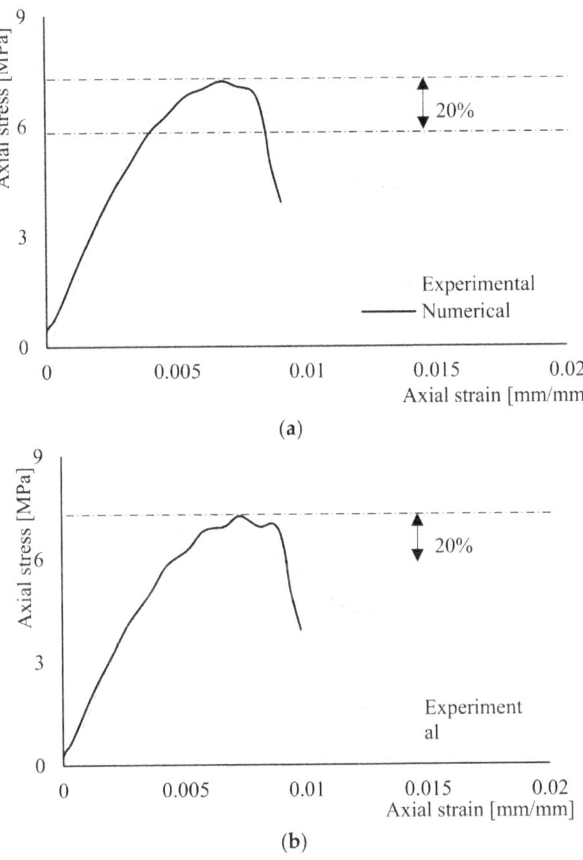

Figure 10. Comparison results for unreinforced column between the numerical and experimental curve reported in Cascardi et al. [65]: (**a**) SIIMI and (**b**) MI approach.

3.3.2. Wrapped Columns

The external reinforcement called FRCM, which the columns reported in [65], were reinforced and modeled in 3D. In particular, the glass fibers were described by a 2D model, while the matrix was by a 3D model. In Figure 11, the comparison between the experimental curves and the numerical curves is shown. It should be noted that the strategy adopted to emphasize the matrix effect of the numerical curve is in good agreement with the experimental ones. In addition, the numerical curves showed good accordance in terms of both peak axial stress and branches until the peak axial stress. The post-peak branch was reduced because of the models used to describe the behavior of the external reinforcement, particularly for the excessive distortion of the *FE* elements used.

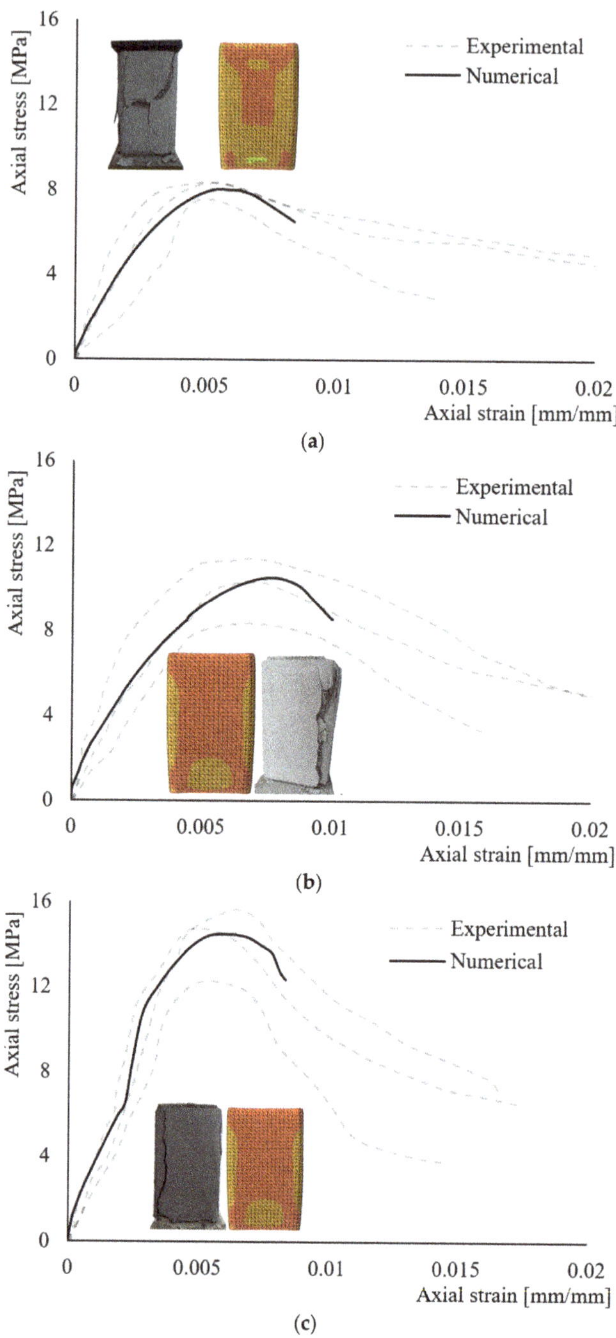

Figure 11. Comparison results for reinforced column between the numerical and average experimental curve reported in Cascardi et al. [65]: (**a**) FRCM_M4, (**b**) FRCM_M7, and (**c**) FRCM_M23.

In Table 8, the error evaluated by Equation (19) is summarized. In Figure 11, the crack pattern at failure is also reported and is compared to that observed experimentally.

The modeling strategy adopted is in good accordance with the experimental in terms of the crack pattern. The error evaluated through Equation (19) is between 1 and 4%. The approach used has a drawback that is due to the excess computational burden. In addition, the matrix effect may be observed, especially for the matrix named FRCM_M23. The columns reinforced with the SRG system [68] were modeled in 2D using the MA approach without distinguishing the matrix and steel fiber. Figure 12 shows the comparison between the experimental curves and that obtained from the numerical model varying the type of matrix and density of steel fibers.

Table 8. Results of comparison.

	$\Delta_{err}[\%]$
FRCM M4 [65]	1.1
FRCM M7 [65]	3.6
FRCM M23 [65]	2.0
Group 1 [68]	5.7
Group 5 [68]	7.1
Group 8 [68]	7.4

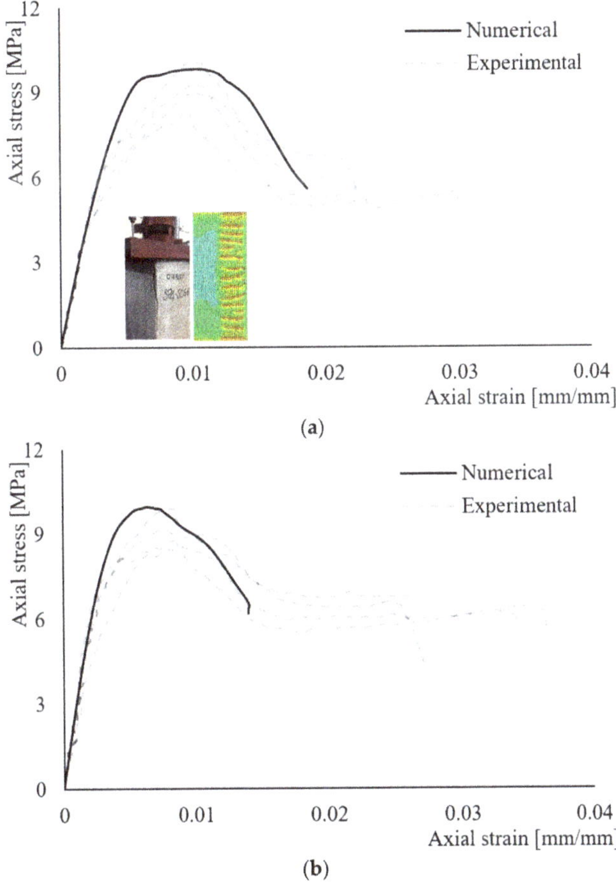

Figure 12. *Cont.*

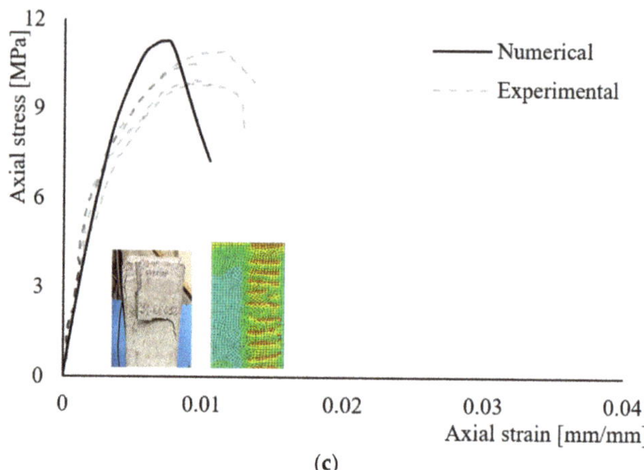

(c)

Figure 12. Comparison results for reinforced column between the numerical and experimental curve reported in Sneed et al. [68]: (**a**) Group 1, (**b**) Group 5, and (**c**) Group 8.

The error is between 5 and 8%. The values are slightly higher than those obtained with the approach used for the FRCM system. This approach has the advantage of a lower computational burden and, consequently, may be used for structures of greater geometric dimensions. The crack pattern at failure is also reported in Figure 12. It is possible to note the detachment along the overlap zone along the entire height of the column. While the one observed experimentally is sometimes localized and does not develop along the entire height, this type has also been noted in [77,92]. Localized detachment is caused by the arrangement of the steel sheet during the casting phase of the steel fibers. The steel fiber sheet is commercialized in 30 cm wide rolls; therefore, multiple sheets of 30 cm wide steel fiber are placed side by side to cover the full height of the column and the several steel cords are joined together by the matrix alone. In Table 8, the errors are summarized.

4. Conclusions

The numerical procedure found, which was based on the finite element, was developed in this paper. Parameters of the numerical model were calibrated on the available experimental results present in the literature. The effectiveness of the model was evaluated by a comparison of test results conducted on clay brick masonry columns confined with FRCM (glass-FRCM) and SRG (with different types of matrix and steel density) systems. Based on the obtained results, the following conclusions can be drawn:

- The different numerical strategies adopted furnish accurate outcomes in terms of axial strength for unconfined masonry columns;
- The proposed strategy adopted to describe the external reinforcement FRCM/SRG by the *MA* and *MI* approaches in terms of axial stress and crack pattern are similar;
- For glass-FRCM- and SRG-confined columns, numerical predictions in terms of axial stress–axial strain curves are in good agreement with experimental results in the ascending branches of the curves, while they are inaccurate for describing the post-peak;
- The approach used for the FRCM system resulted in errors of less than 4%, but with a considerable increase in computational burden;
- The approach used for the SRG system could possibly obtain an error of between 5 and 8%.

Further experimental analysis is needed to confirm the results obtained in the investigation described and discussed in the paper.

Funding: This research received no external funding.

Data Availability Statement: Not applicable.

Conflicts of Interest: The author declares no conflict of interest.

References

1. Dolatshahi, K.M.; Aref, A.J. Multi-directional response of unreinforced masonry walls: Experimental and computational investigations. *Earthq. Eng. Struct. Dyn.* **2016**, *45*, 1427–1449. [CrossRef]
2. Dolatshahi, K.M.; Yekrangnia, M. Out-of-plane strength reduction of unreinforced masonry walls because of in-plane damages. *Earthq. Eng. Struct. Dyn.* **2015**, *44*, 2157–2176. [CrossRef]
3. Dolatshahi, K.M.; Nikoukalam, M.T.; Beyer, K. Numerical study on factors that influence the in-plane drift capacity of unreinforced masonry walls. *Earthq. Eng. Struct. Dyn.* **2018**, *47*, 1440–1459. [CrossRef]
4. Wilding, B.V.; Dolatshahi, K.M.; Beyer, K. Influence of load history on the force displacement response of in-plane loaded unreinforced masonry walls. *Eng. Struct.* **2017**, *152*, 671–682. [CrossRef]
5. Smoljanović, H.; Živaljić, N.; Nikolić, Ž. A combined finite-discrete element analysis of dry stone masonry structures. *Eng. Struct.* **2013**, *52*, 89–100. [CrossRef]
6. Smoljanović, H.; Živaljić, N.; Nikolić, Ž.; Munjiza, A. Numerical analysis of 3D drystone masonry structures by combined finite-discrete element method. *Int. J. Solids Struct.* **2018**, *136*, 150–167. [CrossRef]
7. Petracca, M.; Pelá, L.; Rossi, R.; Zaghi, S.; Camata, G.; Spacone, E. Micro-scale continuous and discrete numerical models for nonlinear analysis of masonry shear walls. *Constr. Build. Mater.* **2017**, *149*, 296–314. [CrossRef]
8. Addessi, D.; Sacco, E. Nonlinear analysis of masonry panels using a kinematic enriched plane state formulation. *Int. J. Solids Struct.* **2016**, *90*, 194–214. [CrossRef]
9. Serpieri, R.; Albarella, M.; Sacco, E. A 3D microstructured cohesivefrictional interface model and its rational calibration for the analysis of masonry panels. *Int. J. Solids Struct.* **2017**, *122*, 110–127. [CrossRef]
10. Orduña, A.; Lourenço, P.B. Three-dimensional limit analysis of rigid blocks assemblages. Part I: Torsion failure on frictional interfaces and limit analysis formulation. *Int. J. Solids Struct.* **2005**, *42*, 5140–5160. [CrossRef]
11. Orduña, A.; Lourenço, P.B. Three-dimensional limit analysis of rigid blocks assemblages. Part II: Load-path following solution procedure and validation. *Int. J. Solids Struct.* **2005**, *42*, 5161–5180. [CrossRef]
12. Portioli, F.; Casapulla, C.; Gilbert, M.; Cascini, L. Limit analysis of 3D masonry block structures with non-associative frictional joints using cone programming. *Comput. Struct.* **2014**, *143*, 108–121. [CrossRef]
13. Milani, G. 3d upper bound limit analysis of multi-leaf masonry walls. *Int. J. Mech. Sci.* **2008**, *50*, 817–836. [CrossRef]
14. Abdulla, K.F.; Cunningham, L.S.; Gillie, M. Simulating masonry wall behaviour using a simplified micro-model approach. *Eng. Struct.* **2017**, *151*, 349–365. [CrossRef]
15. Zhai, C.; Wang, X.; Kong, J.; Li, S.; Xie, L. Numerical simulation of masonry-infilled rc frames using xfem. *J. Struct. Eng.* **2017**, *143*, 04017144. [CrossRef]
16. Roca, P.; Cervera, M.; Gariup, G.; Pela, L. Structural analysis of masonry historical constructions, classical and advanced approaches. *Arch. Comput. Methods Eng.* **2010**, *17*, 299–325. [CrossRef]
17. Minga, E.; Macorini, L.; Izzuddin, B.A. A 3D mesoscale damage-plasticity approach for masonry structures under cyclic loading. *Meccanica* **2018**, *53*, 1591–1611. [CrossRef]
18. D'Altri, A.M.; Messali, F.; Rots, J.; Castellazzi, G.; de Miranda, S. A damaging blockbased model for the analysis of the cyclic behaviour of full-scale masonry structures. *Eng. Fract. Mech.* **2019**, *209*, 423–448. [CrossRef]
19. Bruggi, M. Finite element analysis of no-tension structures as a topology optimization problem. *Struct. Multidiscipl. Optim.* **2014**, *50*, 957–973. [CrossRef]
20. Bartoli, G.; Betti, M.; Vignoli, A. A numerical study on seismic risk assessment of historic masonry towers: A case study in San Gimignano. *Bull. Earthq. Eng.* **2016**, *14*, 1475–1518. [CrossRef]
21. Valente, M.; Milani, G. Seismic assessment of historical masonry towers using simplified approaches and standard FEM. *Constr. Build. Mater.* **2016**, *108*, 74–104. [CrossRef]
22. Castellazzi, G.; D'Altri, A.M.; de Miranda, S.; Chiozzi, A.; Tralli, A. Numerical insights on the seismic behavior of a non-isolated historical masonry tower. *Bull. Earthq. Eng.* **2018**, *16*, 933–961. [CrossRef]
23. Betti, M.; Vignoli, A. Numerical assessment of the static and seismic behaviour of the basilica of Santa Maria all'Impruneta (Italy). *Constr. Build. Mater.* **2011**, *25*, 4308–4324. [CrossRef]
24. Milani, G.; Valente, M. Failure analysis of seven masonry churches severely damaged during the 2012 Emilia-Romagna (Italy) earthquake: Non-linear dynamic analyses vs conventional static approaches. *Eng. Fail. Anal.* **2015**, *54*, 13–56. [CrossRef]
25. Fortunato, G.; Funari, M.F.; Lonetti, P. Survey and seismic vulnerability assessment of the baptistery of San Giovanni in Tumba (Italy). *J. Cult. Herit.* **2017**, *26*, 64–78. [CrossRef]
26. Betti, M.; Galano, L. Seismic analysis of historic masonry buildings: The vicarious palace in Pescia (Italy). *Buildings* **2012**, *2*, 63–82. [CrossRef]
27. Castellazzi, G.; D'Altri, A.M.; de Miranda, S.; Ubertini, F. An innovative numerical modeling strategy for the structural analysis of historical monumental buildings. *Eng. Struct.* **2017**, *132*, 229–248. [CrossRef]

28. Degli Abbati, S.; D'Altri, A.M.; Ottonelli, D.; Castellazzi, G.; Catteri, S.; de Miranda, S. Seismic assessment of interacting structural units in complex historic masonry constructions by nonlinear static analysis. *Comput. Struct.* **2019**, *213*, 51–71. [CrossRef]
29. Pelà, L.; Aprile, A.; Benedetti, A. Seismic assessment of masonry arch bridges. *Eng. Struct.* **2009**, *31*, 1777–1788. [CrossRef]
30. Zampieri, P.; Zanini, M.A.; Modena, C. Simplified seismic assessment of multi-span masonry arch bridges. *Bull. Earthq. Eng.* **2015**, *13*, 2629–2646. [CrossRef]
31. Cecchi, A.; Sab, K. A homogenized reissner-mindlin model for orthotropic periodic plates: Application to brickwork panels. *Int. J. Solids Struct.* **2007**, *44*, 6055–6079. [CrossRef]
32. Mistler, M.; Anthoine, A.; Butenweg, C. In-plane and out-of-plane homogenization of masonry. *Comput. Struct.* **2007**, *85*, 1321–1330. [CrossRef]
33. Taliercio, A. Closed-form expressions for the macroscopic in-plane elastic and creep coefficients of brick masonry. *Int. J. Solids Struct.* **2014**, *51*, 2949–2963. [CrossRef]
34. Cecchi, A.; Milani, G. A kinematic FE limit analysis model for thick english bond masonry walls. *Int. J. Solids Struct.* **2008**, *45*, 1302–1331. [CrossRef]
35. Godio, M.; Stefanou, I.; Sab, K.; Sulem, J.; Sakji, S. A limit analysis approach based on Cosserat continuum for the evaluation of the in-plane strength of discrete media: Application to masonry. *Eur. J. Mech. A Solids* **2017**, *66*, 168–192. [CrossRef]
36. Calderini, C.; Lagomarsino, S. A micromechanical inelastic model for historical masonry. *J. Earthq. Eng.* **2006**, *10*, 453–479. [CrossRef]
37. Marfia, S.; Sacco, E. Multiscale damage contact-friction model for periodic masonry walls. *Comput. Methods Appl. Mech. Eng.* **2012**, *205*, 189–203. [CrossRef]
38. Salerno, G.; de Felice, G. Continuum modeling of periodic brickwork. *Int. J. Solids Struct.* **2009**, *46*, 1251–1267. [CrossRef]
39. De Bellis, M.L.; Addessi, D. A Cosserat based multi-scale model for masonry structures. *Int. J. Multisc. Comput. Eng.* **2011**, *9*, 543. [CrossRef]
40. Greco, F.; Leonetti, L.; Luciano, R.; Blasi, P.N. An adaptive multiscale strategy for the damage analysis of masonry modeled as a composite material. *Compos. Struct.* **2016**, *153*, 972–988. [CrossRef]
41. Reccia, E.; Leonetti, L.; Trovalusci, P.; Cecchi, A. A multi-scale/multi-domain model for the failure analysis of masonry walls: A validation with a combined FEM/DEM approach. *Int. J. Multisc. Comput. Eng.* **2018**, *16*, 325–343. [CrossRef]
42. Lloberas-Valls, O.; Rixen, D.; Simone, A.; Sluys, L. Multiscale domain decomposition analysis of quasi brittle heterogeneous materials. *Int. J. Numer. Methods Eng.* **2012**, *89*, 1337–1366. [CrossRef]
43. Siano, R.; Roca, P.; Camata, G.; Pela', L.; Sepe, V.; Spacone, E. Numerical investigation of non-linear equivalent-frame models for regular masonry walls. *Eng. Struct.* **2018**, *173*, 512–529. [CrossRef]
44. Quagliarini, E.; Maracchini, G.; Clementi, F. Uses and limits of the equivalent frame model on existing unreinforced masonry buildings for assessing their seismic risk: A review. *J. Build. Eng.* **2017**, *10*, 166–182. [CrossRef]
45. Roca, P.; Molins, C.; Marì, A.R. Strength capacity of masonry wall structures by the equivalent frame method. *J. Struct. Eng.* **2005**, *131*, 1601–1610.
46. Pasticier, L.; Amadio, C.; Fragiacomo, M. Non-linear seismic analysis and vulnerability evaluation of a masonry building using the sap2000 v. 10 code. *Earthq. Eng. Struct. Dyn.* **2008**, *37*, 467–485. [CrossRef]
47. Belmouden, Y.; Lestuzzi, P. An equivalent frame model for seismic analysis of masonry and reinforced concrete buildings. *Constr. Build. Mater.* **2009**, *23*, 40–53.
48. Raka, E.; Spacone, E.; Sepe, V.; Camata, G. Advanced frame element for seismic analysis of masonry structures: Model formulation and validation. *Earthq. Eng. Struct. Dyn.* **2015**, *44*, 2489–2506. [CrossRef]
49. Chen, S.Y.; Moon, F.; Yi, T. A macro-element for the nonlinear analysis of in-plane unreinforced masonry piers. *Eng. Struct.* **2008**, *30*, 2242–2252. [CrossRef]
50. Caliò, I.; Marletta, M.; Pantò, B. A new discrete element model for the evaluation of the seismic behaviour of unreinforced masonry buildings. *Eng. Struct.* **2012**, *40*, 327–338. [CrossRef]
51. Xu, H.; Gentilini, C.; Yu, Z.; Wu, H.; Zhao, S. A unified model for the seismic analysis of brick masonry structures. *Constr. Build. Mater.* **2018**, *184*, 733–751. [CrossRef]
52. Block, P. Ochsendorrust network analysis: A new methodology for threedimensional equilibrium. *J. Int. Assoc. Shell Spat. Struct.* **2007**, *48*, 167–173.
53. Fantin, M.; Ciblac, T. Extension of thrust network analysis with joints consideration and new equilibrium states. *Int. J. Space Struct.* **2016**, *31*, 190–202. [CrossRef]
54. Fraternali, F. A thrust network approach to the equilibrium problem of unreinforced masonry vaults via polyhedral stress functions. *Mech. Res. Commun.* **2010**, *37*, 198–204. [CrossRef]
55. Fraddosio, A.; Lepore, N.; Piccioni, M.D. Lower bound limit analysis of masonry vaults under general load conditions. In *Structural Analysis of Historical Constructions*; Springer: Berlin/Heidelberg, Germany, 2019; pp. 1090–1098.
56. Milani, G. Upper bound sequential linear programming mesh adaptation scheme for collapse analysis of masonry vaults. *Adv. Eng. Softw.* **2015**, *79*, 91–110. [CrossRef]
57. Chiozzi, A.; Milani, G.; Tralli, A. A genetic algorithm NURBS-based new approach for fast kinematic limit analysis of masonry vaults. *Comput. Struct.* **2017**, *182*, 187–204.

58. Mezrea, P.E.; Yilmaz, I.A.; Ispir, M.; Binbir, E.; Bal, I.E.; Ilki, A. External jacketing of unreinforced historical masonry piers with open-grid basalt-reinforced mortar. *J. Comp. Constr.* **2017**, *21*, 04016110. [CrossRef]
59. Fossetti, M.; Minafò, G.; Papia, M. Flexural behaviour of glulam timber beams reinforced with FRP cords. *Constr. Build. Mater.* **2015**, *95*, 54–64. [CrossRef]
60. Santandrea, M.; Quartarone, G.; Carloni, C.; Gu, X. Confinement of masonry columns with steel and basalt FRCM composites. In *International Conference on Mechanics of Masonry Structures Strengthened with Composites Materials*; MuRiCo6; Trans Tech Publications Ltd.: Bäch, Switzerland, 2017.
61. Maddaloni, G.; Cascardi, A.; Balsamo, A.; di Ludovico, M.; Micelli, F.; Aiello, M.A.; Prota, A. Confinement of Full-Scale Masonry Columns with FRCM Systems. *Key Eng. Mater.* **2017**, *747*, 374–381.
62. Ombres, L.; Verre, S. Analysis of the Behavior of FRCM Confined Clay Brick Masonry Columns. *Fibers* **2020**, *8*, 11. [CrossRef]
63. Carloni, C.; Mazzotti, C.; Savoia, M.; Subramaniam, K.V. Confinement of Masonry Columns with PBO FRCM Composites. *Key Eng. Mater.* **2014**, *624*, 644–651. [CrossRef]
64. Aiello, M.A.; Cascardi, A.; Ombres, L.; Verre, S. Confinement of masonry columns with the FRCM-system: Theoretical and experimental investigation. *Infrastructures* **2020**, *5*, 101. [CrossRef]
65. Cascardi, A.; Micelli, F.; Aiello, M.A. FRCM-confined masonry columns: Experimental investigation on the effect of the inorganic matrix properties. *Constr. Build. Mater.* **2018**, *186*, 811–825. [CrossRef]
66. Aiello, M.A.; Bencardino, F.; Cascardi, A.; D'Antino, T.; Fagone, M.; Frana, I.; La Mendola, L.; Lignola, G.P.; Mazzotti, C.; Micelli, F.; et al. Masonry columns confined with fabric reinforced cementitious matrix (FRCM) systems: A round robin test. *Constr. Build. Mater.* **2021**, *298*, 123816.
67. Sneed, L.H.; Carloni, C.; Baietti, G.; Fraioli, G. Confinement of Clay Masonry Columns with SRG. *Key Eng. Mater.* **2017**, *747*, 350–357. [CrossRef]
68. Sneed, L.H.; Baietti, G.; Fraioli, G.; Carloni, C. Compressive Behavior of Brick Masonry Columns Confined with Steel-Reinforced Grout Jackets. *J. Comp. Constr.* **2019**, *23*, 04019037.
69. Ombres, L. Confinement Effectiveness in Eccentrically Loaded Masonry Columns Strengthened by Fiber Reinforced Cementitious Matrix (FRCM) Jackets. *Key Eng. Mater.* **2014**, *624*, 551–558. [CrossRef]
70. Theofanis, K.D. Textile reinforced mortar system as a means for confinement of masonry structures. In Proceedings of the 12th International Symposium on Fiber Rein-forced Polymers for Reinforced Concrete Structures (FRPRCS-12), Nanjing, China, 14–16 December 2015.
71. Incerti, A.; Vasiliu, A.; Ferracuti, B.; Mazzotti, C. Uni-Axial compressive tests on masonry columns confined by FRP and FRCM. In Proceedings of the 12th International Symposium on Fiber Reinforced Polymers for Reinforced Concrete Structures & The 5th Asia-Pacific Conference on Fiber Reinforced Polymers in Structures, Joint Conference, Nanjing, China, 14–16 December 2015.
72. Valdés, M.; Concu, G.; De Nicolo, B. FRP strengthening of masonry columns: Experimental tests and theoretical analysis. *Key Eng. Mater.* **2015**, *624*, 603–610. [CrossRef]
73. Witzany, J.; Zigler, R. Stress state analysis and failure mechanisms of masonry columns reinforced with FRP under concentric compressive load. *Polymers* **2016**, *8*, 176. [CrossRef]
74. EN 772-1:2011; Methods of Test for Masonry Units—Part 1: Determination of Compressive Strength. European Union: Brussels, Beligum, 2011.
75. UNI EN 1926:2007; Metodi di Prova per Pietre Naturali—Determinazione Della Resistenza a Compressione Uniassiale. UNI: Milano, Italy, 2007. (In Italian)
76. Ghiassi, B.; Milani, G. (Eds.) *Numerical Modeling of Masonry and Historical Structures: From Theory to Application*; Woodhead Publishing: Sawston, UK, 2019.
77. Ombres, L.; Verre, S. Numerical modeling approaches of FRCMs/SRG confined masonry columns. *Front. Built Environ.* **2019**, *5*, 143.
78. Murgo, F.S.; Mazzotti, C. Masonry columns strengthened with FRCM system: Numerical and experimental evaluation. *Constr. Build. Mater.* **2019**, *202*, 208–222.
79. Ameli, Z.; D'Antino, T.; Carloni, C. A new predictive model for FRCM-confined columns: A reflection on the composite behavior at peak stress. *Constr. Build. Mater.* **2022**, *337*, 127534. [CrossRef]
80. Maroušková, A. Reinforced Masonry Column's Analysis: The Influence of Rounded Corners. *Adv. Mat. Res.* **2017**, *1144*, 34–39. [CrossRef]
81. Maroušková, A. Inelastic material models for numerical analysis of unreinforced compressed masonry columns. *Key Eng. Mater.* **2016**, *677*, 197–202. [CrossRef]
82. Massart, T.J.; Peerlings, R.H.J.; Geers, M.G.D.; Gottcheiner, S. Mesoscopic modeling of failure in brick masonry accounting for three-dimensional effects. *Eng. Fract. Mech.* **2005**, *72*, 1238–1253. [CrossRef]
83. Lü, W.R.; Wang, M.; Liu, X.J. Numerical analysis of masonry under compression via micro-model. *Adv. Mat. Res.* **2011**, *243*, 1360–1365. [CrossRef]
84. Petersen, R.B.; Masia, M.J.; Seracino, R. In-plane shear behavior of masonry panels strengthened with NSM CFRP strips. I: Experimental investigation. *J. Comp. Constr.* **2010**, *14*, 754–763. [CrossRef]
85. Vermeltfoort, A.T.; Martens, D.R.W.; Van Zijl, G.P.A.G. Brick–mortar interface effects on masonry under compression. *Can. J. Civ. Eng.* **2007**, *34*, 1475–1485. [CrossRef]

86. Vasconccelos, G.D.F.M.; Lourenço, P.B.; Alves, C.A.S.; Pamplona, J. Ultrasonic evaluation of the physical and mechanical properties of granites. *Ultrasonics* **2008**, *48*, 453–466. [CrossRef]

87. Stavridis, A.; Shing, P.B. Finite-element modeling of nonlinear behavior of masonry-infilled RC frames. *J. Struct. Eng.* **2010**, *136*, 285–296. [CrossRef]

88. La Mendola, L.; Accardi, M.; Cucchiara, C.; Licata, V. Nonlinear FE analysis of out-of plane behaviour of masonry walls with and without CFRP reinforcement. *Constr. Build. Mater.* **2014**, *54*, 190–196. [CrossRef]

89. Milani, G. Simple homogenization model for the non-linear analysis of in-plane loaded masonry walls. *Comput. Struct.* **2011**, *89*, 1586–1601. [CrossRef]

90. Feenstra, P.H. Computational Aspect of Biaxial Stress in Plain and Reinforced Concrete. Ph.D. Thesis, Delft University of Technology, Delft, The Netherlands, 1993.

91. Bolhassani, M.; Hamid, A.A.; Lau, A.C.; Moon, F. Simplified micro modeling of partially grouted masonry assemblages. *Constr. Build. Mater.* **2015**, *83*, 159–173. [CrossRef]

92. *ABAQUS Theory and User's Manual*; Version 6.12; Hibbitt, Karlsson & Sorensen: Cheshire, UK, 2012.

93. Ombres, L.; Verre, S. Masonry columns strengthened with Steel Fabric Reinforced Cementitious Matrix (S-FRCM) jackets: Experimental and numerical analysis. *Measurement* **2018**, *127*, 238–245. [CrossRef]

94. Verre, S.; Cascardi, A.; Aiello, M.A.; Ombres, L. Numerical modelling of FRCMs confined masonry column. *Key Eng. Mater.* **2019**, *817*, 9–14. [CrossRef]

95. Micelli, F.; Cascardi, A. Structural assessment and seismic analysis of a 14th century masonry tower. *Eng. Fail. Anal.* **2020**, *107*, 104198. [CrossRef]

96. Cascardi, A.; Verre, S.; Sportillo, A.; Giorgio, G. A Multiplex Conversion of a Historical Cinema. *Adv. Civ. Eng.* **2022**, *2022*, 2191315. [CrossRef]

97. Ombres, L.; Verre, S. Shear strengthening of reinforced concrete beams with SRG (Steel Reinforced Grout) composites: Experimental investigation and modelling. *J. Build. Eng.* **2021**, *42*, 103047. [CrossRef]

98. Ombres, L.; Verre, S. Experimental and numerical investigation on the steel reinforced grout (SRG) composite-to-concrete bond. *J. Compos. Sci.* **2020**, *4*, 182. [CrossRef]

99. Jawdhari, A.; Adheem, A.H.; Kadhim, M.M.A. Parametric 3D finite element analysis of FRCM-confined RC column under eccentric loading. *Eng. Struct.* **2020**, *212*, 110504. [CrossRef]

100. Drougkas, A.; Roca, P.; Molins, C. Numerical prediction of the behavior, strength and elasticity of masonry in compression. *Eng. Struct.* **2015**, *90*, 15–28. [CrossRef]

101. Ombres, L.; Verre, S. Flexural strengthening of RC beams with steel-reinforced grout: Experimental and numerical investigation. *J. Comp. Constr.* **2019**, *23*, 04019035. [CrossRef]

102. Bencardino, F.; Condello, A. SRG/SRP—Concrete bond—Slip laws for externally strengthened RC beams. *Comput. Struct.* **2015**, *132*, 804–815. [CrossRef]

103. Cascardi, A.; Dell'Anna, R.; Micelli, F.; Lionetto, F.; Aiello, M.A.; Maffezzoli, A. Reversible techniques for FRP-confinement of masonry columns. *Constr. Build. Mater.* **2019**, *225*, 415–428. [CrossRef]

104. Lu, X.Z.; Teng, J.G.; Ye, L.P.; Jiang, J.J. Bond slip models for FRP sheet/plates bonded to concrete. *Eng. Struct.* **2005**, *27*, 920–937. [CrossRef]

105. Chen, G.M.; Teng, J.G.; Chen, J.F.; Xiao, Q.G. Finite element modeling of debonding failures in FRP-strengthened RC beams: A dynamic approach. *Comput. Struct.* **2015**, *158*, 167–183. [CrossRef]

106. *CNR-DT 215/2018*; Guide for the Design and Construction of Externally Bonded Fibre Reinforced Inorganic Matrix Systems for Strengthening Existing Structures. National Research Council: Rome, Italy, 2020.

107. Akbarzade, A.; Tasnimi, A. Nonlinear analysis and modeling of unreinforced masonry shear walls based on plastic damage model. *J. Seismol. Earthq. Eng.* **2011**, *11*, 189–203.

108. Gumaste, K.S.; Nanjunda Rao, K.S.; Reddy, B.V.V.; Jagadish, K.S. Strength and elasticity of brick masonry prisms and wallettes under compression. *Mater. Struct.* **2007**, *40*, 241–253. [CrossRef]

109. Oliveira, D.V.D.C.; Lourenço, P.B.; Roca, P. Cyclic behaviour of stone and brick masonry under uniaxial compressive loading. *Mater. Struct.* **2006**, *39*, 247–257. [CrossRef]

110. *EN 1992-1-1*; Eurocode 2: Design of Concrete Structures—Part 1-1: General Rules and Rules for Buildings. CEN (European Committee for Standardization): Brussels, Belgium, 2003.

111. Kerakoll, S.p.A. 2018. Available online: http://www.kerakoll.com (accessed on 1 February 2018).

 *buildings*

Article

Application and Practice of Variable Axial Force Cable in Powerhouse Truss Reinforcement System

Zizhen Shen [1,*], Min Hong [2], Xunfeng Li [1], Zigang Shen [3], Lianbo Wang [4,*] and Xueping Wang [1]

1 School of Architectural Engineering, Zhejiang College of Construction, Hangzhou 310053, China
2 School of Materials Science and Engineering, Shanghai Jiao Tong University, Shanghai 200240, China
3 Zhejiang Min An Testing Technology Co., Hangzhou 310000, China
4 School of Materials Science and Engineering, Shanghai Institute of Technology, Shanghai 201418, China
* Correspondence: hzhgjgc123@126.com (Z.S.); lbwabg@sit.edu.cn (L.W.)

Abstract: Long-span steel structure trusses are widely used in factory buildings. However, with the increase in service time and dynamic load fatigue, transverse cracks at the bottom of the middle span and oblique deformation of the abdomen during the operation process may appear in a considerable part of long-span trusses with dynamic load. The U-shaped cracks at the bottom and belly, as well as the mid-span down deflection of the main truss, can also reduce the functionality of the factory building truss structure and limit the original crane load, thus affecting the normal safety and durability of the structure. Therefore, the principle of variable axial force cable system in the long-span factory building truss structure and 3D3S software modelling were applied. Analysing and studying the reinforcement method of large-span powerhouse trusses can provide practical experience for subsequent similar projects. In view of the above phenomenon, the large-span powerhouse trusses of Hongcheng Powerhouse No. 1 and No. 2, located in Tonglu, Zhejiang Province, were used as the research objects, and the variable axial force cable method was proposed to strengthen and lift the load. Considering the span of the powerhouse truss, a cable system with 22 m and a controlling force of 400 kN was proposed for Powerhouse 1, and a cable system with a variable axial force of 24 m was proposed for Powerhouse 2. The force model of large-span trusses was established by using the finite element method, which is commonly used to analyse the force of the truss. The influence of the reinforcement effect was analysed under two working conditions and compared from three aspects: stiffness, bearing capacity and stability. Furthermore, the phenomenon of uneven stress distribution was analysed. The stress distribution characteristics of each node were understood by simulating the most disadvantageous node plates with the greatest internal force before and after reinforcement.

Keywords: long-span steel structure truss; variable axial force cable; 3D3S finite element model; joint plate analysis; variable system reinforcement combination stiffness; load domain

Citation: Shen, Z.; Hong, M.; Li, X.; Shen, Z.; Wang, L.; Wang, X. Application and Practice of Variable Axial Force Cable in Powerhouse Truss Reinforcement System. *Buildings* **2023**, *13*, 1271. https://doi.org/10.3390/buildings13051271

Academic Editor: Hiroshi Tagawa

Received: 21 March 2023
Revised: 27 April 2023
Accepted: 29 April 2023
Published: 12 May 2023

1. Introduction

Many previous researchers have conducted corresponding research on the application of variable axial force cables in reinforcement engineering. For example, the application of variable axial force cables in bridge reinforcement has been widely studied. One study investigated the application of VLM.TS-type outer cable in the Dongming Huanghe Bridge reinforcement project [1]. Gong proposed strengthening the Pu Shan Wan cantilever bridge using a cable system [2]. Hu et al. studied the application of 2000 MPa parallel steel cables in highway and cable bridges [3]. Simultaneously, the application of the variable axial force cable in bridge reinforcement has attracted the attention of researchers [4]. For example, He et al. conducted a stress and reinforcement analysis of steel truss structures considering the influence of global joint stiffness [5]. Pan studied the application of cable installation and construction technology of single-tower suspension bridges [6]. Yu et al. studied the application of cable installation and construction technology of composite beam suspension bridges.

The Variable Axial Load Cable Method is a structural retrofitting technique used to strengthen trusses. This method involves the installation of steel cables with appropriate tension to transfer the loads from weak or damaged members of the truss to stronger ones.

The process of retrofitting through this method starts with the identification of the damaged or weak member(s) and an assessment of the truss system's overall strength. Once identified, steel cables are installed in place of the damaged or weakened member(s), so that the original load-bearing function is restored.

The cables are then pre-tensioned to a specified load and attached to the adjacent members of the truss. The tension in the cables is adjusted to ensure the load is evenly distributed among all members of the truss. This ensures that the strength of the entire truss system is improved without overloading any one member [7].

Additionally, a number of researchers have studied the application of variable axial force cables in concrete bridge reinforcement. For instance, Hu studied damage inversion analysis and variable system reinforcement of concrete bridges based on an equivalent sandwich beam model [8], whilst Zhao et al. applied and investigated the effect of a cable shock absorber in the Jiayu Yangtze River Highway Bridge [9]. Several studies have also cited the variable axial force cable construction and related technologies of the standard specification, such as building structure load calculation code (GB50009-2019) [10], steel structure construction quality acceptance code (GB50205-2014) [11], cable construction technical code (JGJ257-2012) [12] and building structure test technical code (GB/T50344-2019) [13]. Teng et al. determined the axial force on stay cables whilst accounting for their bending stiffness and rotational end restraints by free vibration test [14]. Other researchers have introduced the design method for the overall strength and stability of steel structures [15–18] and used advanced structural inspection and evaluation techniques, such as 3D3S modelling inspection [19–21], and introduced the stress loss and strength failure detection methods of some high-strength bolts and steel [22–24].

Finally, researchers have also used the variable axial force cable application in structural engineering experimental research and finite element (FE) analysis. For example, one study verified the FE analysis method by conducting experiments on reinforced concrete beams [25], whilst another study conducted the stiffness evaluation and FE analysis of fibre-reinforced epoxy resin laminates [26]. In summary, the variable axial force cable has been widely used in reinforcement engineering, with great success. In practical engineering, the construction scheme of the cable with variable axial force should be reasonably designed in accordance with the application research of the cable with variable axial force in reinforcement engineering and relevant standards, combined with the actual situation, thus improving the reinforcement effect. However, studies on the application of the variable axial force cable in reinforcement engineering of steel truss structures have been limited, thereby motivating this research.

As typical representatives of modern industrial buildings, large-span factory buildings have various structural forms, huge spans and weak seismic ability and can easily be affected by natural disasters and human factors. Therefore, how to improve the earthquake resistance and overall stability of large-span plants has always been a concern of researchers. Recently, variable axial force cable technology has gradually been widely used, as it can effectively resist the impact of earthquakes, wind and other abnormal loads; improve the structural stability and seismic ability of long-span factory buildings; and can be widely used in large bridges, high-rise buildings and other fields.

This paper explores the application of cable technology with variable axial forces in the reinforcement of large-span powerhouses. Taking two large-span powerhouses as examples, the influence of cable reinforcement with variable axial force on large-span powerhouse structure is studied through FE analysis and experimental verification. The results of this study can provide a theoretical and practical basis for seismic reinforcement of large-span powerhouses. Simultaneously, this paper also discusses the application prospects and development trends of the variable axial force cable reinforcement technology in other fields. Through this study, we make an important contribution to the structural reinforcement and seismic capacity improvement of large-span factory buildings, as well as provide a useful reference for the sustainable development of modern industrial buildings.

2. Project Profile

Powerhouses 1 and 2 of Zhejiang Hongcheng Industrial Co., Ltd. are double-span gantry rigid steel structures. Industrial Powerhouse 1 was built in 2008, covering a building area of 11,200.3 m^2 and with an eaves elevation of 11.400 m, according to the fire risk classification D, fire resistance grade 2 and the redesigned safety grade 2. Powerhouse 1 has waterproof-grade III roofing and uses a moulding steel plate for defensive protection. As for the seismic fortification of this project category C, its seismic fortification intensity is $6°$, the design basic acceleration is 0.05 g, and the engineering design life of the steel frame main body is good for 50 years.

Powerhouse 2 was also built in 2008, covering a building area of 5952 m^2 and with an eaves elevation of 11.100 m (slightly higher than Powerhouse 1). Its fire risk classification, fire resistance grade, safety grade, waterproof grade, seismic fortification category, seismic fortification intensity, design basic acceleration and engineering design life are the same as those of Powerhouse 1. The photos of the two house trusses are shown in Figures 1 and 2, respectively, whilst the corresponding section views of Powerhouses 1 and 2 are shown in Figures 3 and 4.

Figure 1. Powerhouse 1.

Figure 2. Powerhouse 2.

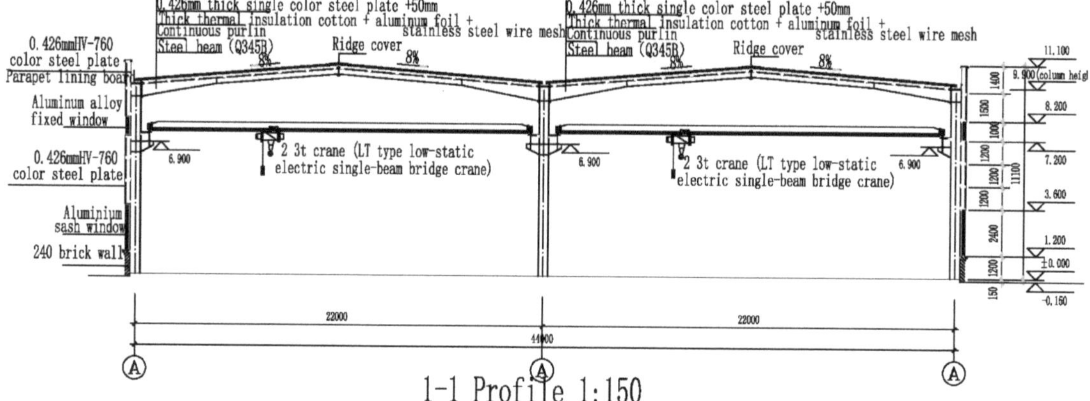

1-1 Profile 1:150

Figure 3. Cross-section view of Powerhouse 2.

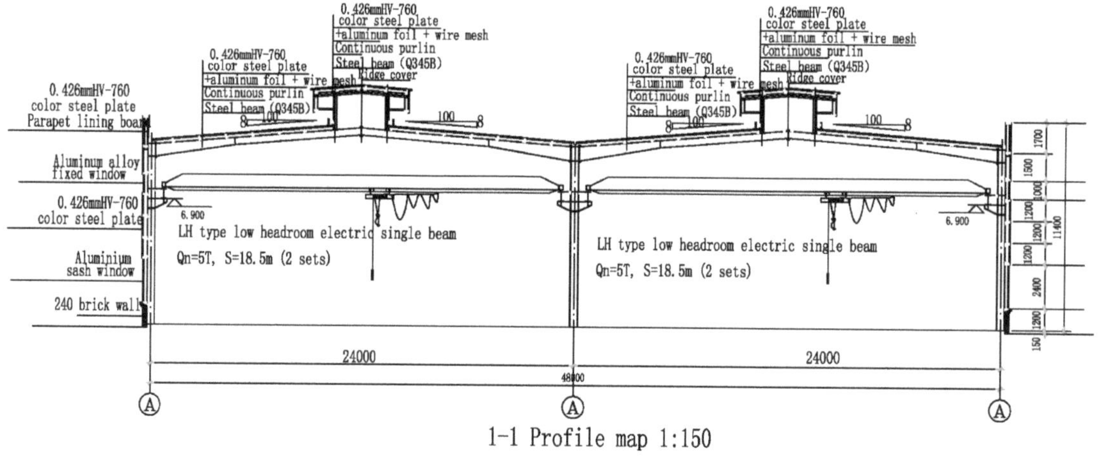

1-1 Profile map 1:150

Figure 4. Cross-section view of Powerhouse 1.

3. Reinforcing Analysis

3.1. Reinforcement Scheme

Due to the addition of roof photovoltaic panels in the two plant buildings, the existing plant buildings cannot meet the requirements of the new code. Therefore, under the influence of the above two factors, the buildings must be strengthened comprehensively. Simultaneously, the roofs of the two factory buildings are cracked, causing water leakage that affects their normal use. Furthermore, the deflection of the truss exceeds the limit, the crane track is seriously worn, and the crane in the lane cannot pass normally. The plant area is reinforced and reformed under the influence of the new regulations.

Due to the large span of the trusses in the powerhouses of Zhejiang Hong Cheng Company, if the conventional increase force surface of the column is arranged between the rigid trusses, the headroom area of the powerhouse will be reduced, and the crane and the vehicles inside the powerhouse cannot be used or passed normally. Therefore, we selected a variable axial force cable for the overall truss reinforcement. The cable is arranged in a radial manner, and the anchor block is arranged in the purlin of the original rigid frame node. Upon reinforcement, the original truss only must bear the dead weight of the truss, whilst the new cable variable axial force bears two parts of the load, mainly the dead weight and tension of the cable, as well as the additional dead load and other live loads of the truss, thereby improving the overall bearing capacity of the truss.

To summarise the prestressing force value of the cable, two kinds of cable calculations and comparisons were selected in this paper. Type 1 simulates the cable reinforcement with a low controlling force. The span of the original truss was 22 m, whilst the controlling force was 400 kN. For the cable reinforcement of the conventional control force in the Type 2 simulation, the span of the original truss was 24 m, and the control force was 600 kN, as shown in Figures 5 and 6. Through theoretical calculations and analysis, we obtained the influence of two kinds of cable reinforcement effects, from which information the whole node plates are designed.

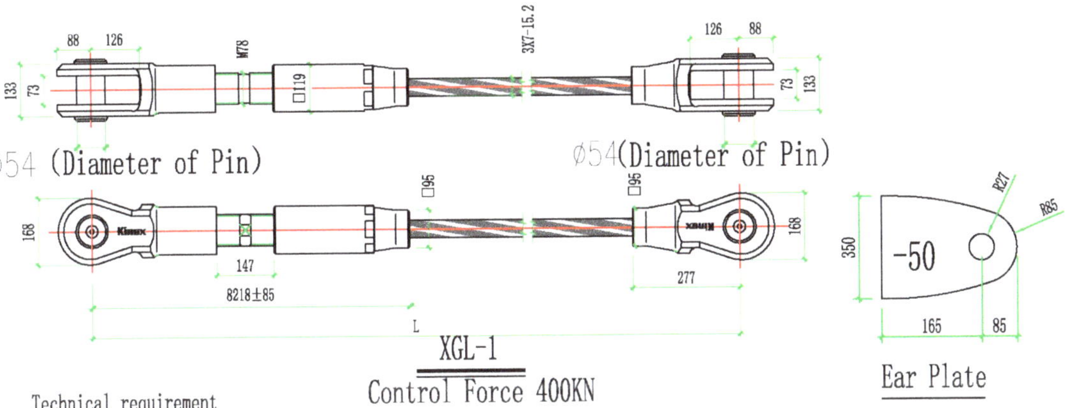

Technical requirement
1. Steel cable adopts steel strand.
2. Steel wire resistance and tensile strength grade is 1860MPa.
3. Theoretical minimum breaking force of cable body is 781kN, and static load of cable is 100KN.
4. Cable making error: ±15mm.
5. Adjustable cable head, adjustable anchor cup, fixed cable head outer surface hot spray zinc after spraying zinc rich primer, pin shaft, pin cover plate, screw surface galvanized. After completion of construction, the construction unit will paint the anchorage of the cable twice, the same as the steel structure.

Figure 5. Detailed drawing of the variable axial force cable in Powerhouse 1.

3.2. Reinforcement Mechanism

In this study, we developed a method of cable reinforcement with variable axial force from a bridge system. The cable is a kind of cable-bearing bridge, in which the force form of the main truss is similar to the continuous beam that is supported by an elastic multi-point position. In truss calculations, the position and controlling force of the cable have a great influence on the whole force of the truss.

The main impact of changing the controlling force and position of the cable is that the height of the cable directly affects the dip angle of the cable. In the general truss design, the cable provides elastic support for the roof structure [1]. Therefore, to obtain a larger vertical component, a cable with a larger dip angle must be selected. A study of the bridge systems found that the ratio between the layout height of the conventional cable and the span of the truss generally ranges from 1/4–1/6, which is the reasonable layout height of a cable. If the height of the truss is short, the ratio between the layout height and the path of the truss generally ranges from 1/8–1/12 [2]. Due to the different layout heights, the stress characteristics of the two are also different. Simultaneously, for the transverse diameter cable tie truss/cover, both ends of the suspension cable can be designed to be equal or unequal height, and the sag should have a range of 1/10–1/20 of the span, according to JGJ257-2012 cable structure technical regulations [3]. Here, we selected the design value of the cable control force and verified that this is actual practice, according to Articles 5.6.1 and 5.6.2 of the JGJ257-2012 Cable Structure Technical Regulations.

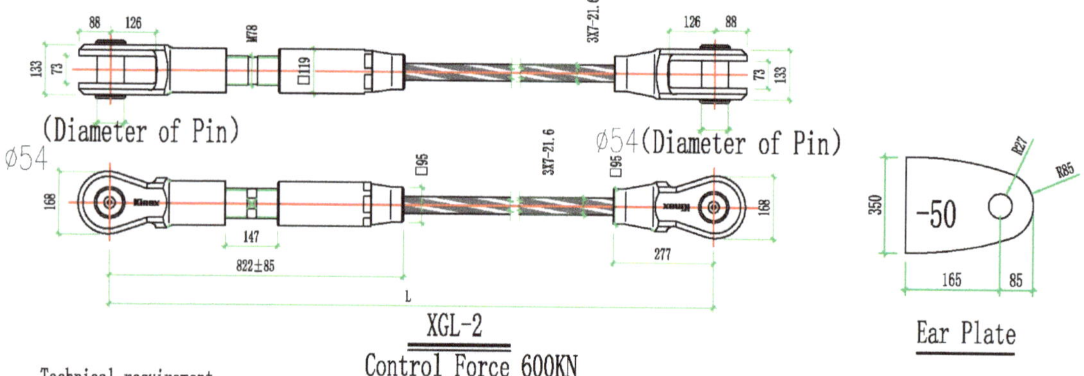

Figure 6. Detailed drawing of the variable axial force cable in Powerhouse 2.

3.3. Computational Analysis and Modelling

(1)   The overall situation is shown in Figures 3 and 4.

(2)   The force analysis of the 3/A axis column and the deformation analysis of the column section after stress are shown in Figure 7. The schematic diagram of the roof truss load G1 and crane beam load G2 is shown in Figure 8.

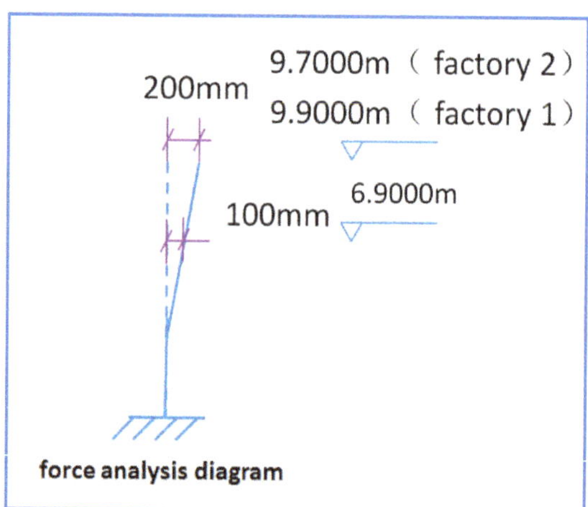

Figure 7. Column stress deformation diagram.

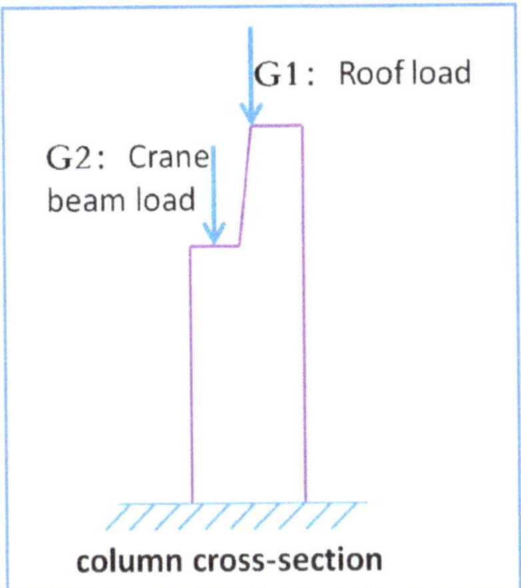

Figure 8. Column cross-section.

(3) Analysis of the upper part: The height of the column section is within the height range of 16.9–9.9 m. The original eccentricity of G1 is 400 mm, which has turned 100 mm to the right—a condition that is favourable to the upper end and ensures that no breakage occurs. The calculation is based on the GB 50009-2019 building structure loading code.

(4) Overall analysis (see Figure 9):

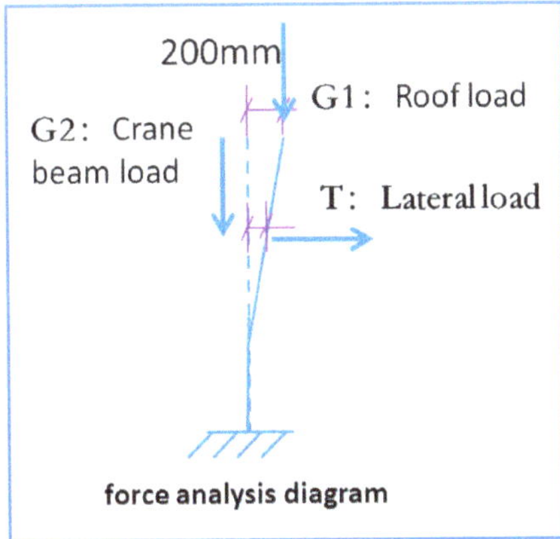

Figure 9. Schematic of the overall mechanical analysis.

G1: Original eccentricity + 0.3 (positive to the left)
G2: original eccentricity − 0.5
Deflection due to the column force:
G1: original eccentricity $e_1 = +0.5$
G2: eccentricity becomes $e_2 = -0.4$
Based on the calculation of a single piece of house truss in Powerhouse 1:

1) Roof plate dead load: 3 kN/m^2; Live load: 0.5 kN/m^2; The span ranges from 22–24 m, with Seta pin spacing of 8 m

$$G_1 = (1.3 \times 3 + 1.5 \times 0.5) \times 8 \times 14 \times \frac{1}{2} = 45.9 \text{ kN}$$

2) Wind load: Using simplified calculation, we take 1 as a uniform load, and then the line load is as follows: $1 \times 8 = 8\frac{kN}{m} = q_w$.

3) Vertical crane load (see Figure 10), because the eccentricity of the crane load is negative, the worst case is 0.

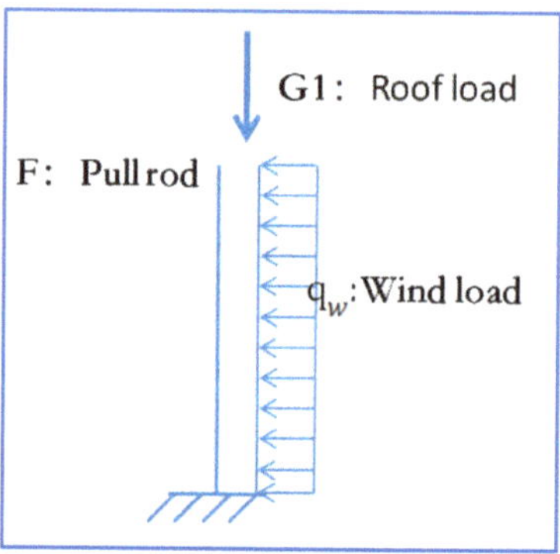

Figure 10. Schematic of vertical crane load mechanics.

4) Dead weight of the wall: $M_4 = 8 \times 0.3 \times 30 \times 22 \times 0.1 \times 1.3 = 205.4 \text{ kN/m}$

5) The horizontal load (transverse) of the crane is two sets of $Q = 20/5$ t soft hook crane A6 and heavy car.

$$\sum T_k = 2 \times 2 \times 0.1 \times \frac{5+20}{4} \times 9.8 = 24.5 \text{ kN}$$

$$M = M_1 + M_2 + M_4 + 0.7 \times M_5$$

For the convenience of calculation, we considered the roof live load, according to the dead load, to meet the guaranteed rate. This includes the following:

$$= 45.9 \times 0.5 + 205.4 + \frac{1}{2} \times 8 \times 30 \times 30 \times 1.5 + 0.7 \times 915.08 = 6268.906 \text{ kN/m}$$

6) Current situation: mm column bending capacity configuration: 10 HRB400 rebar with a diameter of 25 mm:

$$M_{\text{configuration}} = 2425 \text{ kN/m}$$

7) Horizontal tie-bar tension:

$$T = \frac{M}{11.4} = 549.9 \text{ kN}$$

When the current mm can bear part of the bending distance, then:

$$T = \frac{6268.906 - 2425}{11.4} = 337.18 \text{ kN}$$

We considered adding a cable rod between the columns at an elevation of 6.9 m. The cable rod coordinates the bending distance between the two ends [27]. It can either be removed after reinforcement or retained permanently (more suitable) to reduce the eccentricity of the columns.

(5) 3D3S modelling was applied after importing the overall calculation data [28], as shown in Figures 11 and 12.

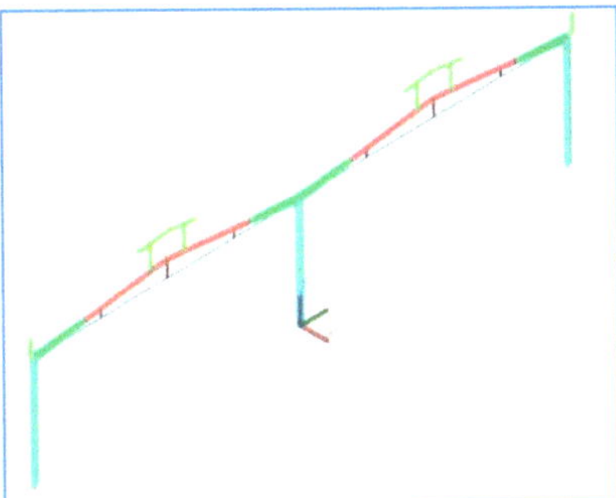

Figure 11. 3D3S model of the southeast axis.

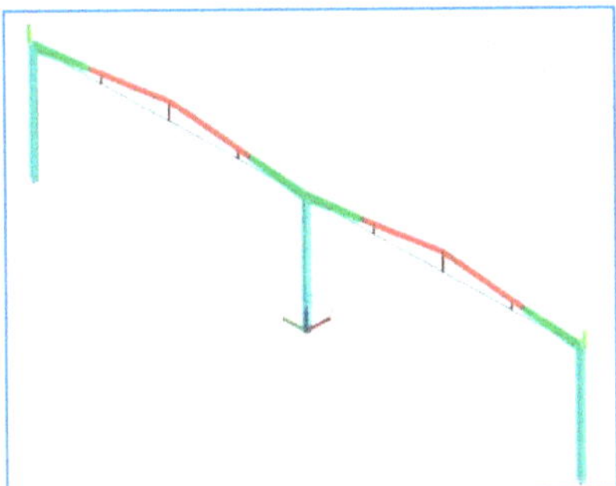

Figure 12. 3D3S model of the southwest axis.

3.3.1. Condition 1: Uniform Load

We compared and analysed the deflection values of different span trusses before and after reinforcement under a uniform load in working condition 1, as shown in Figure 11. The maximum deflection value was 55.42 mm before reinforcement, whilst the maximum deflection value was 12.52 mm after the span of 22 m. Compared with the state before reinforcement, the deflection value of each component decreased by more than 77.4%. The maximum deflection value of the cable span after reinforcement at 24 m was 15.64 mm. Compared with the state before reinforcement, the deflection value of each component decreased by more than 71.8% [29], as shown in Figure 13.

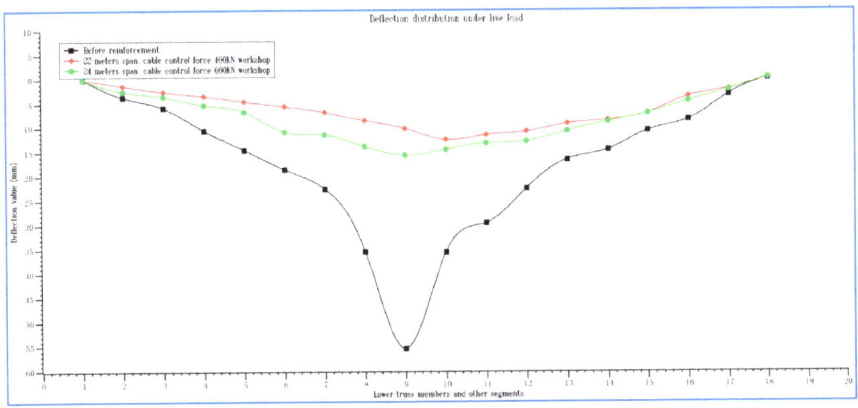

Figure 13. Deflection distribution under live load.

The difference in cable span and axial force led to a change in the cable dip angle. When the span and axial force of the cable changed, the components of the cable force along the X axis and the Z axis also differed, in which the larger the span, the larger the axial force (i.e., the larger the component along the Z axis, the larger the deflection reduction value). Compared with the axial forces of 400 kN with a 22 m span and 600 kN with a 24 m span, the reduction rate of the deflection value of the component strengthened with the latter was greater than that of the former under the same uniform load conditions. When the controlling force exceeded 600 kN, the reduction rate of the deflection value became smaller [30].

3.3.2. Condition 2: Stress Value

Table 1 lists the changes in the maximum stress value of the lower member after cable reinforcement. Combined with the data in the table and the stress distribution trend of the lower member of the truss in Figure 14, it can be seen that the stress system of the original truss changed after reinforcement, after which the internal forces of each truss member changed accordingly. Before the original truss reinforcement, it can be regarded as a simply supported truss structure, with the maximum bending moment value at the mid-span. The reinforced cable was similar to the elastic support, which changed the type of original structure, i.e., the 1-span 22 m/24 m truss changed into a 4-span 5.5 m/6 m continuous beam. Furthermore, the structural type and single span changed, greatly reducing the internal force value of the reinforced truss. After reinforcement, the initial tension of the cable expanded the range of the compression member of the truss under the action of dead load, whilst the tensile stress of the tension rod decreased.

Table 1. Changes in the stress values of lower truss members after cable reinforcement.

Span (m)/Control Force (kN)	Maximum Stress (MPa)	Stress Reduction (MPa)	Reduction Rate (%)
Before reinforcement	310.82	0.0	0.0
22 m/400 kN	112.52	212.54	65.38
24 m/600 kN	70.54	244.70	77.62

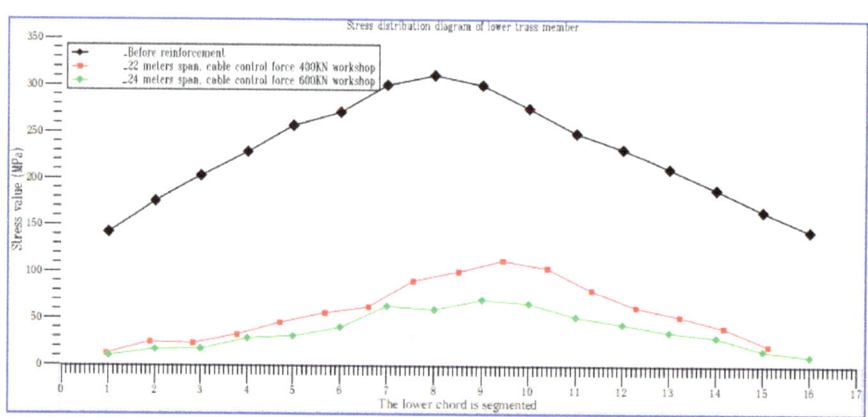

Figure 14. Stress distribution diagram of the lower truss members.

Figure 15 shows that the maximum compressive stress of the upper member of the reinforced front truss exceeded the yield strength of the steel used for the upper member, with the value reaching 155.85 MPa. Some members will be damaged. The compressive stress value of the upper member almost reached 30 MPa, and the stress value was greatly reduced after the reinforcement with a variable axial force cable. When the original truss was a single-span truss that was a simply supported structure, the mid-span maximum compressive stress of the upper member was nearly three times that of the fulcrum. The multi-point elastic support provided by the reinforced cable reduced the span of a single span, thus decreasing the compressive stress difference of the upper component. This resulted in more uniform stress of the component and increased structural life. Due to the influence of the initial tension of the cable, the truss has an upward arch under the action of a dead load, and there is a reserve of tensile stress on the upper member. When the strain is applied to the whole truss, the reserve stress generated by the initial tension can offset part of the load effect, resulting in a substantial reduction in the tensile stress value.

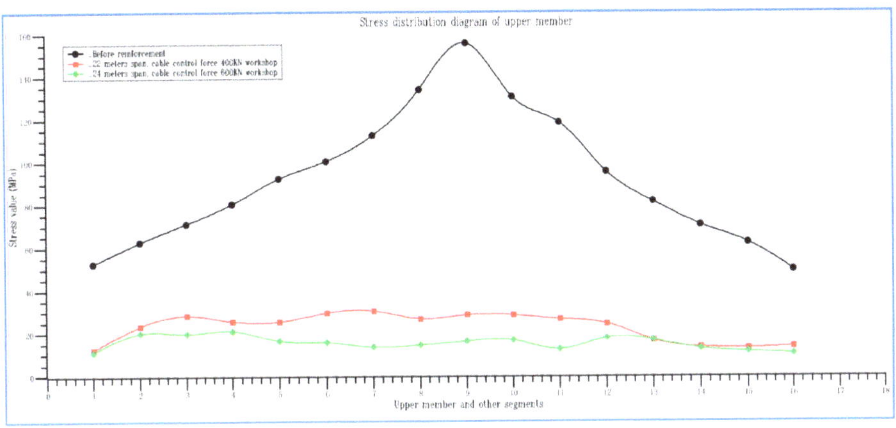

Figure 15. Stress distribution diagram of the upper truss member.

As shown in Figure 16, the stress distribution of the inclined rod of the reinforced front truss consisted of two inclined rods connected to the same node (one under tension and one under pressure), whilst the absolute value of the stress decreased continuously from the fulcrum to the span [30]. After reinforcement, eight of the 24 diagonal rods on each side were placed under tension, whilst 16 were under pressure. Furthermore, the two diagonal rods at the node position of the cable anchorage were under pressure, whilst the tension and pressure of nodes at adjacent positions were placed alternately. The details presented in Table 2 below can also be seen in Figures 16–18.

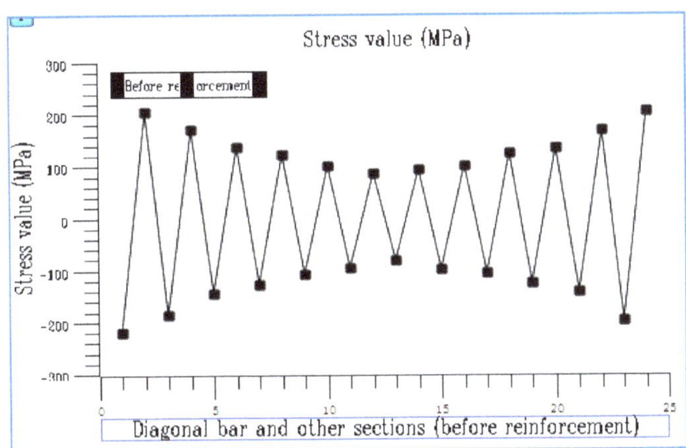

Figure 16. State before reinforcement.

Table 2. The stress changes in the inclined bar of the truss after cable reinforcement.

Component Name	Maximum Compressive Stress (MPa)/Position	Component Name	Maximum Compressive Stress (MPa)/Position	Reduced Value (MPa)
Reinforce the front inclined bar	219/Near-truss	Reinforce the rear diagonal bar	33.8/Near-truss	185.2 ↓
Reinforce the front mid-span diagonal bar	73.5	Reinforce the diagonal bar in the middle of the rear span	38.2	34.6 ↓
The rear inclined rod was reinforced with 22 m controlling force and 400 kN cable	39.9	The rear inclined rod was reinforced with a 24 m controlling force of 600 kN cable	38.2	1.7 ↓

Note: ↓: It means to decrease or decrease.

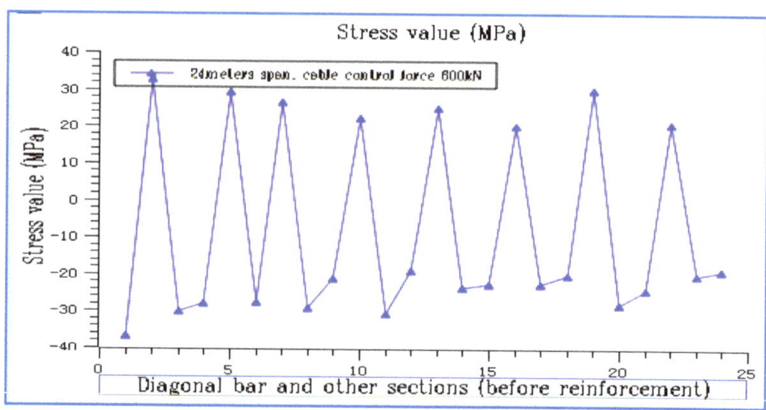

Figure 17. 400 kN control force cable after reinforcement.

Figure 18. 600 kN control cable after reinforcement.

By comparing the stress changes after the control force changes, the maximum stress value of the inclined bar with a 22 m span and 400 kN control force decreased by 1.7 MPa compared with that with a 24 m span and 600 kN control force. Therefore, the maximum stress value of the inclined bar changes with the span and control force.

By comparing the deflection, stress values of the upper and lower truss members and inclined bar of two kinds of plant trusses with variable axial force reinforcement under live load, the results showed that, in terms of the reinforcement of large-span trusses, the cable with a large controlling force within a reasonable span had better performance, and the controlling force increased with the increase in span. According to the standard, the 600 kN control force can be considered the conventional control force. The cable reinforcement produces a certain inclination angle, and the Z-component increases, which greatly unloads the powerhouse truss. Moreover, the vertical support provided by the cable can transform the large-span simply supported structure into a multi-span continuous structure, thus shortening the span of the single-span structure and greatly reducing its internal force value.

4. Nodal Analysis

4.1. FE Modelling

A solid model was established for the lug plate at the joint position of the plant with complex forces (Figure 19). The joint plate was made of Q345B [30] steel, and the thickness was 20 mm. Each side of the roof truss was provided with internal and external node plates, each with a diameter of 20 mm large hexagonal head high-strength bolts 64 and a hole diameter of 22 mm. The bolt strength class was 10.9 S, and the bolt pretension ranged from 155–187 kN, the value of which was specified in Appendix B of the GB50205-2020 Steel Structure Engineering Construction Quality Acceptance Standard [31].

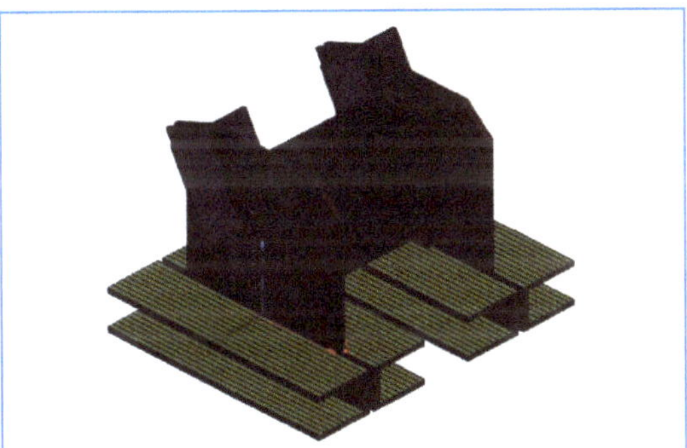

Figure 19. Finite element model of gusset plate.

Here, the force of each member of the node was transferred through welded connections, such as the flange plate and bottom plate, and friction was provided by friction-type high-strength bolts [5]. The ASET software was used to establish the model. The material nonlinearity and geometric linearity were considered in the calculation, and the model was loaded in five steps, according to the GB50344-2004 Technical Standard of Building Structure Inspection, thus ensuring the convergence of the calculation structure. To save computing resources, only bolts were established in this paper (the stress test of side plate materials was omitted). For the simulation test, we selected the most unfavourable joint plate with the greatest internal force before and after reinforcement.

4.2. Node Plate Analysis

As can be seen, the stress value of the bolt group is greatly reduced after adding cable reinforcement, which improves the overall stability of the powerhouse trusses.

5. Conclusions

(1)   After the cable reinforcement, the stiffness of the building truss increased, and the stress distribution trend of each component changed. Under the action of the crane and other main live loads, the reduction rate of the deflection value exceeded 50%. Fur-

thermore, the maximum stress reduction rates of the upper and lower truss members exceeded 60%, whilst the overall load increase rates of the first and second power-house trusses exceeded 70% (Figures 11–21) after being reinforced by the variable axial force cable.

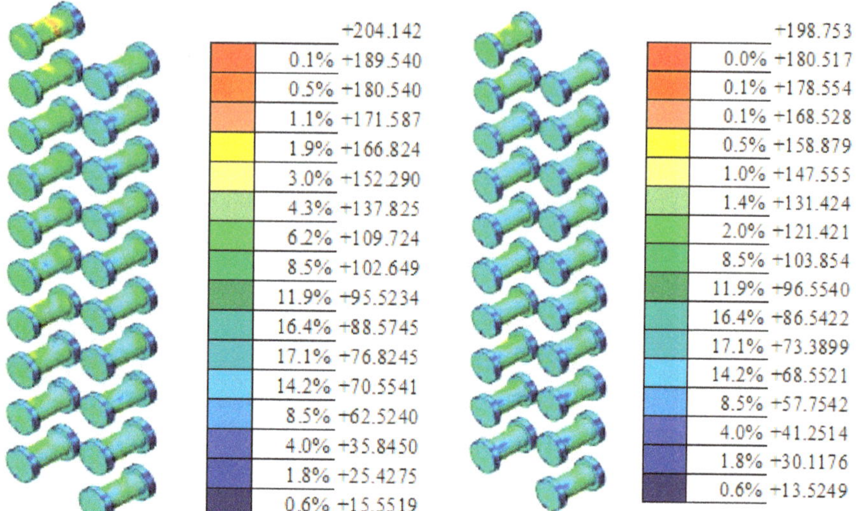

(**a**) Outside connecting plate bolt (**b**) Inner connecting plate bolt

Figure 20. Stress distribution diagram of the bolt before reinforcement.

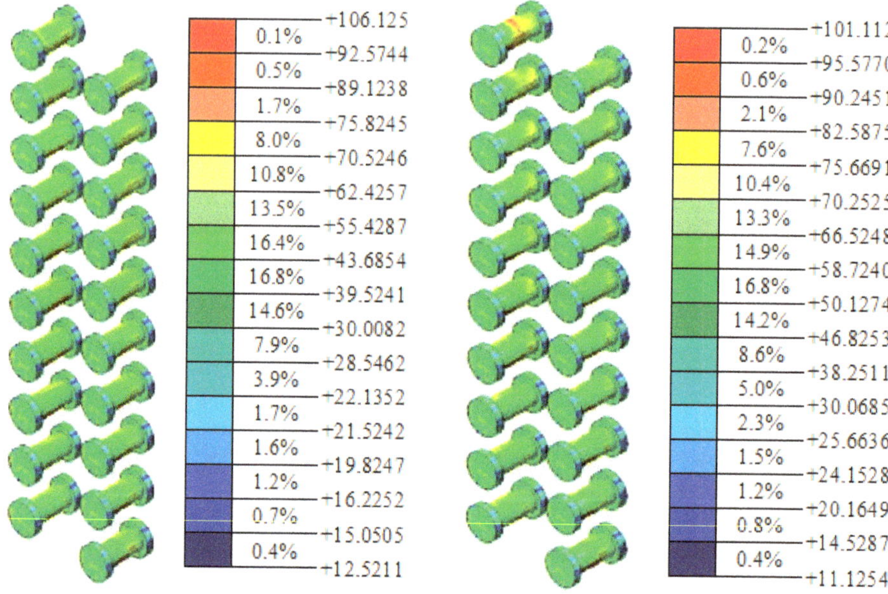

(**a**) Outside connecting plate bolt (**b**) Inner connecting plate bolt

Figure 21. Stress distribution of bolts reinforced with 600 kN cable with 24 m controlling force.

(2) Before reinforcement, the overall stress level of each node was reasonable, but this increased due to the increased load and service life limit after the installation of photovoltaic panels, along with the second-highest stress concentration at the bolt position of the node plate. After the reinforcement, the stress values of the node plates all decreased significantly. Moreover, the stress below 100 MPa accounted for over 90% of the nodes, whilst the higher stress accounted for less than 1% of the nodes (Tables 1 and 2). The node plates after the reinforcement were in the stable stress area without tearing or stress damage.

(3) The high-stress and secondary high-stress areas of the bolt group were mainly distributed near the load position. After the reinforcement, the stress zone area of 50–100 MPa accounted for about 35% of the bolt group, whilst the stress zone area above 100 MPa accounted for less than 1% (Table 3) TA. The problem of stress concentration has been solved, and the requirements of the new specification have been met.

Table 3. Results of the comparative analysis of Figure 20a,b and Figure 21a,b.

Condition	Maximum Stress of the Connecting Plate Bolt (MPa)	The Stress Distribution Area above 100 MPa Accounted for (%)	The Stress Distribution Area between 50 and 100 MPa Accounted for (%)	The Stress Distribution Area below 50 MPa Accounted for (%)
Before reinforcement	204	25.6	68.1	6.3
After using 24 m of control force and 600 kN cable reinforcement	106	0.1 ↓	34.6 ↓	65.3 ↑

Note: ↓: It means to decrease or decrease. ↑: It means to rise or increase.

In summary, the plant truss reinforced using the variable axial force cable method can play a role in lifting and strengthening the whole truss [3]. Two kinds of cables with different spans and controlling forces were used to conduct load-lifting reinforcement application practice in Powerhouses 1 and 2. Through comprehensive comparative analysis, the results revealed that the cable reinforcement of powerhouse trusses plays an active and effective role in the stress system. When the cable reinforcement of Powerhouse 2 was used with a 24 m span and controlling force of 600 kN, the reinforcement effect of the structural system was better than that of Powerhouse 1.

Author Contributions: Conceptualization, Z.S. (Zizhen Shen); methodology, Z.S. (Zizhen Shen); software, Z.S. (Zizhen Shen).; validation, M.H. and X.L.; formal analysis, X.L.; investigation, L.W.; resources, X.W.; data curation, M.H.; writing—original draft preparation, Z.S. (Zizhen Shen); writing—review and editing, X.L.; visualization, Z.S. (Zizhen Shen); supervision, X.W.; project administration, Z.S. (Zigang Shen); funding acquisition, Z.S. (Zigang Shen). All authors have read and agreed to the published version of the manuscript.

Funding: This research was funded by [Research Project of Zhejiang Provincial Department of Housing and Urban-Rural Development] grant number [2022K010] and The APC was funded by [Zhejiang Provincial Department of Housing and Urban-Rural Development].

Conflicts of Interest: The authors declare no conflict of interest.

References

1. Wang, X.; Yang, Y.; Xin, X.; Jia, J.; Xu, G.; Chen, Z. Computer Aided Design and Numerical Simulation of Dongming Yellow River Bridge Strengthened by a Cable-Stayed System. In Proceedings of the 2021 3rd International Conference on Artificial Intelligence and Advanced Manufacture, Manchester, UK, 23–25 October 2021; pp. 2625–2633.
2. Cai, H.; Aref, A.J. On the design and optimization of hybrid carbon fiber reinforced polymer-steel cable system for cable-stayed bridges. *Compos. Part B Eng.* **2015**, *68*, 146–152. [CrossRef]
3. Ceballos, M.A.; Prato, C.A. Determination of the axial force on stay cables accounting for their bending stiffness and rotational end restraints by free vibration tests. *J. Sound Vib.* **2008**, *317*, 127–141. [CrossRef]
4. Simões, L.M.C.; Negrão, J.H.O. Sizing and geometry optimization of cable-stayed bridges. *Comput. Struct.* **1994**, *52*, 309–321. [CrossRef]
5. Song, C.; Xiao, R.; Sun, B. Optimization of cable pre-tension forces in long-span cable-stayed bridges considering the counterweight. *Eng. Struct.* **2018**, *172*, 919–928. [CrossRef]

6. Malekinejad, M.; Rahgozar, R. An analytical approach to free vibration analysis of multi-outrigger-belt truss-reinforced tall buildings. *Struct. Des. Tall Spec. Build.* **2013**, *22*, 382–398. [CrossRef]
7. Khanorkar, A.; Sukhdeve, S.; Denge, S.V.; Raut, S.P. Outrigger and belt truss system for tall building to control deflection: A review. *GRD J. Glob. Res. Dev. J. Eng.* **2016**, *1*, 6–15.
8. Qi, J.; Yang, H.C. Improvement of a truss-reinforced, half-concrete slab floor system for construction sustainability. *Sustainability* **2021**, *13*, 3731. [CrossRef]
9. Hwang, J.S.; Lee, K.S. Seismic strengthening effects based on pseudodynamic testing of a reinforced concrete building retrofitted with a wire-woven bulk kagome truss damper. *Shock. Vib.* **2016**, *2016*, 1–17. [CrossRef]
10. American Society of Civil Engineers. *Minimum Design Loads and Associated Criteria for Buildings and Other Structures*; American Society of Civil Engineers: New York, NY, USA, 2017.
11. Hou, Z.; He, W.; Zhou, J.; Qiu, L. national standard «Code for acceptance of construction quality of steel structure engineering» GB50205 Introduction of special research results in revision. *Stand. Eng. Constr.* **2014**, 49–55. [CrossRef]
12. Keane, B.; Schwarz, G.; Thurnherr, P. Cables and cable glands for hazardous locations. In Proceedings of the 2018 IEEE Petroleum and Chemical Industry Technical Conference (PCIC), Cincinnati, OH, USA, 24–27 September 2018; Volume 42, p. 9.
13. Revanna, N.; Moy, C.K.S.; Krevaikas, T. *Verifying a Finite Element Analysis Methodology with Reinforced Concrete Beam Experiments*; Research Press of America, Ed.; Research Press of America 2020 Paper Collection IV; Scientific Research Publishing: Chengdu, China, 2021; pp. 340–347.
14. Teng, J.G.; Yu, T.; Fernando, D. Strengthening of steel structures with fiber-reinforced polymer composites. *J. Constr. Steel Res.* **2012**, *78*, 131–143. [CrossRef]
15. Gardner, L. Stability and design of stainless steel structures—Review and outlook. *Thin-Walled Struct.* **2019**, *141*, 208–216. [CrossRef]
16. Subramanian, N. *Design of Steel Structures*; Oxford University Press: Oxford, UK, 2008.
17. Trahair, N.S.; Bradford, M.A.; Nethercot, D.A.; Gardner, L. *The Behaviour and Design of Steel Structures to EC3*; CRC Press: Boca Raton, FL, USA, 2017.
18. Ban, H.; Shi, G. A review of research on high-strength steel structures. *Proc. Inst. Civ. Eng. Struct. Build.* **2018**, *171*, 625–641. [CrossRef]
19. Marshall, J.D.; Jaiswal, K.; Gould, N.; Turner, F.; Lizundia, B.; Barnes, J.C. Post-earthquake building safety inspection: Lessons from the Canterbury, New Zealand, earthquakes. *Earthq. Spectra* **2013**, *29*, 1091–1107. [CrossRef]
20. Bortolini, R.; Forcada, N. Building inspection system for evaluating the technical performance of existing buildings. *J. Perform. Constr. Facil.* **2018**, *32*, 04018073. [CrossRef]
21. Ferraz, G.T.; De Brito, J.; De Freitas, V.P.; Silvestre, J.D. State-of-the-art review of building inspection systems. *J. Perform. Constr. Facil.* **2016**, *30*, 04016018. [CrossRef]
22. Wen, J.; Liu, L.; Jiao, Q.; Yang, J.; Liu, Q.; Chen, L. Failure analysis on 20MnTiB steel high-strength bolts in steel structure. *Eng. Fail. Anal.* **2020**, *118*, 104820. [CrossRef]
23. Caccese, V.; Mewer, R.; Vel, S.S. Detection of bolt load loss in hybrid composite/metal bolted connections. *Eng. Struct.* **2004**, *26*, 895–906. [CrossRef]
24. Schauwecker, F.; Moncayo, D.; Middendorf, P. Characterization of high-strength bolts and the numerical representation method for an efficient crash analysis. *Eng. Fail. Anal.* **2022**, *137*, 106249. [CrossRef]
25. Ahmad, F.; Bajpai, P.K. Evaluation of stiffness in a cellulose fiber reinforced epoxy laminates for structural applications:Experimental and finite element analysis. *Def. Technol.* **2018**, *14*, 278–286. [CrossRef]
26. Jake, F.; Fidelis, M. Finite Element Analysis of Telecommunication Structure Reinforcement. *ce/papers* **2022**, *5*, 1084–1091. [CrossRef]
27. Chhushyabaga, B.; Karki, S.; Khadka, S.S. Effect of Mechanical Vibration in a Power House Located in the Nepal Himalaya. In *IOP Conference Series: Earth and Environmental Science*; IOP Publishing: Bristol, UK, 2022; Volume 1037, p. 012065.
28. Kiarasi, F.; Babaei, M.; Sarvi, P.; Asemi, K.; Hosseini, M.; Omidi Bidgoli, M. A review on functionally graded porous structures reinforced by graphene platelets. *J. Comput. Appl. Mech.* **2021**, *52*, 731–750.
29. Lin, Y.H.; Lin, Z.H.; Chen, Q.T.; Lei, Y.P.; Fu, H.G. Laser in-situ synthesis of titanium matrix composite coating with TiB–Ti network-like structure reinforcement. *Trans. Nonferrous Met. Soc. China* **2019**, *29*, 1665–1676. [CrossRef]
30. López, D.L.; Roca, P.; Liew, A.; Echenagucia, T.M.; Van Mele, T.; Block, P. A three-dimensional approach to the Extended Limit Analysis of Reinforced Masonry. In *Structures*; Elsevier: Amsterdam, The Netherlands, 2022; Volume 35, pp. 1062–1077.
31. Koshcheev, A.A.; Roshchina, S.I.; Naichuk, A.Y.; Vatin, N.I. The effect of eccentricity on the strength characteristics of glued rods made of steel cable reinforcement in solid wood. In *IOP Conference Series: Materials Science and Engineering*; IOP Publishing: Bristol, UK, 2020; Volume 896, p. 012059.

 *buildings*

Article

A Continuously Derivable Uniaxial Tensile Stress-Strain Model of Cold-Formed Circular Steels

Chang Yang, Ling Ying, Binbin Wang and Qi Li *

College of Architecture and Urban-Rural Planning, Sichuan Agricultural University, Chengdu 611830, China; 41479@sicau.edu.cn (C.Y.); yingling9898@163.com (L.Y.); wangbinbin0806@163.com (B.W.)
* Correspondence: 41471@sicau.edu.cn

Abstract: Promoting prefabricated steel structures is considered one of the crucial approaches to meeting the objectives of "carbon peak" and "carbon neutrality" in the construction industry. Due to insufficient practical experience and incomplete fine engineering techniques in civil construction, the sustainable development of prefabricated building systems in China faces many challenges. Taking steel components as an example, the design process of tubular columns does not pay enough attention to the influence of the cold-working effect on material mechanical properties, and the constitutive relationship of cold-formed steels is not clear, which will cause an engineering economic burden and may affect the judgment of catastrophic problems. To serve the refined design and meet the intelligent construction technology using the computer platform, a modified Menegotto-Pinto model using a continuously derivable function is proposed in the paper. The proposed model can successfully describe the complete stress-strain curve of cold-formed circular mild steels as long as the basic mechanical parameters of the parent material are determined. Taking into account the influence of the strength and thickness of the parent steel sheets, as well as the internal bending radius r, on the cold-rolling effect, the model can also flexibly track the elastic-plastic nonlinearity of the cold-formed materials. In addition, the research shows that the cold-rolling effect will weaken with the increase of the yield strength $f_{sy,0}$ of the parent steels and r/t ratio, and may disappear when $f_{sy,0}$ reaches 1748 MPa or the r/t ratio is approximately 60, which can be used as economic indicators during the design process.

Keywords: cold-formed; mild steels; circular hollow sections; uniaxial tensile stress-strain model; material property; high-strength

Citation: Yang, C.; Ying, L.; Wang, B.; Li, Q. A Continuously Derivable Uniaxial Tensile Stress-Strain Model of Cold-Formed Circular Steels. *Buildings* **2024**, *14*, 36. https://doi.org/10.3390/buildings14010036

Academic Editors: Zechuan Yu and Dongming Li

Received: 21 November 2023
Revised: 14 December 2023
Accepted: 19 December 2023
Published: 22 December 2023

1. Introduction

To accomplish the objectives of "carbon peak" and "carbon neutrality" in the building and construction industries, the promotion and application of prefabricated steel structure buildings are advocated to meet the requirements of the circular economy and sustainable development in China [1,2]. The measures to promote prefabricated steel structure buildings can be divided into the following broad categories: construction technology, the use of high-strength steel, intelligent construction technology using the computer platform, and so on [3]. In recent years, due to their high capacity, ease of construction, and recyclable utilization, cold-formed steel structures have shown great potential in intelligent design. While experiencing cold-forming, the uniaxial tensile stress-strain curve of mild steels exhibits a more rounded stress-strain response which is no longer suitable to be described by an ideal elastic-plastic model or simple broken line models. However, the most popular structural design is still inclined to use such models to simulate cold-formed steels. On the other side, the guidance contents in most of the current specifications [4–6] are on the basis of early experimental work, which limits the range of material strength and geometry parameters. In fact, there are significant differences in the strength improvement of different steel section types formed through cold-rolling. Those specifications primarily concentrate on the strength enhancement for the corner sections but neglect the cold-rolling effect for the circular hollow sections (CHSs for short). The above issues may lead to some key technical issues being vague and increasing the economic burden of engineering, which contradicts the intention of green building and intelligent design.

Determining a material constitutive model is one of the essential parts of structural analysis. Establishing an efficient computational numerical model in intelligent design requires defined material strength and a continuous function to deal with the stress-strain relationship of materials. Many studies have been carried out to study the uniaxial tensile stress-strain models of cold-formed steels as well as the strength enhancement due to cold work. Li et al. [7] have contrasted several available predictive methods of strength enhancement and found that current specifications, such as AISI [5] and Eurocode 3 [6], overestimate the cold-formed effect on corner yield strength, while the empirical methods proposed by various researchers have limitations in terms of the r/t ratio. Liu et al. [8] proposed a new method to predict the corner strength of cold-formed conventional steels based on measured data, and the method can be applicable to the yield strength $f_{sy,0}$ of parent materials, which ranges from 256 MPa to 497 MPa and the r/t ranges from 0.57 to 7.54. Masoud Kalani [9] investigated the cold work effect on the tensile behavior of thick steel plates and verified the accuracy of several available equations for predicting the average yield stress of the experimental specimens. Pham [10] investigated the $G450$ channel steels to better predict the strength enhancement of high-strength steels. Chen [11] conducted an investigation into the material properties of high-strength CHS steels with r/t ranging from 11.5 to 32.3 and found that the yield strength improvement rate $f_{sy}/f_{sy,0}$ of the $Q460$ was 1.09, while the $Q960$ section exhibited no increase in strength. Meng [12] discovered that tensile coupons of high-strength CHS ($f_{sy,0}$ = 799 MPa, r/t = 7~29) exhibit lower f_{su}/f_{sy} as the r/t ratio decreases. Generally, the use of high-strength material leads to a longer life span of the structure and brings cost-effectiveness [13]. However, based on experimental investigations, Chan et al. [14] found that the measured failure strain of high-strength steels cannot meet the requirements of Eurocode 3 due to the press-braking process.

Given that most research has focused on the effect of cold-forming on material strength enhancement, the influence of cold-forming on the deformation ability of steels should be given more attention and carefully introduced into the relationship between stress and strain. Based on the differences in the fabrication process, Yao et al. [15] established a finite element-based method for plastic strains, as well as residual stresses, in cold-formed steel hollow sections, but the stress-strain relationships of cold-formed steels used in the finite element model were transformed from experimental curves. Gardner [16] proposed a method to predict the strength enhancement in the corner regions of cold-formed sections by considering the plastic strains associated with the dominant stages in the fabrication process. Further, Gardner [17] improved the Ramberg-Osgood model [18] by using piecewise functions to describe the stress-strain curve of cold-formed steels, and the improved model was confirmed to have good accuracy. Similarly, Quach and Huang [19] also raised a modified Ramberg-Osgood model to describe the uniaxial tensile stress-strain curve of cold-formed steels. Based on detailed experimental testing, Li et al. [20] established a material model to simulate cold-formed high-strength steels. Note that although the above models can reflect the rounded tensile curve characteristics of cold-formed steels, their mathematical carriers are all piecewise functions, and the steel yield strength f_{sy} after cold work is required before using these models.

In summary, the cold-rolling effect of steel is generally affected by the yield strength $f_{sy,0}$, and thickness t of the parent steel sheets, the internal bending radius r, and the section shape [21,22]. The available prediction expressions used to reveal the strength enhancement and stress-strain curves of cold-formed steels were more for corner sections than for CHSs. In order to better promote the efficient, intelligent design of prefabricated steel structures, the objective of this paper is to propose a continuously derivable uniaxial tensile stress-strain model of cold-formed circular steels. The paper focuses on the influence of cold-rolling on the material properties of CHS mild steels based on the design parameters of the parent steels. More specifically, based on the collected experimental data of CHS mild steels with a wide range of yield strength and r/t ratio, a systematic analysis was conducted on the influence of cold-rolling on the strength and deformation of different sections of slenderness. The proposed constitutive model was established by modifying the Megenetto-Pinto model [23] since the function image can reflect a complete nonlinear characteristic of cold-formed steels under uniaxial tension and exhibits a clear physical meaning. Finally, in order to demonstrate the superiority of the proposed model, comparative studies of the full range of stress-strain curves and corresponding absorption capacity were carried out with the measured curves and available models.

2. Uniaxial Tensile Stress-Strain Model Based on the Menegotto-Pinto Model

Establishing a complete stress-strain model with a unified function will enhance the design and process of intelligent platforms. Through detailed comparison and investigation, a modified Menegotto-Pinto model was raised to describe the uniaxial tensile stress-strain relationship of cold-formed CHS mild steels. In fact, the Menegotto-Pinto model was first raised to deal with the nonlinear responses of reinforced concrete members under earthquakes, and the material law of mild steels is a four-parameter model, which is described by a continuously derivable composite function [23]. In previous studies, the model was successfully improved to fit the ascending stage of the equivalent stress-strain relationship of cold-formed steel stub columns [24], and its mathematical expression and corresponding figure are shown in Equation (1) and Figure 1, respectively. Note that the parameter S, as shown in Figure 1, represents the ratio of ultimate tensile strength to yield strength. Combining Equation (1) and Figure 1, it can be seen that the modified model is composed of a derivable function and can flexibly track the material nonlinearity by adjusting the value of N.

$$f_s = E_s \varepsilon_s \left(Q + \frac{1 - Q}{\left(1 + \left(\frac{\varepsilon_s}{\varepsilon_{sy}}\right)^N\right)^{\frac{1}{N}}} \right), \quad \varepsilon_s \leq \varepsilon_{su} \tag{1}$$

where Q is the strain-hardening coefficient; the exponent N mainly controls the roundness of the yield stage; ε_{sy} is the nominal yield strain; ε_{su} is the ultimate strain corresponding to the ultimate tensile strength f_{su}.

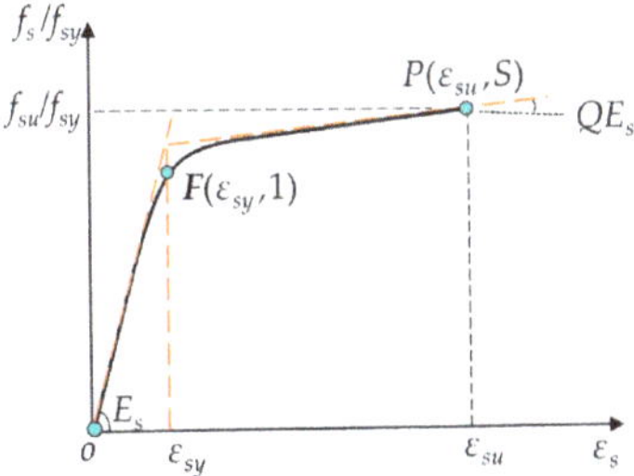

Figure 1. Outline of the improved uniaxial stress-strain model for cold-formed CHS steels based on the Menegotto-Pinto model.

To improve the efficiency of intelligent design and to highlight the relationship between the constitutive model of cold-formed mild steels and design parameters ($f_{sy,0}$ and t of the parent material and the r/t ratio of CHS), the paper intends to continue improving the modified Megenetto-Pinto model for fitting the material constitutive model of cold-formed CHS mild steels.

3. Tensile Coupon Details of Cold-Formed CHS Steels

A comprehensive collection of 74 experimental results from the available literature is assembled. Figure 2 displays the labels assigned to the tensile coupon, while specimens from the weld area are excluded from the analysis. Additionally, specimens failing to meet the ductility requirements of the specifications are also eliminated. A summary of the variables of specimens as well as necessary experimental results is provided in Table 1, where $f_{sy,0}$ and f_{sy} are the yield strengths of the same material before and after cold-rolling, respectively. Additionally, f_{su} represents the ultimate tensile strength of cold-formed CHS steel specimens. For convenience, based on the parent steel, coupons with $f_{sy,0}$ exceeding 460 MPa are considered high-strength steels [25]. As shown in Table 1, the range of $f_{sy,0}$ is from 400 MPa to 1400 MPa, corresponding to the thickness t of steel sheets varying from 1.5 mm to 10 mm, while the yield strength improves from 357 MPa to 1402 MPa for cold-formed CHSs with a r/t ratio of 6~32.

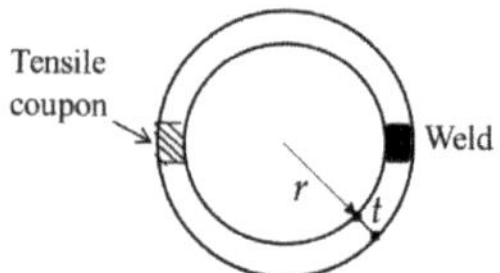

Figure 2. Location of the tensile coupon.

Table 1. Summary of the detail of tensile coupons.

Ref.	Specimens	r/t	E_s (Gpa)	$f_{sy,0}$ (MPa)	f_{sy} (MPa)	f_{su} (MPa)
[11]	4C200 × 3	32.3	207.4	546.5	571.7	632.8
	4C150 × 3	24	205.4	546.5	574.4	623.2
	4C150 × 6	11.5	217.2	580.7	623.9	694.8
	4C200 × 6	15.7	216	580.7	630.3	698.5
	4C250 × 6	19.8	217.8	580.7	603.5	685.2
	6C150 × 6	11.5	198.8	756.3	765.1	808.6
	6C200 × 6	15.7	208	756.3	758.3	808
	6C350 × 6	28.2	207.7	756.3	755.6	804.1
	9C150 × 6	11.5	205.6	973.3	959	1045.2
	9C200 × 6	19	208.3	973.3	964.7	1040.5
	9C300 × 6	29	207.7	973.3	969.9	1037.1
[12]	CHS139.7 × 4	16.6	213.3	700	742.4	842.3
	CHS168.3 × 4	20.3	211.7	700	720	823.4
	CHS139.7 × 5	13.3	212.5	700	729.7	843.3
	CHS139.7 × 6	10.6	207.9	700	779	866.7
	CHS139.7 × 8	7.9	205.7	700	784.8	866.8
	CHS139.7 × 10	6.1	205.6	700	787.6	877.5
[26]	89 × 4	10.1	209	1100	1084	1242
	108 × 4	12.5	208	1100	1233	1327
	133 × 4	15.6	210	1100	1164	1278
	89 × 3	13.8	203	900	980	1093
[27]	V89 × 4	10.4	210	900	1054	1108
	S89 × 4	10.4	205	1100	1180	1317
	S108 × 4	12.9	215	1100	1180	1292
	S133 × 4	16.1	204	1100	1159	1291
	S139 × 6	10.8	194	1100	1014	1382
	V89 × 3	14.03	209	900	1053	1124
Total	21 coupons	6.1~32.3	198.8~217.8	546.5~1100	571.7~1233	623.2~1382
[28]	CHS01	11.5	203	690	746	811
	CHS02	15.7	204	690	747	816
	CHS03	9	202	690	757	837
	CHS04	11.5	201	690	767	827

Table 1. *Cont.*

Ref.	Specimens	r/t	E_s (Gpa)	$f_{sy,0}$ (MPa)	f_{sy} (MPa)	f_{su} (MPa)
[29]	193.7 × 8	11.1	198.6	355	404	480
[30]	C1	8.7	191	350	454	520
	C2	11.3	220	350	416	484
	C3	15.5	204	350	453	521
	C4	18.3	200	350	430	514
	C5	19.4	204	350	379	440
	C6	22.8	207	350	357	474
	C7	23	193	350	433	479
	C8	27.5	206	350	395	481
[31]	CBC1	19.1	200	350	365	469
	CBC2	14.9	210	350	432	538
	CBC3	14.6	218	350	415	534
	CBC4	11.4	211	350	433	508
	CBC5	10.8	205	350	456	548
	CBC6	9.1	204	350	408	503
	CBC7	7.1	207	350	442	511
	CBC8	5.4	209	350	460	568
[32]	TS1A	10.7	190.9	1350	1402	1558
	TS1B	10.8	195.1	1350	1392	1533
	TS1C	10.6	190.7	1350	1400	1550
	TS2A	9.3	198.3	1350	1361	1513
	TS2B	9.4	204.4	1350	1360	1507
	TS2C	9.2	197.6	1350	1362	1499
	TS3A	8.3	195.6	1350	1328	1477
	TS3B	8.4	197.1	1350	1329	1495
	TS3C	8.3	200.2	1350	1332	1487
	TS4A	16.8	203	1350	1346	1506
	TS4B	16.6	194.2	1350	1365	1519
	TS4C	16.9	197	1350	1368	1540
	TA5A	16.9	195.2	1350	1363	1540
	TS5B	16.9	196.7	1350	1370	1568
	TS5C	22.7	203.7	1350	1399	1520
[33]	1	22.6	201.6	355	456.8	527
	2	22.9	203.6	355	451.7	534.2
	3	14	200.2	355	455.6	529.2
	4	14	195.4	355	392	503.7
	5	17.9	196.6	355	405.2	511.8
	6	17.9	198.5	355	443.9	508.1
	7	21.2	196.7	355	385	500
	8	21.2	197.3	355	397.4	511.1
	9	21.2	196.7	355	436.4	502.1
[34]	CHS139.7 × 8	8	202.2	700	856.8	893.7
	CHS139.7 × 10	6	203.1	700	762	804.9
Total	53 coupons	5.4~27.5	190.7~220	350~1350	357~1402	440~1568

Figure 3 summarizes the range of values for several key parameters of the collected tensile coupons. The coupons are divided into three groups, with the boundaries of $f_{sy,0}$ being 460 MPa and 690 MPa, respectively. As shown in Figure 3, the number of coupons for the three groups is relatively uniform, as well as the range of r/t.

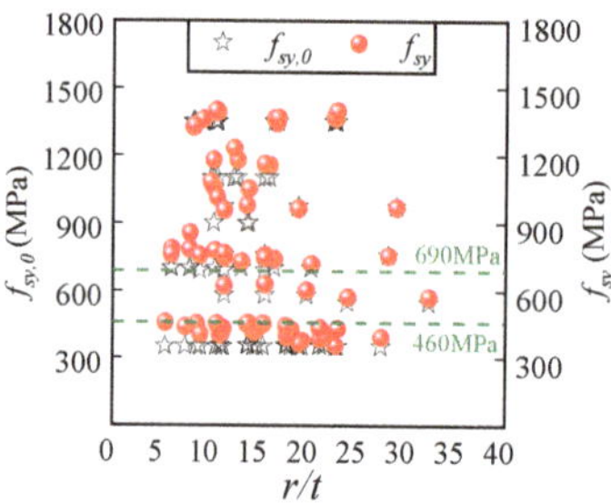

Figure 3. Distribution of the main measured variables of the collected experimental data.

All the gathered measured data are utilized to establish a predictive expression for the yield strength f_{sy} and ultimate tensile strength f_{su} considering the cold work of materials. 54 sets of experimental values of ultimate strain ε_{su} from cold-formed CHS steel specimens are used to assess the corresponding empirical formula. Furthermore, a total of 26 stress-strain curves spanning the full range are examined and analyzed to derive suitable prediction formulas for the curvature coefficient N and the strain-hardening exponent Q. The mathematical expressions of these parameters will be derived in the following sections.

4. Analysis of Results and Recommendations

4.1. Yield Strength f_{sy} of Cold-Formed CHSs

Based on the collected data of Table 1, Figure 4 presents the relationship between the measured yield strength f_{sy} and the r/t ratio, and f_{sy} has been normalized by the measured $f_{sy,0}$. According to the trend line of the red dotted line shown in Figure 4, it can be found that the $f_{sy}/f_{sy,0}$ ratio decreases with the r/t ratio increasing, and the cold-rolling effect seems to disappear when the value of the r/t ratio reaches 60 based on the trend line of Figure 4.

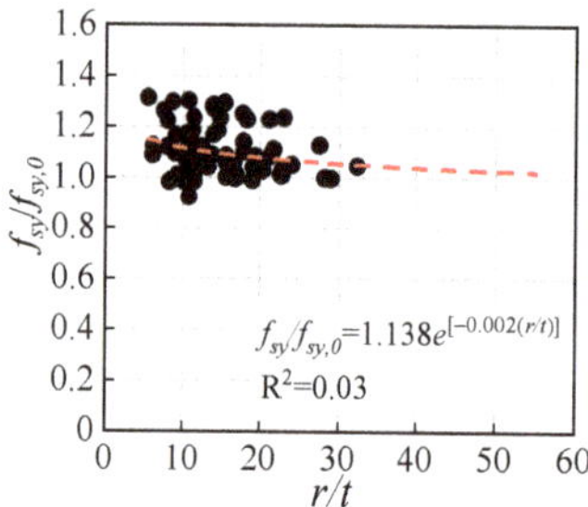

Figure 4. Relationship between the measured $f_{sy}/f_{sy,0}$ ratio and r/t.

The yield strength of the cold-formed CHS steels differs from that of the parent material. To examine the influence of the cold-rolling effect, a sensitivity analysis is conducted on the r/t ratio and the yield strengths $f_{sy,0}$ and f_{sy}. By comparing the relationship between f_{sy} and three different physical quantities shown in Figure 5, it can be found that the independent r/t is different in establishing connections with f_{sy}, while the compound parameter $f_{sy,0}/(r/t)^{0.5}$ exhibits a better correlation with f_{sy} as the R^2 reaches 0.87. The optimal group is the relationship between $f_{sy,0}$ and f_{sy} because the measured f_{sy} has the strongest correlation with $f_{sy,0}$, which can be expressed by a linear function.

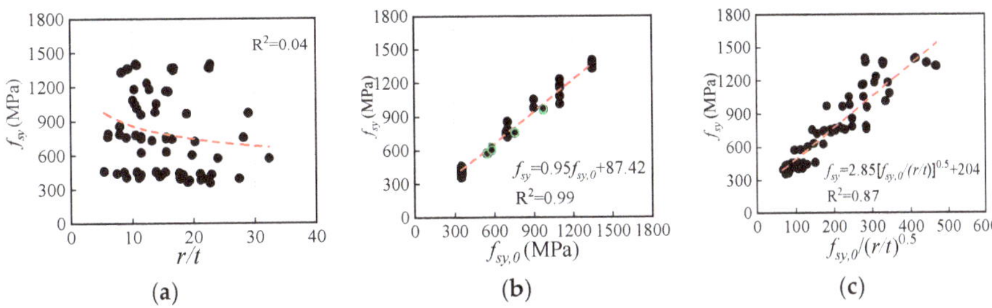

Figure 5. Relationships between the yield strength f_{sy} of CHS and the main variables. (**a**) r/t. (**b**) $f_{sy,0}$. (**c**) combined parameters based on r/t and $f_{sy,0}$.

The red trend line in Figure 5b shows that the relationship between $f_{sy,0}$ and f_{sy} exhibits a strong linear correlation. The difference between f_{sy} and $f_{sy,0}$ decreases and even disappears with the increase of $f_{sy,0}$, and the results is also confirmed by previous research [11], the relevant experimental data of which is highlighted in green dots in Figure 5b. Furthermore, if setting the $f_{sy}/f_{sy,0}$ ratio to 1.0, it can be obtained that the value of f_{sy} is 1748 MPa, and the value suggests that the influence of material strength on the cold-rolling effect has an upper limit. Based on the results shown above, a more physically meaningful way of determining f_{sy} can be obtained, as shown in Equation (2). When $f_{sy,0}$ exceeds 1748 MPa, the value of f_{sy} can be considered consistent with $f_{sy,0}$.

$$\frac{f_{sy}}{1748} = 0.95\frac{f_{sy,0}}{1748} + 0.05 \ (f_{sy,0} \leq 1748 \text{ MPa}) \tag{2}$$

4.2. Curvature Coefficient N

The curvature coefficient N is the critical roundness variable at the yield stages of the stress-strain curve. In this section, the analytical curve is required to pass through the coordinate origin and the point $(\varepsilon_{su}, f_{su})$ shown in Figure 1. The ideal value of N is evaluated based on two criteria: (1) the envelope area ratio of the calculated results A_{cal} to the experimental results A_{exp} (corresponding to Figure 6a), and (2) the coincidence degree of the transition curvature in the elastic-plastic stage (corresponding to Figure 6b). Through debugging and analysis, as shown in Figure 6, the empirical values of N are almost in the range of 4 to 8.

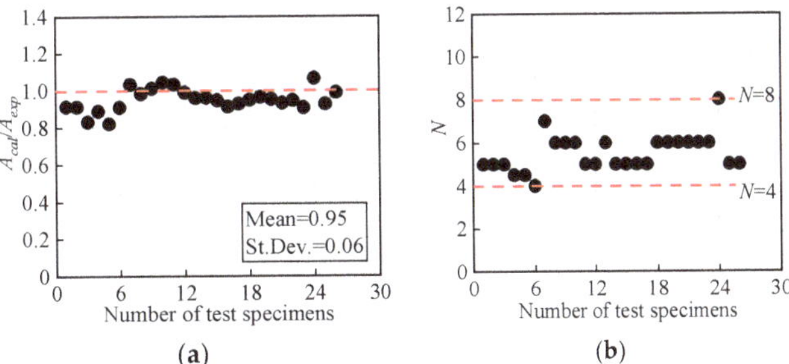

Figure 6. The value of the curvature coefficient N. (**a**) Optimal envelope area. (**b**) Optimal elastic-plastic curvature.

The relationships between the curvature coefficient N and the main variables are shown in Figure 7. It is clear that the r/t ratio and the combined parameter $f_{sy,0}/(r/t)^{0.5}$ are not the primary factors affecting the curvature coefficient N, while $f_{sy,0}$ and the corresponding strain $\varepsilon_{sy,0}$ have the same correlation coefficient relationship with N. Considering the influence of the elastic modulus *Es* of tensile coupons, $\varepsilon_{sy,0}$ is chosen to be the primary factor, and Equation (3) is proposed to calculate the curvature coefficient *N*.

$$N = 0.33\varepsilon_{sy,0}^{-0.05} = 0.33\left(f_{sy,0}/E_s\right)^{-0.5} \tag{3}$$

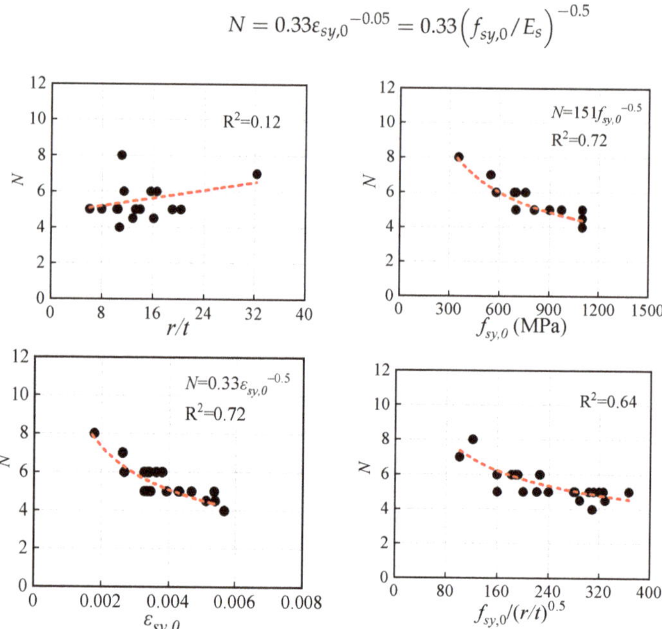

Figure 7. Relationships between the curvature coefficient *N* and the main variables.

4.3. Strain-Hardening Exponent *Q*

The strain-hardening exponent *Q* characterizes the strain-hardening behavior of the strengthening segment. The value of *Q* can be determined by defining two points $(\varepsilon_{sy}, f_{sy})$ and $(\varepsilon_{su}, f_{su})$, as shown in Figure 1. To identify the critical factor for *Q*, the relationships between the measured *Q* and the main variables are shown in Figure 8. It can be concluded that the correlation coefficient R^2 between the combination parameter $(r/t)^2/f_{sy,0}$ and *Q* is 0.65, which is higher than that of the other variables. Therefore, the strain-hardening coefficient *Q* can be calculated using Equation (4).

$$Q = 0.0053\left[(r/t)^2/f_{sy,0}\right]^{-0.13} \tag{4}$$

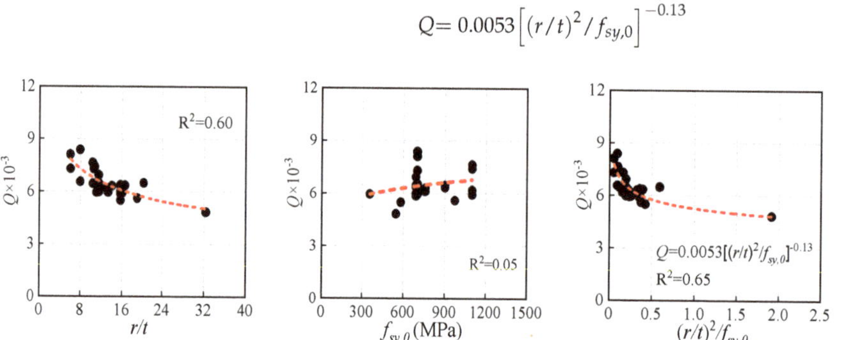

Figure 8. Relationships between the strain-hardening exponent *Q* and the main variables.

4.4. Ultimate Strain ε_{su}

The relationships between the ultimate strain ε_{su} and the main variables are shown in Figure 9. There is no obvious relationship between the measured ultimate strain $\varepsilon_{su,cal}$, and the *r/t* ratio. The correlation coefficient between ε_{su} and the $\varepsilon_{sy,0}$ is 0.82. It can be observed that the ultimate strain ε_{su} shows a stronger correlation with the combined parameter $f_{sy,0}/(r/t)^{0.5}$ compared with $\varepsilon_{sy,0}$, whose correlation coefficient is 0.9.

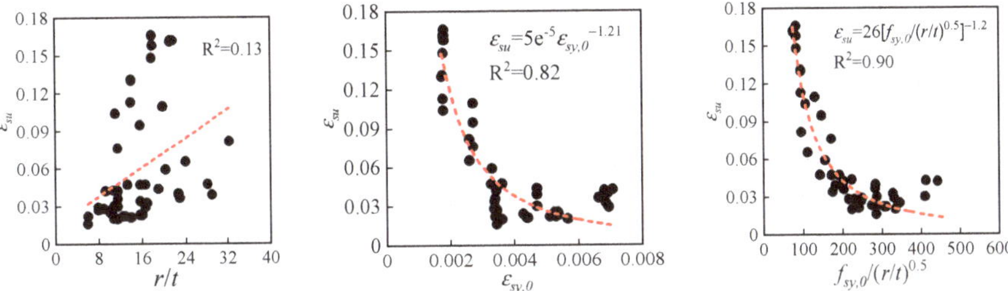

Figure 9. Relationships between the ultimate strain ε_{su} and the main variables.

To summarize, the ultimate strain ε_{su} can be calculated by the following formula:

$$\varepsilon_{su} = 26\left[f_{sy,0}/(r/t)^{0.5}\right]^{-1.2} \tag{5}$$

4.5. Ultimate Strength f_{su}

Figure 10 shows the relationship between ultimate tensile strength f_{su} and three main factors. There is no obvious relationship between the *r/t* ratio and f_{su}. The correlation coefficient between f_{su} and the combined parameter is 0.87, while the strongest correlation can be observed between $f_{sy,0}$ and f_{su} with a correlation coefficient as high as 0.99, which implies an extremely strong linear correlation.

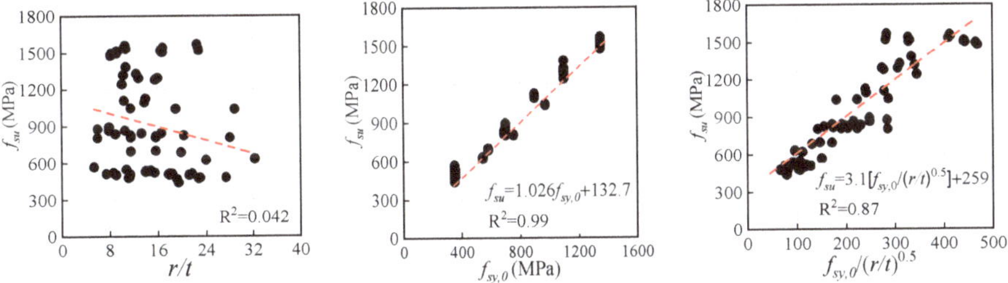

Figure 10. Relationships between the ultimate strength f_{su} and the main variables.

By comparing the relationship between f_{su} and three different physical quantities shown in Figure 10, it can be found that the optimal group is the linear relationship between $f_{sy,0}$, and f_{su}. This indicates that there is a significant correlation between the yield strength of the parent steels and f_{sy} compared with other factors, and the following equation is derived to calculate the ultimate tensile strength:

$$f_{su} = 1.026f_{sy,0}+132.7 \tag{6}$$

5. Verification of the Proposed Model

5.1. Comparison of the Calculated Results and Measured Results

Comparisons between the experimental and calculated results are conducted, including ultimate tensile strength f_{su}, ultimate strain ε_{su}, and uniaxial tensile stress-strain curves of cold-formed steels. Firstly, the experimental ultimate strain of 54 groups of specimens is compared with the calculated results $\varepsilon_{su,cal}$ based on both Equation (5) and the method by Gardner [17]. As shown in Figure 11, the proposed method by Equation (5) can better predict the actual results by Gardner than the method proposed by Gardner (the corresponding equations can be found in Appendix A), because its mean $\varepsilon_{su,cal}/\varepsilon_{su,exp}$ of 1.079 and moderate standard deviation of 0.3400 are both smaller than the other.

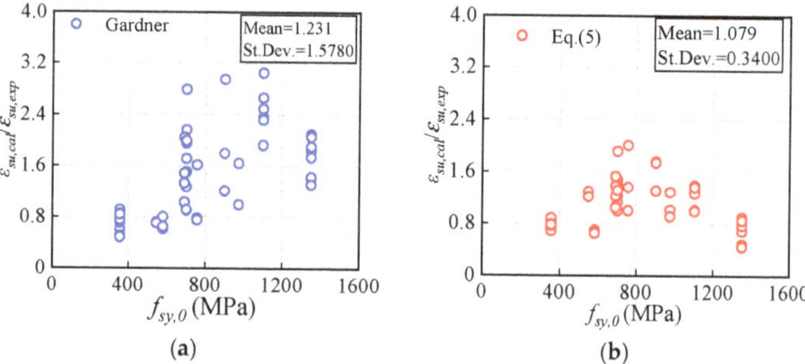

(a)

(b)

Figure 11. Comparison between the measured and calculated ultimate strain ε_{su}. (**a**) Available method raised by Gardner. (**b**) Proposed method in the paper

Secondly, the comparison between the experimental ultimate tensile strength $f_{su,exp}$ (from 74 test groups) and the calculated results $f_{su,cal}$ (obtained using Equation (6)) is conducted, as illustrated in Figure 12. The predictive expression of ultimate strength f_{su} proposed by Gardner is also plotted in the figure. Both expressions proposed demonstrate relatively high accuracy in predicting the ultimate tensile strength with errors of approximately 15%, while Equation (6) proposed by the authors is better suited for high-strength steels.

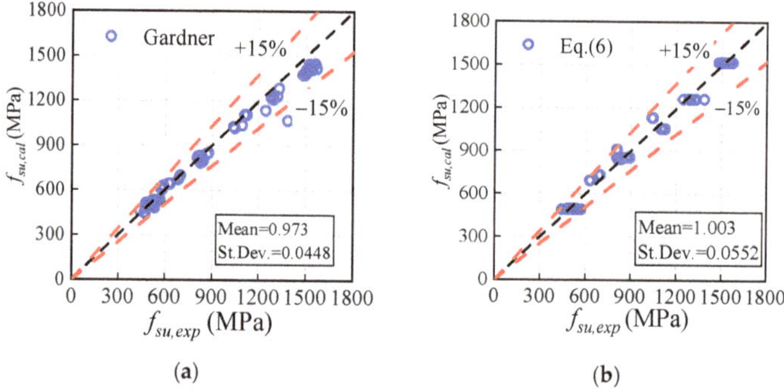

(a)

(b)

Figure 12. Comparisons between the measured and calculated ultimate strength f_{su}. (**a**) Available method raised by Gardner. (**b**) Proposed method in the paper

It is crucial to elucidate the difference between the stress-strain relationship prediction models. In order to evaluate the overall accuracy of the proposed model, a comparison is made between the measured stress-strain curves of tensile coupons and the corresponding calculated ones. Similarly, as a representative of existing models, the improved Ramberg-Osgood model proposed by Gardner is still being compared. Table 2 lists the main variables of the chosen specimens. As shown in Figure 13, both of the predictive models for cold-formed CHS steels have good consistency. Concretely, when $f_{sy,0}$ is within 600 MPa, as shown in Figure 13a,b, the proposed model based on the Megenetto-Pinto model is almost better than the one based on the Ramberg-Osgood model. With the value of $f_{sy,0}$ increasing, as shown in Figure 13c, the two models are seen to have consistent accuracy. However, it should be noted that the improved Megenetto-Pinto model proposed in the paper is directly based on the most basic design parameters, including $f_{sy,0}$ and t of the parent steels and internal bending radius r, while the model proposed by Gardner needs to know the yield strength of cold-formed steel.

Table 2. Main variables of the test specimens for comparison.

Notation	193.7 × 8	4C200 × 3	4C200 × 6	CHS168.3 × 4	CHS139.7 × 5	6C200 × 6	6C150 × 6	9C200 × 5
$f_{sy,0}$ (MPa)	355	546.5	580	700	700	756.3	756.3	973
r/t	11.1	32.2	15.7	20.3	13.3	15.7	15.7	15
Ref.	[29]	[11]		[12]			[11]	

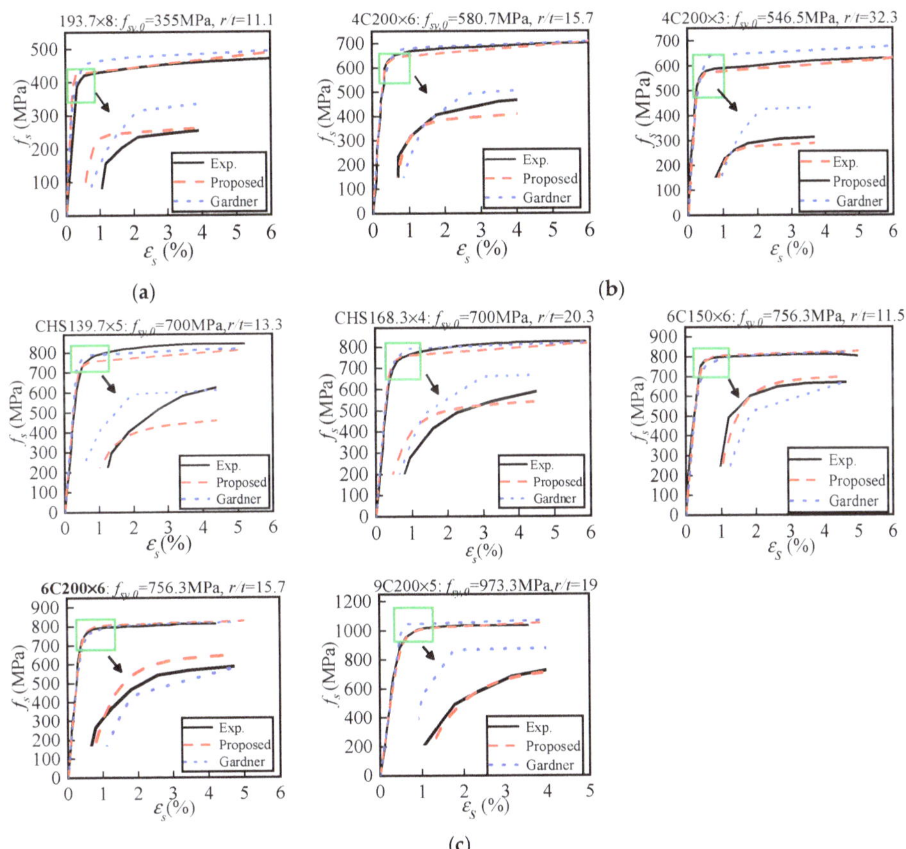

Figure 13. Comparison of experimental and calculated uniaxial tensile stress-strain curves. (**a**) Conventional strength steels. (**b**) High-strength steels with an average $f_{sy,0}$ of 565 MPa. (**c**) Ultra-high strength steels with $f_{sy,0}$ greater than 690 MPa

5.2. Case Application Analysis

Choosing the cold-formed CHS steel stub columns under axial load as simulated subjects, a comparison of energy absorption, which is gained from the load-bearing curves, are carried out based on 0material constitutive models of the ideal elastoplastic model and the modified model, respectively. The energy absorption can be obtained by Equation (7). The higher the value, the stronger the ability of the component to resist deformation will become.

$$E = \int_0^D F(s)\,ds \tag{7}$$

where E represents the energy absorbed by the component through the entire ascending stage; $F(s)$ is the load; and D is the displacement corresponding to the load.

Table 3 shows the main variables of the numerical examples, in which the thickness t is 3 mm, the r/t ratio varies from 10 to 60, and the range of $f_{sy,0}$ is 235 MPa to 1900 MPa. Figure 14 shows the ultimate energy absorption of the total numerical examples. It can be seen that the energy absorption is often underestimated for the specimens without considering the cold-rolling effect. For the example specimens with $f_{sy,0}$ of 235 MPa shown in Figure 14a, the energy absorption capacity obtained by using the proposed model is twice that of using the ideal elastoplastic model. However, the difference in the energy absorption capacity caused by the two models decreases with $f_{sy,0}$ increasing. When the value of $f_{sy,0}$ is 960 MPa and the r/t ratio is about 60, the cold-rolling effect has little influence on the energy absorption. As $f_{sy,0}$ reaches 1800 MPa, the influence of material models seems to disappear regardless of the value of r/t.

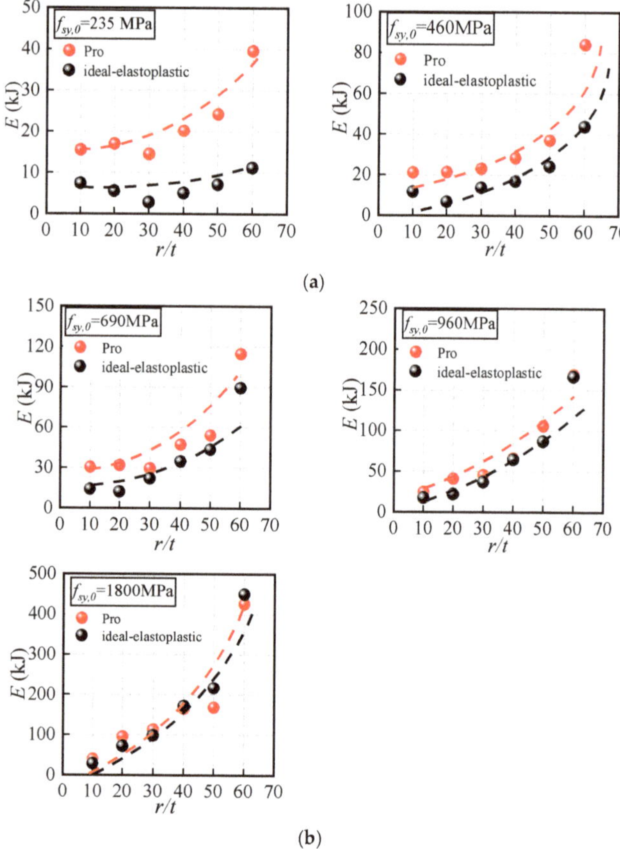

Figure 14. Energy absorption of numerical examples using different material constitutive models. (**a**) Conventional strength steels within 460 MPa. (**b**) High-strength steels exceeding 690 Mpa.

Overall, the consistency between the calculated results and measured results shows that the proposed model can be used as a digital platform for intelligent construction. In addition, compared with the ideal elastoplastic model, the proposed model established in this paper has more distinct material properties and can more accurately evaluate the ultimate bearing capacity and corresponding deformation of the components to achieve energy saving, which is beneficial to the sustainable development of prefabricated steel buildings.

Table 3. Main variables of the numerical examples.

Sets	$f_{sy,0}$ (MPa)	t (mm)	r/t	L (mm)
1	235, 460, 690, 960, 1800	3	10	198
2	235, 460, 690, 960, 1800	3	20	378
3	235, 460, 690, 960, 1800	3	30	558
4	235, 460, 690, 960, 1800	3	40	738
5	235, 460, 690, 960, 1800	3	50	918
6	235, 460, 690, 960, 1800	3	60	1098

6. Conclusions

In order to provide a unified and efficient material constitutive model for the digital intelligent design, a continuously derivable function based on the Menegotto-Pinto model is proposed to describe a complete uniaxial tensile stress-strain relationship for cold-formed circular mild steel. The key physical quantities and auxiliary parameters of the proposed model can be calibrated once the r/t ratio of CHSs, thickness t and yield strength $f_{sy,0}$ of parent steels are determined.

To verify the validity and accuracy of the proposed model, 74 sets of experimental data on mild steel have been collected. Through comparisons between the measured results and the calculated ones, the following conclusions can be drawn:

(1) The proposed model can predict the complete uniaxial tensile stress-strain behavior of cold-formed circular steels with high accuracy. Considering the wide varying range of the collected experimental variables such as $f_{sy,0}$ (400~1400 MPa), and r/t (5.4~32.3), the good agreement observed between the predictive and measured stress-strain curves indicates that the improved Menegotto-Pinto model proposed in this paper has a wide application scope.

(2) The ultimate tensile strain ε_{su} of cold-formed circular steels can be predicted by Equation (5) with more improved accuracy than the model proposed by Gardner, due to the comprehensive consideration of the influence of $f_{sy,0}$ and r/t.

(3) The cold-rolling effect that causes strength enhancement will weaken with $f_{sy,0}$ and r/t increasing and seems to be neglected when $f_{sy,0}$ reaches 1748 MPa or the r/t ratio is approximately 60.

(4) Compared with the ideal elastoplastic model, the proposed model can more accurately estimate the load-bearing capacity of the components under extreme loads, which reduces the economic burden of engineering.

Based on the mathematical and statistical analysis process presented in the paper, the proposed material constitutive model has reparability to a certain degree, which can also be a helpful tool to develop more models for the CHS and CFST members for analyzing catastrophic engineering problems. The related studies will be reported in the near future.

Author Contributions: Conceptualization, C.Y., L.Y. and Q.L.; methodology, C.Y. and L.Y. software, C.Y. and L.Y.; analysis, C.Y., L.Y., B.W. and Q.L.; writing—original draft preparation, L.Y.; writing—reviewing and editing, C.Y. and L.Y.; supervision, C.Y., B.W. and Q.L. All authors have read and agreed to the published version of the manuscript.

Funding: This work is supported by the Opening Project of Sichuan Province University Key Laboratory of Bridge Non-destruction Detecting and Engineering Computing (Grant No. 2021QZJ02 and 2019OYJ01).

Data Availability Statement: Data is contained within the article.

Conflicts of Interest: The authors declare no conflict of interest.

Appendix A

Uniaxial tensile stress-strain model of cold-formed steels including the mathematical expressions for the ultimate tensile strength f_{su} and the corresponding strain ε_{su} proposed by Gardner [17] are expressed by Equations (A1)–(A3), respectively. Note that the models are based on the available experimental specimens with $f_{sy,0}$ of 235~1100 MPa.

$$\varepsilon_s = \begin{cases} \frac{f_s}{E_s} + 0.002\left(\frac{f_s}{f_{sy}}\right)^n, \ f_s \leq f_{sy} \\ \frac{f_s - f_{sy}}{E_{0.2}} + \left(\varepsilon_{su} - \varepsilon_{0.2} - \frac{f_{su} - f_{sy}}{E_{0.2}}\right)\left(\frac{f_s - f_{sy}}{f_{su} - f_{sy}}\right)^m + \varepsilon_{0.2}, \ f_{sy} < f_s \leq f_{su} \end{cases} \tag{A1}$$

$$\varepsilon_{su} = 0.6\left(1 - f_{sy}/f_{su}\right) \tag{A2}$$

$$f_{su} = 1 + \left(130/f_{sy}\right)^{1.4} \tag{A3}$$

References

1. *Guiding Opinions of the General Office of the State Council on Vigorously Developing Prefabricated Buildings*; State Council Bulletin of the People's Republic of China; The State Council: Beijing, China, 2016; pp. 24–26. (In Chinese)
2. *GB/T 51232-2016*; Prefabricated Steel Structure Building Technical Standards. China Construction Industry Press: Beijing, China, 2016. (In Chinese)
3. Li, Q.W.; Yue, Q.R.; Feng, P.; Xie, N.; Liu, Y. Development Status and Prospect of Steel Structure Industry Based on Carbon Peak and Carbon Neutrality Target. *Prog. Steel Build. Struct.* **2022**, *24*, 1–6+23. (In Chinese)
4. *GB 50017-2017*; Standard for Design of Steel Structures. Ministry of Housing Urban-Rural Construction of the People's Republic of China: Beijing, China, 2017.
5. *AISI S100-16*; North American Specification for the Design of Cold-Formed Steel Structural Members. American Iron and Steel Institute: Washington, DC, USA, 2016.
6. *EN 1993-1-12*; Eurocode 3: Design of Steel Structures-Part 1-12: Additional Rules for the Extension of EN 1993 up to Steel Grades S 700. European Committee for Standardization (CEN): Brussels, Belgium, 2007.
7. Li, C.L.; Yuan, H.; Hong, H.P. Predicting yield strength of cold-formed carbon steel: A review and new approaches. *J. Constr. Steel. Res.* **2023**, *206*, 107926. [CrossRef]
8. Liu, H.X.; Chen, J.; Chan, T.M. Predictive methods for material properties of cold-formed conventional steels in the corner region. *Thin-Walled Struct.* **2023**, *187*, 110740. [CrossRef]
9. Kalani, M.; Bakhshi, A. Investigation of cold work effect, Karren (AISI) and ECCS (Euro Code3) equations validity and tension failure modes in thick plate cold-formed steel angle sections. *Structures* **2023**, *57*, 105092. [CrossRef]
10. Pham, C.H.; Trinh, H.N.; Proust, G. Effect of manufacturing process on microstructures and mechanical properties, and design of cold-formed G450 steel channels. *Thin-Walled Struct.* **2021**, *162*, 107620. [CrossRef]
11. Chen, J.B.; Chan, T.M. Material properties and residual stresses of cold-formed high-strength-steel circular hollow sections. *J. Constr. Steel Res.* **2020**, *170*, 106099. [CrossRef]
12. Meng, X.; Gardner, L. Cross-sectional behaviour of cold-formed high strength steel circular hollow sections. *Thin. Wall. Struct.* **2020**, *156*, 106822. [CrossRef]
13. Javidan, F.; Heidarpour, A.; Zhao, X.L.; Minkkinen, J. Application of high strength and ultra-high strength steel tubes in long hybrid compressive members: Experimental and numerical investigation. *Thin-Walled Struct.* **2016**, *102*, 273–285. [CrossRef]
14. Liu, J.Z.; Fang, H.; Chan, T.M. Experimental investigations on material properties and stub column behavior of high strength steel irregular hexagonal hollow sections. *J. Constr. Steel Res.* **2022**, *196*, 107343. [CrossRef]
15. Yao, Y.; Quach, W.M.; Young, B. Finite element-based method for residual stresses and plastic strains in cold-formed steel hollow sections. *Eng. Struct.* **2019**, *188*, 24–42. [CrossRef]
16. Rossi, B.; Afshan, S.; Gardner, L. Strength enhancements in cold-formed structural sections—Part II: Predictive models. *J. Construct. Steel Res.* **2013**, *83*, 189–196. [CrossRef]
17. Gardner, L.; Yun, X. Description of stress-strain curves for cold-formed steels. *Constr. Build. Mater.* **2018**, *189*, 527–538. [CrossRef]
18. Ramberg, W.; Osgood, W.R. *Description of Stress-Strain Curves by Three Parameters*; National Advisory Committee for Aeronautics: Washington, DC, USA, 1943.
19. Quach, W.; Huang, J. Stress-strain models for light gauge steels. *Procedia Eng.* **2011**, *14*, 288–296. [CrossRef]
20. Li, G.W.; Wu, Z.X.; Wen, D.H.; Shao, J.F.; Hu, C.X. Experimental investigation on the cold-forming effect of high strength cold-formed steel sections. *Structures* **2023**, *58*, 105615. [CrossRef]
21. Chajes, A.; Britvec, S.; Winter, G. Effects of cold-straining on structural sheet steels. *J. Struct. Div.* **1963**, *89*, 1–32. [CrossRef]
22. Bui, Q.; Ponthot, J. Numerical simulation of cold roll-forming processes. *J. Mater. Process. Technol.* **2008**, *202*, 275–282. [CrossRef]

23. Menegotto, M.; Pinto, R.E. Method of analysis for cyclically loaded RC plane frames including changes in geometry and non-elastic behavior of elements under combined normal force and bending. In *IABSE Symposium on Resistance and Ultimate Deformability of Structures Acted on by Well Defined Repeated Loads*; IABSE: Zurich, Switzerland, 1973; Volume 13, pp. 15–22.
24. Yang, C.; Zhao, H.; Sun, Y.P.; Zhao, S.C. Compressive stress-strain model of cold-formed circular hollow section stub columns considering local buckling. *Thin-Walled Struct.* **2017**, *120*, 495–505. [CrossRef]
25. Ban, H.Y.; Shi, G.; Shi, Y.J.; Wang, Y.Q. Research progress on the mechanical property of high strength structural steels. *Adv. Mater. Res.* **2011**, *250*, 640–648. [CrossRef]
26. Su, M.; Cai, Y.; Chen, X.; Young, B. Behaviour of concrete-filled cold-formed high strength steel circular stub columns. *Thin-Walled Struct.* **2020**, *157*, 107078. [CrossRef]
27. Ma, J.L.; Chan, T.M.; Young, B. Material properties and residual stresses of cold-formed high strength steel hollow sections. *J. Constr. Steel Res.* **2015**, *109*, 152–165. [CrossRef]
28. Hu, Y.F.; Xiao, M.; Chung, K.; Ban, H.; Nethercot, D.A. Experimental investigation into high strength S690 cold-formed circular hollow sections under compression. *J. Constr. Steel Res.* **2022**, *194*, 107306. [CrossRef]
29. McCann, F.; Gardner, L.; Kirk, S. Elevated temperature material properties of cold-formed steel hollow sections. *Thin-Walled Struct.* **2015**, *90*, 84–94. [CrossRef]
30. Elchalakani, M.; Zhao, X.L.; Grzebieta, R. Tests on concrete filled double-skin (CHS outer and SHS inner) composite short columns under axial compression. *Thin-Walled Struct.* **2002**, *40*, 415–441. [CrossRef]
31. Elchalakani, M.; Zhao, X.L.; Grzebieta, R.H. Concrete-filled circular steel tubes subjected to pure bending. *J. Constr. Steel Res.* **2001**, *57*, 1141–1168. [CrossRef]
32. Jiao, H.; Zhao, X.L. Material ductility of very high strength (VHS) circular steel tubes in tension. *Thin-Walled Struct.* **2001**, *39*, 887–906. [CrossRef]
33. Sadowski, A.J.; Wong, W.J.; Li, S.C.S.; Málaga-Chuquitaype, C.; Pachakis, D. Critical buckling strains in thick cold-formed circular-hollow sections under cyclic loading. *J. Struct. Eng.* **2020**, *146*, 04020179. [CrossRef]
34. Zhong, Y.; Zhao, O. Experimental and numerical studies on post-fire behaviour of S700 high strength steel circular hollow sections under combined compression and bending. *Thin-Walled Struct.* **2022**, *181*, 110004. [CrossRef]

MDPI

St. Alban-Anlage 66

4052 Basel

Switzerland

www.mdpi.com

Buildings Editorial Office

E-mail: buildings@mdpi.com

www.mdpi.com/journal/buildings